Lecture Notes in Artificial Intelligence 13199

Subseries of Lecture Notes in Computer Science

More information about this subseries at https://link.springer.com/bookseries/1244

Katsuhiro Honda · Tomoe Entani ·
Seiki Ubukata · Van-Nam Huynh ·
Masahiro Inuiguchi (Eds.)

Integrated Uncertainty in Knowledge Modelling and Decision Making

9th International Symposium, IUKM 2022
Ishikawa, Japan, March 18–19, 2022
Proceedings

 Springer

Editors
Katsuhiro Honda ⓘ
Osaka Prefecture University
Sakai, Osaka, Japan

Seiki Ubukata ⓘ
Osaka Prefecture University
Sakai, Japan

Masahiro Inuiguchi ⓘ
Osaka University
Toyonaka, Osaka, Japan

Tomoe Entani ⓘ
University of Hyogo
Kobe, Japan

Van-Nam Huynh ⓘ
Japan Advanced Institute of Science
and Technology
Nomi, Japan

ISSN 0302-9743   ISSN 1611-3349 (electronic)
Lecture Notes in Artificial Intelligence
ISBN 978-3-030-98017-7   ISBN 978-3-030-98018-4 (eBook)
https://doi.org/10.1007/978-3-030-98018-4

LNCS Sublibrary: SL7 – Artificial Intelligence

This Springer imprint is published by the registered company Springer Nature Switzerland AG
The registered company address is: Gewerbestrasse 11, 6330 Cham, Switzerland

Preface

This volume contains the papers that were presented at the 9th International Symposium on Integrated Uncertainty in Knowledge Modelling and Decision Making (IUKM 2022) held in Ishikawa, Japan, during March 18–19, 2022.

The IUKM symposia aim to provide a forum for exchanges of research results and ideas, and experience of application among researchers and practitioners involved with all aspects of uncertainty modelling and management. Previous editions of the conference were held in Ishikawa, Japan (IUM 2010), Hangzhou, China (IUKM 2011), Beijing, China (IUKM 2013), Nha Trang, Vietnam (IUKM 2015), Da Nang, Vietnam (IUKM 2016), Ha Noi, Vietnam (2018), Nara, Japan (2019), Phuket, Thailand (2020), and their proceedings were published by Springer in AISC 68, LNAI 7027, LNAI 8032, LNAI 9376, LNAI 9978, LNAI 10758, LNAI 11471, and LNAI 12482 respectively.

IUKM 2022 was jointly organized by Osaka University, Osaka Prefecture University, and the Japan Advanced Institute of Science and Technology.

This year the conference received 46 submissions from authors in 11 different countries. Each submission was peer reviewed by at least three members of the Program Committee. After a thorough review process, 36 papers were accepted for presentation at IUKM 2022, of which 30 papers (65.22%) were accepted for publication in the LNAI proceedings. In addition to the regular and short presentations, three keynote lectures by leading researchers on topics ranging from theory and methods to applications in the fields of IUKM were organized.

We express our sincere thanks to Motohide Umano, Salvatore Greco, and Rudolf Felix for providing valuable and stimulating lectures. We are very thankful to the local organizing team from Osaka University, Osaka Prefecture University, and the Japan Advanced Institute of Science and Technology for their hard working, efficient services, and wonderful local arrangements.

We would like to express our appreciation to the members of the Program Committee for their support and cooperation in this publication. We are also thankful to the team at Springer for providing a meticulous service for the timely production of this volume. Last, but certainly not the least, our special thanks go to all the authors who submitted papers and all the attendees for their contributions and fruitful discussions that made this conference a great success.

March 2022

Katsuhiro Honda
Tomoe Entani
Seiki Ubukata
Van-Nam Huynh
Masahiro Inuiguchi

Organization

General Chair

Masahiro Inuiguchi   Osaka University, Japan

Advisory Board

Michio Sugeno   European Center for Soft Computing, Spain
Hung T. Nguyen   New Mexico State University, USA, and Chiang
               Mai University, Thailand
Sadaaki Miyamoto   University of Tsukuba, Japan
Akira Namatame   AOARD/AFRL and National Defense Academy
               of Japan, Japan

Program Chair

Katsuhiro Honda   Osaka Prefecture University, Japan

Program Vice-chairs

Tomoe Entani   University of Hyogo, Japan
Seiki Ubukata   Osaka Prefecture University, Japan

Local Arrangement Co-chairs

Van-Nam Huynh   Japan Advanced Institute of Science and
               Technology, Japan
Hideomi Gokon   Japan Advanced Institute of Science and
               Technology, Japan

Program Committee

Tomoyuki Araki   Hiroshima Institute of Technology, Japan
Yaxin Bi   University of Ulster, UK
Matteo Brunelli   University of Trento, Italy
Tru Cao   University of Texas Health Science Center at
               Houston, USA
Tien-Tuan Dao   Centrale Lille Institut, France

Yong Deng	University of Electronic Science and Technology of China, China
Thierry Denoeux	University of Technology of Compiègne, France
Sebastien Destercke	University of Technology of Compiègne, France
Zied Elouedi	ISG de Tunis, Tunisia
Tomoe Entani	University of Hyogo, Japan
Katsushige Fujimoto	Fukushima University, Japan
Lluis Godo	IIIA - CSIC, Spain
Yukio Hamasuna	Kindai University, Japan
Katsuhiro Honda	Osaka Prefecture University, Japan
Tzung-Pei Hong	National Univesity of Kaohsiung, Taiwan
Jih Cheng Huang	SooChow University, Taiwan
Van-Nam Huynh	JAIST, Japan
Hiroyuki Inoue	University of Fukui, Japan
Atsuchi Inoue	Eastern Washington University, USA
Masahiro Inuiguchi	Osaka University, Japan
Radim Jirousek	Prague University of Economics and Business, Czech Republic
Yuchi Kanzawa	Shibaura Institute of Technology, Japan
Yasuo Kudo	Muroran Institute of Technology, Japan
Yoshifumi Kusunoki	Osaka University, Japan
Anh Cuong Le	Ton Duc Thang University, Vietnam
Bac Le	University of Science, Vietnam National University Ho Chi Minh City, Vietnam
Churn-Jung Liau	Academia Sinica, Taiwan
Marimin Marimin	Bogor Agricultural University, Indonesia
Luis Martinez	University of Jaen, Spain
Radko Mesiar	Slovak University of Technology in Bratislava, Slovakia
Tetsuya Murai	Chitose Institute of Science and Technology, Japan
Michinori Nakata	Josai International University, Japan
Canh Hao Nguyen	Kyoto University, Japan
Duy Hung Nguyen	Sirindhorn International Institute of Technology, Thailand
Akira Notsu	Osaka Prefecture University, Japan
Vilem Novak	University of Ostrava, Czech Republic
Warut Pannakkong	Sirindhorn International Institute of Technology, Thailand
Irina Perfilieva	University of Ostrava, Czech Republic
Zengchang Qin	Beihang University, China
Jaroslav Ramik	Silesian University in Opava, Czech Republic
Hiroshi Sakai	Kyushu Institute of Technology, Japan

Hirosato Seki	Osaka University, Japan
Kao-Yi Shen	Chinese Culture University, Taiwan
Roman Slowinski	Poznan University of Technology, Poland
Martin Stepnicka	University of Ostrava, Czech Republic
Kazuhiro Takeuchi	Osaka Electro-Communication University, Japan
Roengchai Tansuchat	Chiang Mai University, Thailand
Phantipa Thipwiwatpotjana	Chulalongkorn University, Thailand
Vicenc Torra	University of Skovde, Sweden
Seiki Ubukata	Osaka Prefecture University, Japan
Guoyin Wang	Chongqing University of Posts and Telecommunications, China
Woraphon Yamaka	Chiang Mai University, Thailand
Chunlai Zhou	Renmin University of China, China

Keynote Lectures

Partition of Time Series Using Hierarchical Clustering

Motohide Umano

Osaka Prefecture University and Hitachi Zosen Corporation, Japan

Abstract. We understand a long time-series through features and trends and their transitions, for example, "Globally it increases a little, but it starts with a medium value, decreases a little in the beginning and has big oscillations at end." It is often the case where the periods of features and trends are determined by the data themselves. We must, therefore, partition time-series into several periods of different features and trends.

We propose a method to partition time-series data by clustering adjacent data with the total similarity of their values, changes of values and degrees of oscillations of adjacent periods. First we have the initial clusters of line segments of adjacent data in time. Next we get the adjacent clusters that have the maximum total similarity and merge them into one. We repeat this process until the condition of termination. We formulate the total similarity as the weighted average of three similarities of the value, change of values and degree of oscillations. The weights are very important. The fixed weights cannot have the clustering results that fit our sense. We, therefore, propose variable weights with three similarities and sizes of adjacent clusters with the operation of ordered weighted average. Furthermore, in order to exclude small clusters of outliers, we define similarities of two clusters adjacent to the small cluster. We apply this method to actual time-series data and show results. The method can improve linguistic expressions of time-series data and retrieval of similar time-series with linguistic similarity.

The Robust Ordinal Regression: Basic Ideas, Principal Models, Recent Developments

Salvatore Greco

University of Catania, Italy

Abstract. Multiple Criteria Decision Aiding (MCDA) is constituted by a set of concepts, techniques and procedure aiming to provide a recommendation in complex decision contexts. MCDA is based on a constructive approach that aims to build a preference model in cooperation between the analyst and the Decision Maker. A typical MCDA methodology is the ordinal regression aiming to define a decision model in a given class (an additive value function, a Choquet integral, an outranking model such as ELECTRE or PROMETHEE and so on) representing the preference information provided by the DM. Recently ordinal regression has been extended and generalized through Robust Ordinal Regression taking into account the idea that there is a plurality of decision models in a given class compatible with the preferences expressed by the decision maker. Originally, the set of compatible decision models was used to define the necessary and possible preference relations holding when the preference holds for all value functions or for at least one value function, respectively. After, a probability distribution on the set of compatible decision model was introduced to define probabilistic preferences. ROR has been also fruitfully applied to interactive optimization procedures.

In this talk I shall present the basic concepts, the principal models, the main applications and the recent developments of Robust Ordinal Regression taking into consideration its advantages in the context of an MCDA constructive approach.

Decision Making and Optimization in Context of Inconsistently Interacting Goals and its Relation to Machine Learning

Rudolf Felix

PSI FLS Fuzzy Logik & Neuro Systeme GmbH, Germany

Abstract. Many traditional optimization models are limited with respect to the management of inconsistency that frequently appear between decision and optimization goals. As consequence, such models in many cases achieve results that may be optimal for the model but are not for the use case to be managed. In real world use cases both decision and optimization goals are usually partly conflicting and therefore partly inconsistent. Assumptions like independence of goals, additivity or monotonicity as preconditions usually do not hold. Due to this, traditional concepts like integration based on weighted sums, for instance, in many cases do not really help.

In this talk we describe some applications of a decision and optimization model based on (extended fuzzy) interactions between goals (DMIG) to some relevant real-world decision and optimization use cases. After a brief discussion of the basics of the concept of the model, example use cases are presented and advantages of their solutions are shown. The use cases are related to real-world decision and optimization problems in business processes such as management and scheduling of field forces that maintain complex industrial infrastructure, management of resources based on sequencing of production orders in car producing factories and automated management of bus and tram depots. Some additional examples are named. It is also shown how the so-called key performance indicators (KPIs) of such real-world use cases are understood as decision and optimization goals and how interactions between decision and optimization goals build a bridge to the optimization of real-world KPIs. Finally, it is discussed why DMIG may be used for learning of consistent preferences between the KPIs and how the concept is connected to the field of machine learning.

Contents

Pattern Classification and Data Analysis

Machine Learning

Economic Applications

Uncertainty Management and Decision Making

Measuring Quality of Belief Function Approximations

Radim Jiroušek[1,2] and Václav Kratochvíl[1,2(✉)]

[1] Faculty of Management, Prague University of Economics and Business,
Jindřichův Hradec, Czechia
{radim,velorex}@utia.cas.cz
[2] Institute of Information Theory and Automation, Czech Academy of Sciences,
Prague, Czechia

Abstract. Because of the high computational complexity of the respective procedures, the application of belief-function theory to problems of practice is possible only when the considered belief functions are approximated in an efficient way. Not all measures of similarity/dissimilarity are felicitous to measure the quality of such approximations. The paper presents results from a pilot study that tries to detect the divergences suitable for this purpose.

Keywords: Belief functions · Divergence · Approximation · Compositional models

1 Introduction

Modeling practical problems usually requires a fair amount of random variables. Even small and simple applications require tens of variables, which complicates the application of belief-function models because the corresponding space of discernment grows super-exponentially with the number of the considered variables. As we will see, to specify a general belief function just for six binary variables, we need $2^{(2^6)} = 2^{64}$ parameters. To avoid problems arising from the high computational complexity of the respective procedures, one should restrict their attention to belief functions representable with a limited number of parameters. For this purpose, we propose models assembled from a sequence of several low-dimensional belief functions – so-called *compositional models*. In connection with this, the question arises, how to recognize whether a compositional model is an acceptable approximation of the considered multidimensional belief function.

In [7] and [6], we studied some heuristics proposed to control the model learning procedures. Inspired by the processes used in probabilistic modeling, we investigated the employment of entropy of belief functions for this purpose. Unfortunately, no belief functions entropy has the properties of probabilistic

Financially supported by the Czech National Science Foundation under grant no. 19-06569S.

K. Honda et al. (Eds.): IUKM 2022, LNAI 13199, pp. 3–15, 2022.
https://doi.org/10.1007/978-3-030-98018-4_1

Shannon entropy that would enable us to detect the optimal approximation. Even worse, in belief function theory, there is no generally accepted measure of similarity (dissimilarity) that could help recognize which of two approximations is better. And this is the goal of the current paper. We will study which of several dissimilarity measures (divergences) are suitable for the purpose. In this paper, we consider only those divergences meeting the following two conditions:

- the values of the divergence are non-negative and equal zero only for identical belief functions (the divergence is *non-degenerative*);
- the complexity of the necessary computation is polynomial with the number of focal elements of the considered basic assignments.

Let us note at the very beginning that the achieved results depend on the fact that we consider only a specific class of approximations: the approximations of belief functions by compositional models. We admit that if considering different approximating functions, one could detect other measures of divergence as suitable.

The approximations of complex models by compositional models were first suggested for multidimensional probability distribution [15]. Similarly, the authors of some of the considered divergences also took inspiration from probability theory. And this is why we will at times turn our exposition to probability theory.

The paper is organized as follows. In the next section, we introduce basic notation and recall the idea of Perez, from whom we took the inspiration. The notation from belief function theory is briefly recollected in Sect. 3. Section 4 introduces the considered divergences, and Sect. 5 explains the class of approximations considered, i.e., the class of compositional models. The computational experiments and the achieved results are described in Sect. 6.

2   Basic Notation and Motivation

In this paper, we consider a finite set N of random variables, which are denoted by lower-case characters from the end of the Latin alphabet ($N = \{u, v, w, \ldots\}$). All the considered variables are assumed to be finite-valued. $\mathbb{X}_u$, $\mathbb{X}_v$, ... denote the finite sets of values of variables u, v, ... Sets of variables are denoted by upper-case characters K, L, V, ... Thus, K may be, say, $\{u, v, w\}$. By a *state* of variables K we understand any combination of values of the respective variables, i.e., in the considered case $K = \{u, w, w\}$, a state is an element of a Cartesian product $\mathbb{X}_K = \mathbb{X}_u \times \mathbb{X}_v \times \mathbb{X}_w$. For a state $a \in \mathbb{X}_K$ and $L \subset K$, $a^{\downarrow L}$ denote a *projection* of $a \in \mathbb{X}_K$ into $\mathbb{X}_L$, i.e., $a^{\downarrow L}$ is the state from $\mathbb{X}_L$ that is got from a by dropping out all the values of variables from $K \setminus L$.

The original idea of Perez [15] was to approximate a multidimensional probability distribution $\mu(N)$ (i.e., $\mu : \mathbb{X}_N \longrightarrow [0, 1]$, for which $\sum_{a \in \mathbb{X}_N} \mu(a) = 1$) by a simpler probability distribution $\kappa(N)$. To measure the quality of such approximation he used their *relative entropy*, which is often called *Kullback-Leibler divergence*[1]

$$KL(\mu \parallel \kappa) = \begin{cases} \sum\limits_{c \in \mathbb{X}_N : \kappa(c) > 0} \mu(c) \log_2 \left(\frac{\mu(c)}{\kappa(c)} \right) & \text{if } \mu \ll \kappa, \\ +\infty & \text{otherwise,} \end{cases}$$

where symbol $\mu \ll \kappa$ denotes that κ *dominates* μ, which means that for all $c \in \mathbb{X}_N$, if $\kappa(c) = 0$ then also $\mu(c) = 0$.

It is known that the Kullback-Leibler divergence is non-negative and equals 0 if and only if $\mu = \kappa$ [13]. It is also evident that it is not symmetric[2], and therefore some authors measure the non-similarity of two distributions by the arithmetic mean $\frac{1}{2}(KL(\mu \parallel \kappa) + KL(\kappa \parallel \mu))$. A more sophisticated symmetrized version of this distance is so called Jensen-Shannon divergence (JS) defined

$$JS(\mu \parallel \kappa) = \frac{1}{2} \left(KL \left(\mu \parallel \frac{\mu + \kappa}{2} \right) + KL \left(\kappa \parallel \frac{\mu + \kappa}{2} \right) \right),$$

which is, obviously, symmetric and always finite (namely, both μ and κ are dominated by $\frac{\mu + \kappa}{2}$). For more properties of this and other distances between probability measures, the reader is referred to [14], where one can learn that there is also an alternative way of expressing JS divergence using Shannon entropy

$$JS(\mu \parallel \kappa) = H \left(\frac{\mu + \kappa}{2} \right) - \frac{1}{2}(H(\mu) + H(\kappa)).$$

Recall that

$$H(\mu) = - \sum_{c \in \mathbb{X}_N} \mu(c) \log_2(\mu(c)),$$

which is known to be non-negative and less or equal to $\log_2(|\mathbb{X}_N|)$ [17].

3   Belief Functions

A basic assignment m for variables N is a function[3] $m : 2^{\mathbb{X}_N} \longrightarrow [0,1]$, for which

- $\sum_{\mathbf{a} \subseteq \mathbb{X}_N} m(\mathbf{a}) = 1$,
- $m(\emptyset) = 0$.

We say that $\mathbf{a} \subseteq \mathbb{X}_N$ is a focal element of m if $m(\mathbf{a}) \neq 0$. We use symbols Bel_m, Pl_m, Q_m to denote belief, plausibility and commonality functions, respectively. These functions, which are known to carry the same information as the corresponding basic assignment m, are defined by the following formulas [16]

$$Bel_m(\mathbf{a}) = \sum_{\mathbf{b} \subseteq \mathbf{a}} m(\mathbf{b}); \quad Pl_m(\mathbf{a}) = \sum_{\mathbf{b} \subseteq \mathbb{X}_N : \mathbf{a} \cap \mathbf{b} \neq \emptyset} m(\mathbf{b}); \quad Q_m(\mathbf{a}) = \sum_{\mathbf{b} \subseteq \mathbb{X}_N : \mathbf{b} \supseteq \mathbf{a}} m(\mathbf{b}).$$

---

[1] We take $0 \log_2(0) = 0$.

[2] To show asymmetry of the Kullback-Leibler divergence consider $\mu = (\frac{1}{3}, \frac{1}{3}, \frac{1}{3})$, and $\kappa = (\frac{1}{2}, \frac{1}{2}, 0)$.

[3] $2^{\mathbb{X}_N}$ denote the set of all subsets of $\mathbb{X}_N$.

When constructing compositional models, we need marginals of the considered basic assignments. Let m be defined for arbitrary set of variables $L \supseteq K$. Symbol $m^{\downarrow K}$ will denote the marginal of m, which is defined for variables K. Thus,

$$m^{\downarrow K}(\mathbf{b}) = \sum_{\mathbf{a} \subseteq \mathbb{X}_L : \mathbf{a}^{\downarrow K} = \mathbf{b}} m(\mathbf{a}).$$

for all $\mathbf{b} \subseteq \mathbb{X}_K$.

When normalizing the plausibility function on singletons, one gets a probability distribution on $\mathbb{X}_N$ called a *plausibility transform* of basic assignment m [1]. There are several other probabilistic transforms described in literature [2,3]. In this paper we use only the above-mentioned plausibility transform λ_m and the so-called *pignistic transform* π_m strongly advocated by Philippe Smets [18], which are defined for all $a \in \mathbb{X}_N$

$$\lambda_m(a) = \frac{Pl_m(\{a\})}{\sum_{c \in \mathbb{X}_N} Pl_m(\{c\})}, \quad \text{and} \quad \pi_m(a) = \sum_{\mathbf{b} \subseteq \mathbb{X}_N : a \in \mathbf{b}} \frac{m(\mathbf{b})}{|\mathbf{b}|}.$$

Up to now, we have recalled a standard notation used in belief function theory. Rather unusual is that, to make the next exposition as simple as possible, we will sometimes view the basic assignment m also as a probability distribution on $2^{\mathbb{X}_N}$. This enables us to speak about Shannon entropy $H(m)$ of m, to say that m_1 dominates m_2, and to compute Kullback-Leibler divergence between two basic assignments.

4 Divergences

Quite a few papers suggesting different tools to measure similarity/dissimilarity of belief functions were published. The reader can find a good survey in [12]. As indicated in the Introduction, in this paper, we are interested only in those measures, the computation of which is tractable even for multidimensional belief functions if the number of focal elements of the considered basic assignments is not too high. In other words, we are interested in the formulas, the computational complexity of which depends on the number of focal elements, regardless of the number of variables, for which the respective basic assignments are defined. Given the goal of this paper, we also restrict our attention only to *non-degenerative* measures, i.e., the measures which can detect the equality of belief functions because they equal zero only for identical basic assignments. In this pilot study, we consider only the six divergences described below.

In this section, we assume that all the considered basic assignments are defined for the set of variables N.

Jousselme et al. (2001). In [11], the authors define a distance between basic assignments meeting all the metric axioms: non-negativity, non-degeneracy, symmetry, and the triangle inequality. Recall that the Kullback-Leibler divergence introduced in Sect. 2 meets only the first two properties; it is not symmetric, nor the triangle inequality holds for KL.

To be able use the notation of linear algebra, consider a fixed ordering of elements of $2^{|\mathbb{X}_N|}$. Then, m can be interpreted as a vector $\boldsymbol{m}$ of $2^{|\mathbb{X}_N|}$ non-negative real numbers. Jousselme et al. define their distance

$$d_{BPA}(m_1, m_2) = \sqrt{\frac{1}{2}(\boldsymbol{m_1} - \boldsymbol{m_2})^T D (\boldsymbol{m_1} - \boldsymbol{m_2})}, \tag{1}$$

where D is $2^{|\mathbb{X}_N|} \times 2^{|\mathbb{X}_N|}$ matrix defined as follows: let $\mathbf{a}_i$ be an element of $2^{\mathbb{X}_N}$ which corresponds the i-th coordinate of the vector $\boldsymbol{m}$. Then, the elements of matrix $D = (d_{ij})$ are defined

$$d_{ij} = \frac{|\mathbf{a}_i \cap \mathbf{a}_j|}{|\mathbf{a}_i \cup \mathbf{a}_j|}.$$

Note that we allow a situation of $\mathbf{a}_i = \emptyset$. In this case define $d_{ii} = 1$. Knowing the matrix D, the argument of the square root of Eq. (1) can be rewritten into the following form

$$(\boldsymbol{m_1} - \boldsymbol{m_2})^T D (\boldsymbol{m_1} - \boldsymbol{m_2})$$
$$= \sum_{\mathbf{a} \subseteq \mathbb{X}_N} m_1(\mathbf{a}) \sum_{\mathbf{b} \subseteq \mathbb{X}_N} \frac{m_1(\mathbf{b}) |\mathbf{a} \cap \mathbf{b}|}{|\mathbf{a} \cup \mathbf{b}|} + \sum_{\mathbf{a} \subseteq \mathbb{X}_N} m_2(\mathbf{a}) \sum_{\mathbf{b} \subseteq \mathbb{X}_N} \frac{m_2(\mathbf{b}) |\mathbf{a} \cap \mathbf{b}|}{|\mathbf{a} \cup \mathbf{b}|}$$
$$- 2 \sum_{\mathbf{a} \subseteq \mathbb{X}_N} \sum_{\mathbf{b} \subseteq \mathbb{X}_N} \frac{m_1(\mathbf{a}) m_2(\mathbf{b}) |\mathbf{a} \cap \mathbf{b}|}{|\mathbf{a} \cup \mathbf{b}|}.$$

Xiao (2019). To define the divergence between two basic assignments m_1 and m_2, Xiao [21] makes use of the fact that a basic assignment on $\mathbb{X}_N$ is a probability measure on $2^{\mathbb{X}_N}$. Thus, she defines a belief function divergence – she calls it Belief Jensen-Shannon divergence (BJS) – which is the probabilistic Jensen-Shannon divergence of the corresponding probability measures, i.e.,

$$BJS(m_1, m_2) = \frac{1}{2} \left[KL \left(m_1 \, \| \, \frac{m_1 + m_2}{2} \right) + KL \left(m_2 \, \| \, \frac{m_1 + m_2}{2} \right) \right], \tag{2}$$

or, equivalently

$$BJS(m_1, m_2) = H \left(\frac{m_1 + m_2}{2} \right) - \frac{H(m_1) + H(m_2)}{2} \tag{3}$$

(recall, H denotes the Shannon entropy).

Song-Deng (2019a). As the authors say in [20], being inspired by Eq. (2), they replaced the arithmetic mean in Eq. (2) by the geometric mean, suggesting a new divergence BRE (perhaps from Belief Relative Entropy) defined by

$$BRE(m_1, m_2) = \sqrt{KL\left(m_1 \parallel \sqrt{m_1 \cdot m_2}\right) \cdot KL\left(m_2 \parallel \sqrt{m_1 \cdot m_2}\right)}. \qquad (4)$$

In contrast to BJS, which is always finite, BRE equals $+\infty$ whenever there is at least one $\mathbf{a} \subseteq \mathbb{X}_N$, which is a focal element of only one of the two basic assignments m_1, m_2.

Song-Deng (2019b). The same pair of authors suggested also another belief function divergence related to the relative Deng entropy D_d, which is defined by the following formula

$$D_d(m_1 \parallel m_2) = \sum_{\mathbf{a} \subseteq \mathbb{X}_N : m_2(\mathbf{a}) > 0} \frac{1}{2^{|\mathbf{a}|} - 1} m_1(\mathbf{a}) \log\left(\frac{m_1(\mathbf{a})}{m_2(\mathbf{a})}\right). \qquad (5)$$

Assume that $D_d(m_1 \parallel m_2) = +\infty$ in case that there is a $\mathbf{a} \subseteq \mathbb{X}_N$ for which $m_1(\mathbf{a}) > 0 = m_2(\mathbf{a})$. In [19], the authors define the divergence D_{SDM} symmetrizing the relative Deng entropy

$$D_{SDM}(m_1, m_2) = \frac{1}{2}\left(D_d(m_1 \parallel m_2) + D_d(m_2 \parallel m_1)\right). \qquad (6)$$

Assuming that for $m_1 \ll m_2$, Eq. (5) defines the relative entropy, and that it equals $+\infty$ in opposite case. Then it is not difficult to show [19] that measure D_{SDM} is non-negative, non-degenerative, and symmetric.

Simple Divergences. With the goal to test also some computationally cheap divergences, we, being inspired by the entropy defined in [9], consider also functions

$$Div_\lambda(m_1, m_2) = KL(\lambda_{m_1} \parallel \lambda_{m_2}) + \sum_{\mathbf{a} \subseteq \mathbb{X}_N} |m_1(\mathbf{a}) - m_2(\mathbf{a})| \cdot \log(|\mathbf{a}|), \qquad (7)$$

and

$$Div_\pi(m_1, m_2) = KL(\pi_{m_1} \parallel \pi_{m_2}) + \sum_{\mathbf{a} \subseteq \mathbb{X}_N} |m_1(\mathbf{a}) - m_2(\mathbf{a})| \cdot \log(|\mathbf{a}|), \qquad (8)$$

where λ and π are plausibility and pignistic transforms introduced in Sect. 2.

Proposition 1. *Both divergences Div_λ and Div_π are non-negative and non-degenerative.*

Proof. The non-negativity of the considered divergences follows directly from the non-negativity of Kullback-Leibler divergence.

To show their non-degenerativity, i.e., $Div_\lambda(m_1, m_2) = 0 \iff m_1 = m_2$, and $Div_\pi(m_1, m_2) = 0 \iff m_1 = m_2$, consider two basic assignments m_1 and m_2. If $m_1 = m_2$, then, trivially, $Div_\lambda(m_1, m_2) = Div_\pi(m_1, m_2) = 0$.

To show the other side of the equivalence, assume that $m_1 \neq m_2$, and

$$\sum_{\mathbf{a} \subseteq \mathbb{X}_N} |m_1(\mathbf{a}) - m_2(\mathbf{a})| \cdot \log(|\mathbf{a}|) = 0. \tag{9}$$

This equality holds if and only if $m_1(\mathbf{a}) = m_2(\mathbf{a})$ for all non-singletons $\mathbf{a} \subseteq \mathbb{X}_N$. Since,

$$\sum_{c \in \mathbb{X}_N} Pl_{m_i}(c) = \sum_{c \in \mathbb{X}_N} m_i(c) + \sum_{c \in \mathbb{X}_N} \left(\sum_{\mathbf{a} \subseteq \mathbb{X}_N : c \in \mathbf{a} \,\&\, |\mathbf{a}| > 1} m_i(\mathbf{a}) \right)$$

$$= \left(1 - \sum_{\mathbf{a} \subseteq \mathbb{X}_N : |\mathbf{a}| > 1} m_i(\mathbf{a}) \right) + \sum_{c \in \mathbb{X}_N} \left(\sum_{\mathbf{a} \subseteq \mathbb{X}_N : c \in \mathbf{a} \,\&\, |\mathbf{a}| > 1} m_i(\mathbf{a}) \right),$$

we can see that $\sum_{c \in \mathbb{X}_N} Pl_{m_1}(c) = \sum_{c \in \mathbb{X}_N} Pl_{m_2}(c)$.

Since we assume that for $m_1 \neq m_2$ Eq. (9) holds, then there exists $c \in \mathbb{X}_N$, for which $m_1(c) \neq m_2(c)$, and therefore also

$$Pl_{m_1}(c) = \sum_{\mathbf{a} \subseteq \mathbb{X}_N : c \in \mathbf{a}} m_1(\mathbf{a}) \neq \sum_{\mathbf{a} \subseteq \mathbb{X}_N : c \in \mathbf{a}} m_2(\mathbf{a}) = Pl_{m_2}(c).$$

Thus,

$$\lambda_{m_1}(c) = \frac{Pl_{m_1}(c)}{\sum_{x \in \mathbb{X}_N} Pl_{m_1}(x)} \neq \frac{Pl_{m_2}(c)}{\sum_{x \in \mathbb{X}_N} Pl_{m_2}(x)} = \lambda_{m_2}(c),$$

and therefore $KL(\lambda_{m_1} \| \lambda_{m_2}) > 0$. This proves that Div_λ is non-degenerative because we have showed that either $KL(\pi_{m_1} \| \pi_{m_2})$ is positive, or Eq. (9) does not hold, whenever $m_1 \neq m_2$.

Similarly, for the considered $c \in \mathbb{X}_N$, for which $m_1(c) \neq m_2(c)$,

$$\pi_{m_1}(c) = m_1(c) + \sum_{\mathbf{a} \subseteq \mathbb{X}_N : c \in \mathbf{a} \,\&\, |\mathbf{a}| > 1} \frac{m_1(\mathbf{a})}{|\mathbf{a}|}$$

$$\neq m_2(c) + \sum_{\mathbf{a} \subseteq \mathbb{X}_N : c \in \mathbf{a} \,\&\, |\mathbf{a}| > 1} \frac{m_2(\mathbf{a})}{|\mathbf{a}|} = \pi_{m_2}(c),$$

and therefore also $KL(\pi_{m_1} \| \pi_{m_2}) > 0$, which proves that also Div_π is non-degenerative. □

5   Compositional Models

The definition of compositional models for belief functions is analogous to that in probability theory [4]. A basic assignment of a multidimensional compositional model is assembled from a system of low-dimensional basic assignments. To do it, one needs a tool to create a more-dimensional basic assignment from two or more low-dimensional ones. In this paper, we use an *operator of composition* ▷. By this term, we understand a binary operator meeting the following four axioms (basic assignments m_1, m_2, m_3 are assumed to be defined for K, L, M, respectively):

A1 *(Domain):* $m_1 ▷ m_2$ is a basic assignment for variables $K \cup L$.
A2 *(Composition preserves first marginal):* $(m_1 ▷ m_2)^{\downarrow K} = m_1$.
A3 *(Commutativity under consistency):* If m_1 and m_2 are consistent, i.e., $m_1^{\downarrow K \cap L} = m_2^{\downarrow K \cap L}$, then $m_1 ▷ m_2 = m_2 ▷ m_1$.
A4 *(Associativity under special condition):* If $K \supset (L \cap M)$, or, $L \supset (K \cap M)$ then $(m_1 ▷ m_2) ▷ m_3 = m_1 ▷ (m_2 ▷ m_3)$.

Because of space limit we cannot discuss these axioms in details (for this we refer the reader to [5]), but roughly speaking, axioms A1, A3, A4 guarantee that the operator of composition uniquely reconstruct basic assignment $m^{\downarrow K \cup L}$ from its marginals $m^{\downarrow K}$ and $m^{\downarrow L}$, if there exists a lossless decomposition of $m^{\downarrow K \cup L}$ into $m^{\downarrow K}$ and $m^{\downarrow L}$. Surprisingly, it is axiom A4, which guarantees that no necessary information from $m^{\downarrow L}$ is lost. Axiom A2 solves the problem arising when non-consistent basic assignments are composed. Generally, there are two ways of coping with this problem. Either find a compromise (a mixture of inconsistent pieces of knowledge) or give preference to one of the sources. The solution expressed by axiom A2 is superior to the other two possibilities from the computational point of view.

By a *compositional model*, we understand a multidimensional belief function, the basic assignment of which is assembled from a sequence of low-dimensional basic assignments with the help of the operator of composition: $m_1 ▷ m_2 ▷ \ldots ▷ m_n$. Since the operator of composition is not associative, this expression is ambiguous. To avoid this ambiguity, we omit the parentheses only if the operators are to be performed from left to right, i.e.,

$$m_1 ▷ m_2 ▷ \ldots ▷ m_n = (\ldots ((m_1 ▷ m_2) ▷ m_3) ▷ \ldots ▷ m_{n-1}) ▷ m_n. \qquad (10)$$

Let $m^\star = m_1 ▷ m_2 ▷ \ldots ▷ m_n$, and let each m_i be defined for variables K_i. Due to axiom A2, m_1 is a marginal of $m^\star$. Similarly, $m_1 ▷ m_2 = m^{\star \downarrow K_1 \cup K_2}$. This, however, does not mean that m_2 is also a marginal of $m^\star$. If all m_i are marginals of $m^\star$, then we say that $m^\star$ is defined by a *perfect compositional model*. The following assertion summarizes the relevant properties that were proved in [8,10].

Proposition 2. *Let* $m^\star = m_1 ▷ m_2 ▷ \ldots ▷ m_n$*, and let each* m_i *be defined for the set of variables* K_i*.*

- *(Compositional models can be perfectized.) There exists a perfect model $m^\star = \bar{m}_1 \triangleright \bar{m}_2 \triangleright \ldots \triangleright \bar{m}_n$ such that each $\bar{m}_i$ is defined for K_i.*
- *(Uniqueness of compositional models.) Let $m^\star = m_1 \triangleright m_2 \triangleright \ldots \triangleright m_n$ be perfect. If there is a permutation $j_1, j_2, \ldots, j_n$ such that $m_{j_1} \triangleright m_{j_2} \triangleright \ldots \triangleright m_{j_n}$ is also perfect, then $m_{j_1} \triangleright m_{j_2} \triangleright \ldots \triangleright m_{j_n} = m^\star$.*
- *(Consistent decomposable models are perfect.) If all m_i are pairwise consistent (i.e., for all $1 \leq i,j \leq n$, $m_i^{\downarrow K_i \cap K_j} = m_j^{\downarrow K_i \cap K_j}$), and the sequence $K_1, K_2, \ldots, K_n$ meets the running intersection property[4], then $m_1 \triangleright m_2 \triangleright \ldots \triangleright m_n$ is perfect.*

Now, let us express the original idea of Perez [15] in the language of compositional models: He proposed to approximate multidimensional probability distributions by compositional models and, as said above, to measure the quality of such approximations using the Kullback-Leibler divergence. He proved that if a perfect model exists, it minimizes the KL divergence (due to the *uniqueness of compositional models*, all perfect models define the identical approximation). This fact fully corresponds with our intuition. When knowing only a system of marginals of an approximated distribution, the best approximation is a distribution having all of them for its marginals.

How to employ this idea within the framework of belief functions? Not having a generally accepted "Kulback-Leibler divergence" for belief functions at our disposal, we try to solve a problem, which is, in a sense, *inverse* to that of Perez. We accept the paradigm that the best approximation of a multidimensional basic assignment is, if it exists, a perfect compositional model assembled from the marginals of the approximated basic assignment. Based on this we test, which belief function divergences detect the optimal approximation. The corresponding computational experiments, as well as the achieved results, are described in the next section. First, however, we owe the reader a specification of the used operator of composition.

In the literature, two operators of composition meeting axioms A1–A4 were introduced. Historically, the first was defined in [10]. Its disadvantage is that it does not comply with the Dempster-Shafer interpretation of belief function theory. The other operator, derived from Dempster's rule of combination, was designed by Shenoy in [8]. Nevertheless, because of its high computational complexity, we did not include it in the described pilot computational experiments. In the experiments described below, we used only the first operator. To present its definition, we need an additional notion.

Consider two arbitrary sets of variables K and L. By a *join* of $\mathbf{a} \subseteq \mathbb{X}_K$ and $\mathbf{b} \subseteq \mathbb{X}_L$ we understand a set

$$\mathbf{a} \bowtie \mathbf{b} = \{ c \in \mathbb{X}_{K \cup L} : c^{\downarrow K} \in \mathbf{a} \ \& \ c^{\downarrow L} \in \mathbf{b} \}.$$

---

[4] $K_1, K_2, \ldots, K_n$ meets the running intersection property if

$$\forall i = 2, 3, \ldots, n \ \exists j \ (1 \leq j < i) \ \ K_i \cap (K_1 \cup \ldots \cup K_{i-1}) \subseteq K_j.$$

Realize that if K and L are disjoint, then $\mathbf{a} \bowtie \mathbf{b} = \mathbf{a} \times \mathbf{b}$, if $K = L$, then $\mathbf{a} \bowtie \mathbf{b} = \mathbf{a} \cap \mathbf{b}$, and, generally, for $\mathbf{c} \subseteq \mathbb{X}_{K \cup L}$, $\mathbf{c}$ is a subset of $\mathbf{c}^{\downarrow K} \bowtie \mathbf{c}^{\downarrow L}$, which may be proper. Notice that the sets, for which $\mathbf{c} = \mathbf{c}^{\downarrow K} \bowtie \mathbf{c}^{\downarrow L}$, were called *Z-layered rectangles* in [22,23].

Definition 1. *Factorizing operator of composition*
Consider two arbitrary basic assignments, m_1 and m_2 defined for sets of variables K and L, respectively. A factorizing composition $m_1 \triangleright m_2$ is defined for each nonempty $\mathbf{c} \subseteq \mathbb{X}_{K \cup L}$ by one of the following expressions:

(i) if $m_2^{\downarrow K \cap L}(\mathbf{c}^{\downarrow K \cap L}) > 0$ and $\mathbf{c} = \mathbf{c}^{\downarrow K} \bowtie \mathbf{c}^{\downarrow L}$, then

$$(m_1 \triangleright m_2)(\mathbf{c}) = \frac{m_1(\mathbf{c}^{\downarrow K}) \cdot m_2(\mathbf{c}^{\downarrow L})}{m_2^{\downarrow K \cap L}(\mathbf{c}^{\downarrow K \cap L})};$$

(ii) if $m_2^{\downarrow K \cap L}(\mathbf{c}^{\downarrow K \cap L}) = 0$ and $\mathbf{c} = \mathbf{c}^{\downarrow K} \times \mathbb{X}_{L \setminus K}$, then

$$(m_1 \triangleright m_2)(\mathbf{c}) = m_1(\mathbf{c}^{\downarrow K});$$

(iii) in all other cases, $(m_1 \triangleright m_2)(\mathbf{c}) = 0$.

6   Computational Experiments

As indicated in the Introduction, the goal of the described experiments is to examine which of the considered divergences can be used to (heuristically) detect the best approximations of basic assignments. To do it, we take into account only the approximations by compositional models and accept the intuitively rational and theoretically well-grounded fact that the perfect model, if it exists, is the best approximation.

In the experiments, we considered 14 binary variables ($|N| = 14$), for which we randomly generated 900 basic assignments[5] (denote them m) with 30 focal elements. For each basic assignment we randomly generated a cover of N, i.e., sets $K_1, K_2, \ldots, K_n$, and $N = K_1 \cup K_2 \cup \ldots \cup K_n$ ($5 \leq n \leq 11$, $2 \leq |K_i| \leq 4$). To assure that we can identify the best approximation, we guaranteed that this sequence met the running intersection property (RIP). Due to Proposition 2, we know that $m^{\downarrow K_1} \triangleright m^{\downarrow K_2} \triangleright \ldots \triangleright m^{\downarrow K_n}$ is perfect, and therefore it is the best approximation of m that can be composed of these marginals. To avoid misunderstanding, recall that we study the behavior of the considered divergences, and therefore, we do not mind that most of the considered approximations were much more complex (in the sense of the number of parameters defining the respective belief functions) than the approximated basic assignment.

For each perfect model, we set up also non-perfect models by randomly permuting the marginals in the sequence. Thus, for each of the 900 randomly generated 14-dimensional basic assignments, we had one RIP and several (on average

---

[5] We generated basic assignments of three types: 300 of them were *nested*, 300 were *quasi-bayesian*, and the remaining 300 basic assignments had 29 fully randomly selected focal elements and the thirties one was $\mathbb{X}_N$.

about 6) non-RIP compositional models[6]. The achieved results are summarized in Table 1. From this, the reader can see that for the 900 basic assignments, we considered 6 458 approximating compositional models, 900 of which were perfect, and the remaining 5 558 were non-perfect. On the right-hand side of Table 1, the behavior of the considered distances is described. As *wrongly detected* we considered those perfect approximations $m^{\downarrow K_1} \triangleright \ldots \triangleright m^{\downarrow K_n}$, for which there was generated non-RIP model (defined by a permutation $m^{\downarrow K_{j_1}} \triangleright \ldots \triangleright m^{\downarrow K_{j_n}}$) such that

Table 1. Numbers of wrongly detected approximations.

	Total	wrongly detected by			
		d_{BPA}	BJS	Div_λ	Div_π
Number of perfect approximations	900	348	216	97	9
Number of non-perfect approximations	5 558	1 613	1 007	167	15

$$Div(m, (m^{\downarrow K_{j_1}} \triangleright \ldots \triangleright m^{\downarrow K_{j_n}})) < Div(m, (m^{\downarrow K_1} \triangleright \ldots \triangleright m^{\downarrow K_n})), \quad (11)$$

where Div stands for the respective divergence from Table 1. Analogously, *wrongly detected* non-perfect models are those non-perfect models $m^{\downarrow K_{j_1}} \triangleright \ldots \triangleright m^{\downarrow K_{j_n}}$, for which Eq. (11) holds true. It means that there is a correspondence between wrongly detected perfect and non-perfect models, however, this correspondence is not a bijection. Each wrongly detected perfect model corresponds with at least one (but often more than one) wrongly detected non-perfect model. Notice that if a perfect approximation $m^{\downarrow K_1} \triangleright \ldots \triangleright m^{\downarrow K_n}$ and its non-perfect permutation $m^{\downarrow K_{j_1}} \triangleright \ldots \triangleright m^{\downarrow K_{j_n}}$ were generated, such that the equality $Div(m, (m^{\downarrow K_{j_1}} \triangleright \ldots \triangleright m^{\downarrow K_{j_n}})) = Div(m, (m^{\downarrow K_1} \triangleright \ldots \triangleright m^{\downarrow K_n}))$ hold, none of these two approximations was recognized as wrongly detected.

Though we said in Sect. 4 that we would study six divergences, only four of them appear in Table 1. It is because the remaining divergences BRE and D_{SDM} (defined by Eq. (4) and Eq. (6), respectively) equal $+\infty$ whenever there is a focal element of the approximation, which is not a focal element of the originally randomly generated basic assignment. This, however, cannot be avoided for any multidimensional basic assignment and its compositional-model approximation.

So, it is not surprising that all divergences computed for BRE and D_{SDM} were $+\infty$, which means that they are useless for the purpose of this study.

7   Conclusions

From Table 1 one can deduce that the simple divergences Div_λ and mainly Div_π may be recommended to identify the best approximations of multidimensional

---

[6] Precisely speaking, we know that all RIP models are perfect, but, theoretically, it may happen that also non-RIP model is perfect. However, this happens very rarely, and when assessing the results, we took that all non-RIP models were non-perfect.

basic assignments. However, let us recall that we have achieved this conclusion when considering only approximations by f-compositional models. We have not yet, achieved any results in the case of experiments with the operator of composition derived from Dempster's rule of combination (d-composition). The main reason is the computational complexity of the operator of d-composition, the calculation of which requires conversions of low-dimensional basic assignments, from which the model is set up, from/to the respective commonality functions.

References

1. Cobb, B.R., Shenoy, P.P.: On the plausibility transformation method for translating belief function models to probability models. Int. J. Approximate Reasoning **41**(3), 314–340 (2006)
2. Cuzzolin, F.: On the relative belief transform. Int. J. Approximate Reasoning **53**(5), 786–804 (2012)
3. Daniel, M.: On transformations of belief functions to probabilities. Int. J. Intell. Syst. **21**(3), 261–282 (2006)
4. Jiroušek, R.: Foundations of compositional model theory. Int. J. Gen. Syst. **40**(6), 623–678 (2011)
5. Jiroušek, R.: A short note on decomposition and composition of knowledge. Int. J. Approximate Reasoning **120**, 24–32 (2020)
6. Jiroušek, R., Kratochvíl, V.: Approximations of belief functions using compositional models. In: Vejnarová, J., Wilson, N. (eds.) ECSQARU 2021. LNCS (LNAI), vol. 12897, pp. 354–366. Springer, Cham (2021). https://doi.org/10.1007/978-3-030-86772-0_26
7. Jiroušek, R., Kratochvíl, V., Shenoy, P.P.: Entropy-based learning of compositional models from data. In: Denœux, T., Lefèvre, E., Liu, Z., Pichon, F. (eds.) BELIEF 2021. LNCS (LNAI), vol. 12915, pp. 117–126. Springer, Cham (2021). https://doi.org/10.1007/978-3-030-88601-1_12
8. Jiroušek, R., Shenoy, P.P.: Compositional models in valuation-based systems. In: Denoeux, T., Masson, M.H. (eds.) Belief Functions: Theory and Applications. Advances in Intelligent and Soft Computing, pp. 221–228. Springer, Berlin, Heidelberg (2012). https://doi.org/10.1007/978-3-642-29461-7_26
9. Jiroušek, R., Shenoy, P.P.: A new definition of entropy of belief functions in the Dempster-Shafer theory. Int. J. Approximate Reasoning **92**(1), 49–65 (2018)
10. Jiroušek, R., Vejnarová, J., Daniel, M.: Compositional models for belief functions. In: de Cooman, G., Vejnarová, J., Zaffalon, M. (eds.) Proceedings of the Fifth International Symposium on Imprecise Probability: Theories and Applications (ISIPTA 2007), pp. 243–252 (2007)
11. Jousselme, A.L., Grenier, D., Bossé, É.: A new distance between two bodies of evidence. Inform. Fusion **2**(2), 91–101 (2001)
12. Jousselme, A.L., Maupin, P.: Distances in evidence theory: comprehensive survey and generalizations. Int. J. Approximate Reasoning **53**(2), 118–145 (2012)
13. Kullback, S., Leibler, R.A.: On information and sufficiency. Ann. Math. Stat. **22**, 76–86 (1951)
14. Österreicher, F., Vajda, I.: A new class of metric divergences on probability spaces and its applicability in statistics. Ann. Inst. Stat. Math. **55**(3), 639–653 (2003). https://doi.org/10.1007/BF02517812

15. Perez, A.: ε-admissible simplifications of the dependence structure of a set of random variables. Kybernetika **13**(6), 439–449 (1977)
16. Shafer, G.: A Mathematical Theory of Evidence. Princeton University Press, Princeton (1976)
17. Shannon, C.E.: A mathematical theory of communication. Bell Syst. Tech. J. **27**(379–423), 623–656 (1948)
18. Smets, P.: Constructing the Pignistic probability function in a context of uncertainty. In: Henrion, M., Shachter, R., Kanal, L.N., Lemmer, J.F. (eds.) Uncertainty in Artificial Intelligence 5, pp. 29–40. Elsevier (1990)
19. Song, Y., Deng, Y.: Divergence measure of belief function and its application in data fusion. IEEE Access **7**, 107465–107472 (2019)
20. Song, Y., Deng, Y.: A new method to measure the divergence in evidential sensor data fusion. Int. J. Distrib. Sens. Netw. **15**(4), 1550147719841295 (2019)
21. Xiao, F.: Multi-sensor data fusion based on the belief divergence measure of evidences and the belief entropy. Inf. Fusion **46**, 23–32 (2019)
22. Yaghlane, B.B., Smets, P., Mellouli, K.: Belief function independence: I. the marginal case. Int. J. Approximate Reasoning **29**(1), 47–70 (2002)
23. Yaghlane, B.B., Smets, P., Mellouli, K.: Belief function independence: II. the conditional case. Int. J. Approximate Reasoning **31**(1–2), 31–75 (2002)

The Lattice Structure of Coverings in an Incomplete Information Table with Value Similarity

Michinori Nakata[1(✉)], Norio Saito[1], Hiroshi Sakai[2], and Takeshi Fujiwara[3]

[1] Faculty of Management and Information Science, Josai International University, 1 Gumyo, Togane, Chiba 283-8555, Japan
nakatam@ieee.org, saitoh_norio@jiu.ac.jp
[2] Department of Mathematics and Computer Aided Sciences, Faculty of Engineering, Kyushu Institute of Technology, Kitakyushu, Tobata 804-8550, Japan
sakai@mns.kyutech.ac.jp
[3] Faculty of Informatics, Tokyo University of Information Sciences, 4-1 Onaridai, Wakaba-ku, Chiba 265-8501, Japan
fujiwara@rsch.tuis.ac.jp

Abstract. Based on Lipski's approach dealing with incomplete information tables, we describe lower and upper approximations using coverings under incomplete information and similarity of values. Lots of coverings, called possible coverings, on a set of attributes are derived in an incomplete information table with similarity of values, although the covering is unique in a complete information table. The family of possible coverings has a lattice structure with the minimum and maximum elements. This is true for the family of maximal descriptions, but is not for the family of minimal descriptions and the family of sets of close friends. As was shown by Lipski, what we can obtain from an information table with incomplete information is the lower and upper bounds of information granules. Using only two coverings: the minimum and maximum possible ones, we obtain the lower and upper bounds of lower and upper approximations. Therefore, there is no difficulty of the computational complexity in our approach.

Keywords: Rough sets · Incomplete information · Possible coverings · Possibly indiscernible classes · Lower and upper approximations

1 Introduction

Rough sets, constructed by Pawlak [1], are based on equality of values characterizing objects. The rough sets are used as an effective method for data mining and so on. The framework is usually used under complete information tables with no similarity of objects and creates significant results in various fields. However, value similarity often appears in the real world. Also, incomplete information ubiquitously occurs in the real world. By dealing with value similarity and

K. Honda et al. (Eds.): IUKM 2022, LNAI 13199, pp. 16–28, 2022.
https://doi.org/10.1007/978-3-030-98018-4_2

incomplete information, we can make better use of information obtained from the real world. Therefore, rough sets need to be extended to deal with incomplete information tables with value similarity.

Lipski showed that we can obtain the lower and upper bounds of the answer set of a query to an information table with incomplete information, although we cannot obtain the precise answer set [2]. This means that when trying to extract information granules from an incomplete information table, what is obtained without information loss is the lower and upper bounds of the information granules. This is true for lower and upper approximations that are the core of rough sets. Therefore, what we can obtain is the lower and upper bounds of these approximations.

It is the process proposed by Kryszkiewicz [3] that most authors use to handle incomplete information. The process a priori gives indiscernibility between an object with incomplete information and another object. Using the given indiscernibility, unique approximations are derived. Clearly, the process produces information loss from Lipski's point of view. As a result, the approach creates poor results [4–6].

We develop an approach using possible coverings without a priori giving indiscernibility between objects. First, we describe a structure of possible coverings. We will show that the lower and upper bounds of lower and upper approximations are obtained without the difficulty of computational complexity under the structure.

Lipski used a possible table as a possible world in possible world semantics. Unfortunately, we cannot use the possible table in an incomplete information table with continuous values. So, we showed a way that does not use the possible table under continuous values [7]. Using a similar way, we deal with categorical values. This means that we can deal with categorical and numerical values in the same framework.

2   Coverings in a Complete Information Table

A complete information table is constructed with $(U, \mathcal{A}, \{V(a) \mid a \in \mathcal{A}\})$, where U is the universe that consists of objects. $\mathcal{A}$ is a non-empty finite set of attributes such that $a : U \to V(a)$ for every $a \in \mathcal{A}$ where $V(a)$ is the set of values that attribute a takes.

Binary relation $R_a^{\delta 1}$ expressing indiscernibility of objects on attribute $a \in \mathcal{A}$ is called the indiscernibility relation for a under threshold δ_a.

$$R_a^\delta = \{(o, o') \in U \times U \mid SIM_a(o, o') \geq \delta_a\}, \tag{1}$$

where $SIM_a(o, o')$ is the similarity degree between objects o and o' for attribute a and δ_a is a threshold fixed for attribute a.

$$SIM_a(o, o') = sim(a(o), a(o')), \tag{2}$$

---

[1]   Unless confused, symbols without subscripts or superscripts are used.

where $sim(a(o), a(o'))$ is the similarity degree between $a(o)$ and $a(o')$. $sim(a(o), a(o'))$ is given whose values are reflexive, symmetric, and not transitive. The indiscernibility relation is a tolerance relation[2].

From indiscernibility relation R_a^δ, the indiscernible class $C(o)_a^\delta$ of object o on a is defined:

$$C_a^\delta(o) = \{o' \mid (o, o') \in R_a^\delta\}. \tag{3}$$

$C_a^\delta(o)$ is not an equivalence class.

Family $\mathcal{C}_a^\delta$ of indiscernible classes on attribute a is:

$$\mathcal{C}_a^\delta = \{C \mid C = C_a^\delta(o) \wedge o \in U\}. \tag{4}$$

Clearly, $\cup_{C \in \mathcal{C}_a^\delta} C = U$. Based on Zakowski [9], $\mathcal{C}_a^\delta$ is a covering, which is unique for a. Under $\mathcal{C}_a^\delta$, minimal description $MdC_a^\delta(o)$ of object o, formulated by [10], is:

$$MdC_a^\delta(o) = \{C \in \mathcal{C}_a^\delta \mid o \in C \wedge \forall C' \in \mathcal{C}_a^\delta (o \in C' \wedge C' \subseteq C \Rightarrow C = C')\}. \tag{5}$$

Set $CFriend_{\mathcal{C}_a^\delta}(o)$ of close friends of o with respect to $\mathcal{C}_a^\delta$, proposed by [11], is:

$$CFriend_{\mathcal{C}_a^\delta}(o) = \cup_{C \in MdC_a^\delta(o)} C. \tag{6}$$

Also, maximal description $MDC_a^\delta(o)$ of object o, described by [11,12], is:

$$MDC_a^\delta(o) = \{C \in \mathcal{C}_a^\delta \mid o \in C \wedge \forall C' \in \mathcal{C}_a^\delta (o \in C' \wedge C' \supseteq C \Rightarrow C = C')\}. \tag{7}$$

Using covering $\mathcal{C}_a^\delta$, lower approximation $\underline{apr}_a^\delta(\mathcal{O})$ and upper approximation $\overline{apr}_a^\delta(\mathcal{O})$ for a of set $\mathcal{O}$ of objects are:

$$\underline{apr}_a^\delta(\mathcal{O}) = \{o \in U \mid C_a^\delta(o) \subseteq \mathcal{O} \wedge C_a^\delta(o) \in \mathcal{C}_a^\delta\}, \tag{8}$$

$$\overline{apr}_a^\delta(\mathcal{O}) = \{o \in U \mid C_a^\delta(o) \cap \mathcal{O} \neq \emptyset \wedge C_a^\delta(o) \in \mathcal{C}_a^\delta\}. \tag{9}$$

3   Coverings in an Incomplete Information Table

An incomplete information table has $a : U \to s_a$ for every $a \in \mathcal{A}$ where s_a is the family of disjunctive sets of values over $V(a)$. So, value $v \in a(o)$ is a possible value that may be the actual one of attribute a in object o.

A covering on a is unique in a complete information table, but lots of coverings, called possible coverings, are derived in an incomplete information table [13,14], although some authors deal with only a covering [15–17]. A possible covering is derived from a possible indiscernibility relation. Many possible indiscernibility relations is derived in an incomplete information table. The number of possible indiscernibility relations may grow exponentially as the number of values with incomplete information increases. However, this does not cause any

---

[2] See [8] for properties of tolerance relations.

difficulties due to computational complexity in obtaining the lower and upper bounds of approximations, as is shown later.

Family FPR_a^δ of possible indiscernibility relations, as is shown in [7,18], is constructed using certain pairs and possible pairs of objects. The certain pair surely has the same characteristic value, while the possible pair may have the same characteristic value. Set SR_a^δ of certain pairs on attribute a is:

$$SR_a^\delta = \{(o, o') \in U \times U \mid (o = o') \vee (\forall u \in a(o) \forall v \in a(o')sim(u, v) \geq \delta_a)\}. \quad (10)$$

Set MPR_a^δ of possible pairs on attribute a is:

$$MPR_a^\delta = \{(o, o') \in U \times U \mid \exists u \in a(o) \exists v \in a(o')sim(u, v) \geq \delta_a\} \backslash SR_a^\delta. \quad (11)$$

Using these two sets, family FPR_a^δ of possible indiscernibility relations is:

$$FPR_a^\delta = \{PR \mid PR = SR_a^\delta \cup e \wedge e \in \mathcal{P}(MPPR_a^\delta)\}, \quad (12)$$

where each element is a possible indiscernibility relation and $\mathcal{P}(MPPR_a^\delta)$ is the power set of $MPPR_a^\delta$ that is:

$$MPPR_a^\delta = \{\{(o', o), (o, o')\} \mid (o', o) \in MPR_a^\delta\}. \quad (13)$$

Clearly, FPR_a^δ is a lattice for set inclusion. SR_a is the minimum possible indiscernibility relation in FPR_a^δ, whereas $SR_a^\delta \cup MPR_a^\delta$ is the maximum possible indiscernibility relation. All the possible indiscernibility relations do not correspond to the indiscernibility relation derived from a possible table where every attribute value is replaced by a possible value in the original information table. The possible indiscernibility relation without a corresponding possible table is artificial. However, the minimum and the maximum possible indiscernibility relations are equal to the intersection and the union of indiscernibility relations derived from possible tables, respectively. The minimum possible indiscernibility relation contains only the pairs of objects that are surely indiscernible with each other, while the maximum possible indiscernibility relation contains all the pairs that are possibly indiscernible. Only the two possible indiscernibility relations are used to derive the lower and upper bounds of approximations, as is shown later. Artificially possible indiscernibility relations are rather useful to derive the lower and upper bounds of approximations.

Example 1. Let similarity degree $sim(u, v)$ on $V(a_1) = \{a, b, c, d, e, f\}$ and incomplete information table IT be as follows:

$$sim(u, v) = \begin{pmatrix} 1 & 0.9 & 0.9 & 0.6 & 0.2 & 0.4 \\ 0.9 & 1 & 0.8 & 0.8 & 0.1 & 0.5 \\ 0.9 & 0.8 & 1 & 0.3 & 0.2 & 0.4 \\ 0.6 & 0.8 & 0.3 & 1 & 0.9 & 0.6 \\ 0.2 & 0.1 & 0.2 & 0.9 & 1 & 0.7 \\ 0.4 & 0.5 & 0.4 & 0.6 & 0.7 & 1 \end{pmatrix}.$$

IT

U	a_1	a_2
o_1	$< a >$	$< x >$
o_2	$< b, e >$	$< x, y >$
o_3	$< c >$	$< x >$
o_4	$< d >$	$< y >$
o_5	$< e >$	$< z >$
o_6	$< f >$	$< z >$

In incomplete information table IT with $U = \{o_1, o_2, o_3, o_4, o_5, o_6\}$, let threshold δ_{a_1} be 0.75 on attribute a_1. Expression $< b, e >$ of a disjunctive set means that the actual value is b or e. The set of certain pairs of indiscernible objects on a_1 under the above $sim(u, v)$ is $\{(o_1, o_1), (o_1, o_3), (o_2, o_2), (o_2, o_4), (o_3, o_3), (o_3, o_1),$ $(o_4, o_4), (o_4, o_2), (o_4, o_5), (o_5, o_5), (o_5, o_4), (o_6, o_6)\}$. The set of possible pairs of indiscernible objects is $\{(o_1, o_2), (o_2, o_1), (o_2, o_3), (o_3, o_2), (o_2, o_5), (o_5, o_2)\}$. Using formulae (10)–(13), the family of possible indiscernibility relations is obtained: $PR_{a_1}^{0.75} = \{PR_1, \cdots, PR_8\}$, and 8 possible indiscernibility relations are:

$PR_1 = \{(o_1, o_1), (o_1, o_3), (o_2, o_2), (o_2, o_4), (o_3, o_3), (o_3, o_1), (o_4, o_4), (o_4, o_2), (o_4, o_5),$
$\qquad (o_5, o_5), (o_5, o_4), (o_6, o_6)\},$

$PR_2 = \{(o_1, o_1), (o_1, o_3), (o_2, o_2), (o_2, o_4), (o_3, o_3), (o_3, o_1), (o_4, o_4), (o_4, o_2), (o_4, o_5),$
$\qquad (o_5, o_5), (o_5, o_4), (o_6, o_6), (o_1, o_2), (o_2, o_1)\},$

$PR_3 = \{(o_1, o_1), (o_1, o_3), (o_2, o_2), (o_2, o_4), (o_3, o_3), (o_3, o_1), (o_4, o_4), (o_4, o_2), (o_4, o_5),$
$\qquad (o_5, o_5), (o_5, o_4), (o_6, o_6), (o_2, o_3), (o_3, o_2)\},$

$PR_4 = \{(o_1, o_1), (o_1, o_3), (o_2, o_2), (o_2, o_4), (o_3, o_3), (o_3, o_1), (o_4, o_4), (o_4, o_2), (o_4, o_5),$
$\qquad (o_5, o_5), (o_5, o_4), (o_6, o_6), (o_2, o_5), (o_5, o_2)\},$

$PR_5 = \{(o_1, o_1), (o_1, o_3), (o_2, o_2), (o_2, o_4), (o_3, o_3), (o_3, o_1), (o_4, o_4), (o_4, o_2), (o_4, o_5),$
$\qquad (o_5, o_5), (o_5, o_4), (o_6, o_6), (o_1, o_2), (o_2, o_1), (o_2, o_3), (o_3, o_2)\},$

$PR_6 = \{(o_1, o_1), (o_1, o_3), (o_2, o_2), (o_2, o_4), (o_3, o_3), (o_3, o_1), (o_4, o_4), (o_4, o_2), (o_4, o_5),$
$\qquad (o_5, o_5), (o_5, o_4), (o_6, o_6), (o_1, o_2), (o_2, o_1), (o_2, o_5), (o_5, o_2)\},$

$PR_7 = \{(o_1, o_1), (o_1, o_3), (o_2, o_2), (o_2, o_4), (o_3, o_3), (o_3, o_1), (o_4, o_4), (o_4, o_2), (o_4, o_5),$
$\qquad (o_5, o_5), (o_5, o_4), (o_6, o_6), (o_2, o_3), (o_3, o_2), (o_2, o_5), (o_5, o_2)\},$

$PR_8 = \{(o_1, o_1), (o_1, o_3), (o_2, o_2), (o_2, o_4), (o_3, o_3), (o_3, o_1), (o_4, o_4), (o_4, o_2), (o_4, o_5),$
$\qquad (o_5, o_5), (o_5, o_4), (o_6, o_6), (o_1, o_2), (o_2, o_1), (o_2, o_3), (o_3, o_2), (o_2, o_5), (o_5, o_2)\}.$

From each possible indiscernibility relation $PR_{a,j}^{\delta}$ in FPR_a^{δ}, possible indiscernible class $C(o)_{a,j}^{\delta}$ on attribute a for object o is:

$$C(o)_{a,j}^{\delta} = \{o' \mid (o, o') \in PR_{a,j}^{\delta} \wedge PR_{a,j}^{\delta} \in FPR_a^{\delta}\}. \tag{14}$$

Proposition 1. If $PR_{a,k}^{\delta} \subseteq PR_{a,l}^{\delta}$, then $C(o)_{a,k}^{\delta} \subseteq C(o)_{a.l}^{\delta}$.

From this proposition, the family of possible indiscernible classes for an object is a lattice for set inclusion.

Example 2. (continuation from Example 1). For object o_1, $C(o_1)_{a_1,j}^{0.75} = \{o_1, o_3\}$ $for\, j = 1, 3, 4, 7$, $C(o_1)_{a_1,j}^{0.75} = \{o_1, o_2, o_3\}$ $for\, j = 2, 5, 6, 8$. For object o_2, $C(o_2)_{a_1,1}^{0.75} = \{o_2, o_4\}$, $C(o_2)_{a_1,2}^{0.75} = \{o_1, o_2, o_4\}$, $C(o_2)_{a_1,3}^{0.75} = \{o_2, o_3, o_4\}$, $C(o_2)_{a_1,4}^{0.75} = \{o_2, o_4, o_5\}$, $C(o_2)_{a_1,5}^{0.75} = \{o_1, o_2, o_3, o_4\}$, $C(o_2)_{a_1,6}^{0.75} = \{o_1, o_2, o_4, o_5\}$, $C(o_2)_{a_1,7}^{0.75} = \{o_2, o_3, o_4, o_5\}$, $C(o_2)_{a_1,8}^{0.75} = \{o_1, o_2, o_3, o_4, o_5\}$. For object o_3, $C(o_3)_{a_1,j}^{0.75} = \{o_1, o_3\}$ $for\, j = 1, 2, 4, 6$, $C(o_3)_{a_1,j}^{0.75} = \{o_1, o_2, o_3\}$ $for\, j = 3, 5, 7, 8$.

For object o_4, $C(o_4)^{0.75}_{a_1,j} = \{o_2, o_4, o_5\}$ *for* $j = 1, \ldots, 8$. For object o_5, $C(o_5)^{0.75}_{a_1,j} =$ $\{o_4, o_5\}$ *for* $j = 1, 2, 3, 5$, $C(o_5)^{0.75}_{a_1,j} = \{o_2, o_4, o_5\}$ *for* $j = 4, 6, 7, 8$, For object o_6, $C(o_6)^{0.75}_{a_1,j} = \{o_6\}$ *for* $j = 1, \ldots, 8$.

A possible covering is derived from a possible indiscernibility relation. Possible covering $PC^\delta_{a,j}$ obtained from possible indiscernibility relation $PR^\delta_{a,j}$ is:

$$PC^\delta_{a,j} = \{e \mid e = C(o)^\delta_{a,j} \wedge o \in U\}. \tag{15}$$

One of possible coverings is the actual covering, although we cannot know it without additional information.

Proposition 2. If $PR^\delta_{a,k} \subseteq PR^\delta_{a,l}$, then $PC^\delta_{a,k} \sqsubseteq PC^\delta_{a,l}$.[3]

From Proposition 2 family FPC^δ_a of possible coverings is a lattice for $\sqsubseteq$.

Proposition 3. If $PR^\delta_{a,k} \subseteq PR^\delta_{a,l}$, then $\forall o \in U \ MDC^\delta_{a,k}(o) \subseteq MDC^\delta_{a,l}(o)$ where $MDC^\delta_{a,k}(o)$ is the maximal description of o with respect to PC^δ_a in $PR^\delta_{k,a}$.

From Proposition 3 family $FMDC^\delta_a(o)$ of maximal descriptions is a lattice for $\subseteq$.

Example 3. Possibly indiscernible classes of objects are obtained in each possible indiscernibility relation PR_i with $i = 1, \ldots, 8$ of Example 1.

In PR_1, $C(o_1)_{a_1} = \{o_1, o_3\}$, $C(o_2)_{a_1} = \{o_2, o_4\}$, $C(o_3)_{a_1} = \{o_1, o_3\}$, $C(o_4)_{a_1} = \{o_2, o_4, o_5\}$, $C(o_5)_{a_1} = \{o_4, o_5\}$, and $C(o_6)_{a_1} = \{o_6\}$.
In PR_2, $C(o_1)_{a_1} = \{o_1, o_2, o_3\}$, $C(o_2)_{a_1} = \{o_1, o_2, o_4\}$, $C(o_3)_{a_1} = \{o_1, o_3\}$, $C(o_4)_{a_1} = \{o_2, o_4, o_5\}$, $C(o_5)_{a_1} = \{o_4, o_5\}$, and $C(o_6)_{a_1} = \{o_6\}$.
In PR_3, $C(o_1)_{a_1} = \{o_1, o_3\}$, $C(o_2)_{a_1} = \{o_2, o_3, o_4\}$, $C(o_3)_{a_1} = \{o_1, o_2, o_3\}$, $C(o_4)_{a_1} = \{o_2, o_4, o_5\}$, $C(o_5)_{a_1} = \{o_4, o_5\}$, and $C(o_6)_{a_1} = \{o_6\}$.
In PR_4, $C(o_1)_{a_1} = \{o_1, o_3\}$, $C(o_2)_{a_1} = \{o_2, o_4, o_5\}$, $C(o_3)_{a_1} = \{o_1, o_3\}$, $C(o_4)_{a_1} = \{o_2, o_4, o_5\}$, $C(o_5)_{a_1} = \{o_2, o_4, o_5\}$, and $C(o_6)_{a_1} = \{o_6\}$.
In PR_5, $C(o_1)_{a_1} = \{o_1, o_2, o_3\}$, $C(o_2)_{a_1} = \{o_1, o_2, o_3, o_4\}$, $C(o_3)_{a_1} = \{o_1, o_2, o_3\}$, $C(o_4)_{a_1} = \{o_2, o_4, o_5\}$, $C(o_5)_{a_1} = \{o_4, o_5\}$, and $C(o_6)_{a_1} = \{o_6\}$.
In PR_6, $C(o_1)_{a_1} = \{o_1, o_2, o_3\}$, $C(o_2)_{a_1} = \{o_1, o_2, o_4, o_5\}$, $C(o_3)_{a_1} = \{o_1, o_3\}$, $C(o_4)_{a_1} = \{o_2, o_4, o_5\}$, $C(o_5)_{a_1} = \{o_2, o_4, o_5\}$, and $C(o_6)_{a_1} = \{o_6\}$.
In PR_7, $C(o_1)_{a_1} = \{o_1, o_3\}$, $C(o_2)_{a_1} = \{o_2, o_3, o_4, o_5\}$, $C(o_3)_{a_1} = \{o_1, o_2, o_3\}$, $C(o_4)_{a_1} = \{o_2, o_4, o_5\}$, $C(o_5)_{a_1} = \{o_2, o_4, o_5\}$, and $C(o_6)_{a_1} = \{o_6\}$.
In PR_8, $C(o_1)_{a_1} = \{o_1, o_2, o_3\}$, $C(o_2)_{a_1} = \{o_1, o_2, o_3, o_4, o_5\}$, $C(o_3)_{a_1} = \{o_1, o_2, o_3\}$, $C(o_4)_{a_1} = \{o_2, o_4, o_5\}$, $C(o_5)_{a_1} = \{o_2, o_4, o_5\}$, and $C(o_6)_{a_1} = \{o_6\}$.

---

[3] $\sqsubseteq$ is defined as $\mathcal{E} \sqsubseteq \mathcal{E}'$ if $\forall E \in \mathcal{E} \exists E' \in \mathcal{E}' \wedge E \subseteq E'$.

Using these possibly indiscernible classes, possible coverings are obtained as follows:

$$PC_1 = \{\{o_1, o_3\}, \{o_2, o_4\}, \{o_2, o_4, o_5\}, \{o_4, o_5\}, \{o_6\}\},$$
$$PC_2 = \{\{o_1, o_2, o_3\}, \{o_1, o_2, o_4\}, \{o_1, o_3\}, \{o_2, o_4, o_5\}, \{o_4, o_5\}, \{o_6\}\},$$
$$PC_3 = \{\{o_1, o_3\}, \{o_2, o_3, o_4\}, \{o_1, o_2, o_3\}, \{o_2, o_4, o_5\}, \{o_4, o_5\}, \{o_6\}\},$$
$$PC_4 = \{\{o_1, o_3\}, \{o_2, o_4, o_5\}, \{o_6\}\},$$
$$PC_5 = \{\{o_1, o_2, o_3\}, \{o_1, o_2, o_3, o_4\}, \{o_2, o_4, o_5\}, \{o_4, o_5\}, \{o_6\}\},$$
$$PC_6 = \{\{o_1, o_2, o_3\}, \{o_1, o_2, o_4, o_5\}, \{o_1, o_3\}, \{o_2, o_4, o_5\}, \{o_6\}\},$$
$$PC_7 = \{\{o_1, o_3\}, \{o_2, o_3, o_4, o_5\}, \{o_1, o_2, o_3\}, \{o_2, o_4, o_5\}, \{o_6\}\},$$
$$PC_8 = \{\{o_1, o_2, o_3\}, \{o_1, o_2, o_3, o_4, o_5\}, \{o_2, o_4, o_5\}, \{o_6\}\}.$$

Minimal descriptions, sets of close friends, and maximal descriptions are as follows:

For PC_1, $MdC(o_1) = \{\{o_1, o_3\}\}$, $MdC(o_2) = \{\{o_2, o_4\}\}$, $MdC(o_3) = \{\{o_1, o_3\}\}$,
$\quad MdC(o_4) = \{\{o_2, o_4\}, \{o_4, o_5\}\}$, $MdC(o_5) = \{\{o_4, o_5\}\}$,
$\quad MdC(o_6) = \{\{o_6\}\}$,
$CFriend_C(o_1) = \{o_1, o_3\}$, $CFriend_C(o_2) = \{o_2, o_4\}$,
$CFriend_C(o_3) = \{o_1, o_3\}$, $CFriend_C(o_4) = \{o_2, o_4, o_5\}$,
$CFriend_C(o_5) = \{o_4, o_5\}$, $CFriend_C(o_6) = \{o_6\}$,
$\quad MDC(o_1) = \{\{o_1, o_3\}\}$, $MDC(o_2) = \{\{o_2, o_4, o_5\}\}$, $MDC(o_3) = \{\{o_1, o_3\}\}$,
$\quad MDC(o_4) = \{\{o_2, o_4, o_5\}\}$, $MDC(o_5) = \{\{o_2, o_4, o_5\}\}$,
$\quad MDC(o_6) = \{\{o_6\}\}$,
For PC_2, $MdC(o_1) = \{\{o_1, o_2, o_4\}, \{o_1, o_3\}\}$,
$\quad MdC(o_2) = \{\{o_1, o_2, o_3\}, \{o_1, o_2, o_4\}, \{o_2, o_4, o_5\}\}$,
$\quad MdC(o_3) = \{\{o_1, o_3\}\}$, $MdC(o_4) = \{\{o_1, o_2, o_4\}, \{o_4, o_5\}\}$,
$\quad MdC(o_5) = \{\{o_4, o_5\}\}$, $MdC(o_6) = \{\{o_6\}\}$,
$CFriend_C(o_1) = \{o_1, o_2, o_3, o_4\}\}$, $CFriend_C(o_2) = \{o_1, o_2, o_3, o_4, o_5\}$,
$CFriend_C(o_3) = \{o_1, o_3\}$, $CFriend_C(o_4) = \{o_1, o_2, o_4, o_5\}$,
$CFriend_C(o_5) = \{o_4, o_5\}$, $CFriend_C(o_6) = \{o_6\}$,
$\quad MDC(o_1) = \{\{o_1, o_2, o_4\}, \{o_1, o_2, o_3\}\}$,
$\quad MDC(o_2) = \{\{o_1, o_2, o_3\}, \{o_1, o_2, o_4\}, \{o_2, o_4, o_5\}\}$,
$\quad MDC(o_3) = \{\{o_1, o_2, o_3\}\}$, $MDC(o_4) = \{\{o_1, o_2, o_4\}, \{o_2, o_4, o_5\}\}$,
$\quad MDC(o_5) = \{\{o_2, o_4, o_5\}\}$, $MDC(o_6) = \{\{o_6\}\}$,
For PC_3, $MdC(o_1) = \{\{o_1, o_3\}\}$, $MdC(o_2) = \{\{o_2, o_3, o_4\}, \{o_1, o_2, o_3\}, \{o_2, o_4, o_5\}\}$,
$\quad MdC(o_3) = \{\{o_1, o_3\}, \{o_2, o_3, o_4\}\}$, $MdC(o_4) = \{\{o_2, o_3, o_4\}, \{o_4, o_5\}\}$,
$\quad MdC(o_5) = \{\{o_4, o_5\}\}$, $MdC(o_6) = \{\{o_6\}\}$,
$CFriend_C(o_1) = \{o_1, o_3\}$, $CFriend_C(o_2) = \{o_1, o_2, o_3, o_4, o_5\}$,
$CFriend_C(o_3) = \{o_1, o_2, o_3, o_4\}$, $CFriend_C(o_4) = \{o_2, o_3, o_4, o_5\}$,
$CFriend_C(o_5) = \{o_4, o_5\}$, $CFriend_C(o_6) = \{o_6\}$,
$\quad MDC(o_1) = \{\{o_1, o_2, o_3\}\}$, $MDC(o_2) = \{\{o_2, o_3, o_4\}, \{o_1, o_2, o_3\}, \{o_2, o_4, o_5\}\}$,
$\quad MDC(o_3) = \{\{o_2, o_3, o_4\}, \{o_1, o_2, o_3\}\}$,

$$MDC(o_4) = \{\{o_2, o_3, o_4\}, \{o_2, o_4, o_5\}\},$$
$$MDC(o_5) = \{\{o_2, o_4, o_5\}\}, MDC(o_6) = \{\{o_6\}\},$$

For PC_4, $Md\mathcal{C}(o_1) = \{\{o_1, o_3\}\}$, $Md\mathcal{C}(o_2) = \{\{o_2, o_4, o_5\}\}$, $Md\mathcal{C}(o_3) = \{\{o_1, o_3\}\}$,
$$Md\mathcal{C}(o_4) = \{\{o_2, o_4, o_5\}\}, Md\mathcal{C}(o_5) = \{\{o_2, o_4, o_5\}\}, Md\mathcal{C}(o_6) = \{\{o_6\}\},$$
$$CFriend_{\mathcal{C}}(o_1) = \{o_1, o_3\}, CFriend_{\mathcal{C}}(o_2) = \{o_2, o_4, o_5\},$$
$$CFriend_{\mathcal{C}}(o_3) = \{o_1, o_3\}, CFriend_{\mathcal{C}}(o_4) = \{o_2, o_4, o_5\},$$
$$CFriend_{\mathcal{C}}(o_5) = \{o_2, o_4, o_5\}, CFriend_{\mathcal{C}}(o_6) = \{o_6\},$$
$$MDC(o_1) = \{\{o_1, o_3\}\}, MDC(o_2) = \{\{o_2, o_4, o_5\}\}, MDC(o_3) = \{\{o_1, o_3\}\},$$
$$MDC(o_4) = \{\{o_2, o_4, o_5\}\}, MDC(o_5) = \{\{o_2, o_4, o_5\}\}, MDC(o_6) = \{\{o_6\}\},$$

For PC_5, $Md\mathcal{C}(o_1) = \{\{o_1, o_2, o_3\}\}$, $Md\mathcal{C}(o_2) = \{\{o_1, o_2, o_3\}, \{o_2, o_4, o_5\}\}$,
$$Md\mathcal{C}(o_3) = \{\{o_1, o_2, o_3\}\}, Md\mathcal{C}(o_4) = \{\{o_1, o_2, o_3, o_4\}, \{o_4, o_5\}\},$$
$$Md\mathcal{C}(o_5) = \{\{o_4, o_5\}\}, Md\mathcal{C}(o_6) = \{\{o_6\}\},$$
$$CFriend_{\mathcal{C}}(o_1) = \{o_1, o_2, o_3\}, CFriend_{\mathcal{C}}(o_2) = \{o_1, o_2, o_3, o_4, o_5\},$$
$$CFriend_{\mathcal{C}}(o_3) = \{o_1, o_2, o_3\}, CFriend_{\mathcal{C}}(o_4) = \{o_1, o_2, o_3, o_4, o_5\},$$
$$CFriend_{\mathcal{C}}(o_5) = \{o_4, o_5\}, CFriend_{\mathcal{C}}(o_6) = \{o_6\},$$
$$MDC(o_1) = \{\{o_1, o_2, o_3, o_4\}\}, MDC(o_2) = \{\{o_1, o_2, o_3, o_4\}, \{o_2, o_4, o_5\}\},$$
$$MDC(o_3) = \{\{o_1, o_2, o_3, o_4\}\}, MDC(o_4) = \{\{o_1, o_2, o_3, o_4\}, \{o_2, o_4, o_5\}\},$$
$$MDC(o_5) = \{\{o_2, o_4, o_5\}\}, MDC(o_6) = \{\{o_6\}\},$$

For PC_6, $Md\mathcal{C}(o_1) = \{\{o_1, o_3\}, \{o_1, o_2, o_4, o_5\}\}$,
$$Md\mathcal{C}(o_2) = \{\{o_1, o_2, o_3\}, \{o_2, o_4, o_5\}\}, Md\mathcal{C}(o_3) = \{\{o_1, o_3\}\},$$
$$Md\mathcal{C}(o_4) = \{\{o_2, o_4, o_5\}\}, Md\mathcal{C}(o_5) = \{\{o_2, o_4, o_5\}\}, Md\mathcal{C}(o_6) = \{\{o_6\}\},$$
$$CFriend_{\mathcal{C}}(o_1) = \{o_1, o_2, o_3, o_4, o_5\}, CFriend_{\mathcal{C}}(o_2) = \{o_1, o_2, o_3, o_4, o_5\},$$
$$CFriend_{\mathcal{C}}(o_3) = \{o_1, o_3\}, CFriend_{\mathcal{C}}(o_4) = \{o_2, o_4, o_5\},$$
$$CFriend_{\mathcal{C}}(o_5) = \{o_2, o_4, o_5\}, CFriend_{\mathcal{C}}(o_6) = \{\{o_6\}\},$$
$$MDC(o_1) = \{\{o_1, o_2, o_3\}, \{o_1, o_2, o_4, o_5\}\},$$
$$MDC(o_2) = \{\{o_1, o_2, o_3\}, \{o_1, o_2, o_4, o_5\}\}, MDC(o_3) = \{\{o_1, o_2, o_3\}\},$$
$$MDC(o_4) = \{\{o_1, o_2, o_4, o_5\}\}, MDC(o_5) = \{\{o_1, o_2, o_4, o_5\}\}, MDC(o_6) = \{\{o_6\}\},$$

For PC_7, $Md\mathcal{C}(o_1) = \{\{o_1, o_3\}\}$, $Md\mathcal{C}(o_2) = \{\{o_1, o_2, o_3\}, \{o_2, o_4, o_5\}\}$,
$$Md\mathcal{C}(o_3) = \{\{o_1, o_3\}, \{o_2, o_3, o_4, o_5\}\}, Md\mathcal{C}(o_4) = \{\{o_2, o_4, o_5\}\},$$
$$Md\mathcal{C}(o_5) = \{\{o_2, o_4, o_5\}\}, Md\mathcal{C}(o_6) = \{\{o_6\}\},$$
$$CFriend_{\mathcal{C}}(o_1) = \{o_1, o_3\}, CFriend_{\mathcal{C}}(o_2) = \{o_1, o_2, o_3, o_4, o_5\},$$
$$CFriend_{\mathcal{C}}(o_3) = \{o_1, o_2, o_3, o_4, o_5\}, CFriend_{\mathcal{C}}(o_4) = \{o_2, o_4, o_5\},$$
$$CFriend_{\mathcal{C}}(o_5) = \{o_2, o_4, o_5\}, CFriend_{\mathcal{C}}(o_6) = \{o_6\},$$
$$MDC(o_1) = \{\{o_1, o_2, o_3\}\}, MDC(o_2) = \{\{o_1, o_2, o_3\}, \{o_2, o_3, o_4, o_5\}\},$$
$$MDC(o_3) = \{\{o_1, o_2, o_3\}, \{o_2, o_3, o_4, o_5\}\}, MDC(o_4) = \{\{o_2, o_3, o_4, o_5\}\},$$
$$MDC(o_5) = \{\{o_2, o_3, o_4, o_5\}\}, MDC(o_6) = \{\{o_6\}\},$$

For PC_8, $Md\mathcal{C}(o_1) = \{\{o_1, o_2, o_3\}\}$, $Md\mathcal{C}(o_2) = \{\{o_1, o_2, o_3\}, \{o_2, o_4, o_5\}\}$,
$$Md\mathcal{C}(o_3) = \{\{o_1, o_2, o_3\}\}, Md\mathcal{C}(o_4) = \{\{o_2, o_4, o_5\}\},$$
$$Md\mathcal{C}(o_5) = \{\{o_2, o_4, o_5\}\}, Md\mathcal{C}(o_6) = \{\{o_6\}\}.$$
$$CFriend_{\mathcal{C}}(o_1) = \{o_1, o_2, o_3\}, CFriend_{\mathcal{C}}(o_2) = \{o_1, o_2, o_3, o_4, o_5\},$$
$$CFriend_{\mathcal{C}}(o_3) = \{o_1, o_2, o_3\}, CFriend_{\mathcal{C}}(o_4) = \{o_2, o_4, o_5\},$$
$$CFriend_{\mathcal{C}}(o_5) = \{o_2, o_4, o_5\}, CFriend_{\mathcal{C}}(o_6) = \{o_6\}.$$

$$MDC(o_1) = MDC(o_2) = MDC(o_3) = MDC(o_4) = MDC(o_5)$$
$$= \{\{o_1, o_2, o_3, o_4, o_5\}\}, MDC(o_6) = \{\{o_6\}\}.$$

The family of possible coverings in Example 3 has the lattice structure for $\sqsubseteq$, which is shown in Fig. 1.

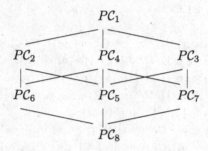

Fig. 1. Lattice structure

PC_1 is the minimum element, whereas PC_8 is the maximum element. On the other hand, the family of minimum descriptions is not a lattice for $\sqsubseteq$; for example, as is clarified for minimum descriptions in PC_6 and PC_8 in Example 3. Also, the family of sets of close friends of an object is not so.

By using possible covering $PC_{a,j}^\delta$, lower and upper approximations of set $\mathcal{O}$ of objects in $PR_{a,j}^\delta$ are:

$$\underline{apr}_{a,j}^\delta(\mathcal{O}) = \{o \in U \mid C_{a,j}^\delta(o) \subseteq \mathcal{O} \wedge C_{a,j}^\delta(o) \in PC_{a,j}^\delta \wedge PC_{a,j}^\delta \in FPC_a^\delta\}, \quad (16)$$

$$\overline{apr}_{a,j}^\delta(\mathcal{O}) = \{o \in U \mid C_{a,j}^\delta(o) \cap \mathcal{O} \neq \emptyset \wedge C_{a,j}^\delta(o) \in PC_{a,j}^\delta \wedge PC_j^\delta \in FPC_a^\delta\}. \quad (17)$$

Proposition 4. If $PC_{a,k} \sqsubseteq PC_{a,l}^\delta$ for possible indiscernibility relations $PC_{a,k}^\delta, PC_{a,l}^\delta \in FPC_a^\delta$, then $\underline{apr}_{a,k}^\delta(\mathcal{O}) \supseteq \underline{apr}_{a,l}^\delta(\mathcal{O})$, and $\overline{apr}_{a,k}^\delta(\mathcal{O}) \subseteq \overline{apr}_{a,l}^\delta(\mathcal{O})$.

This proposition shows that the families of lower and upper approximations under possible coverings are also lattices for set inclusion, respectively. Unfortunately this does not hold in approximations using minimal descriptions and sets of close friends, although various types of covering-based approximation are proposed [11, 19–21].

We aggregate the lower and upper approximations under possible coverings. Certain lower approximation $\underline{Sapr}_a^\delta(\mathcal{O})$ of set $\mathcal{O}$ of objects, the lower bound of the lower approximation, is:

$$\underline{Sapr}_a^\delta(\mathcal{O}) = \{o \in U \mid \forall PC_{a,j}^\delta \in FPC_a^\delta \; o \in \underline{apr}_{a,j}^\delta(\mathcal{O})\}. \quad (18)$$

Possible lower approximation $\underline{Papr}_a^\delta(\mathcal{O})$, the upper bound of the lower approximation, is:

$$\underline{Papr}_a^\delta(\mathcal{O}) = \{o \in U \mid \exists PC_{a,j}^\delta \in FPC_a^\delta \; o \in \underline{apr}_{a,j}^\delta(\mathcal{O})\}. \quad (19)$$

Certain upper approximations $\overline{Sapr}_a^\delta(\mathcal{O})$, the lower bound of the upper approximation, is:

$$\overline{Sapr}_a^\delta(\mathcal{O}) = \{o \in U \mid \forall PC_{a,j}^\delta \in FPC_a^\delta \ o \in \overline{apr}_{a,j}^\delta(\mathcal{O})\}. \tag{20}$$

Possible upper approximation $\overline{Papr}_a^\delta(\mathcal{O})$, the upper bound of the upper approximation, is:

$$\overline{Papr}_a^\delta(\mathcal{O}) = \{o \in U \mid \exists PC_{a,j}^\delta \in FPC_a^\delta \ o \in \overline{apr}_{a,j}^\delta(\mathcal{O})\}. \tag{21}$$

Using Proposition 4, these approximations are transformed into the following formulae:

$$\underline{Sapr}_a^\delta(\mathcal{O}) = \underline{apr}_{a,max}^\delta(\mathcal{O}), \ \underline{Papr}_a^\delta(\mathcal{O}) = \underline{apr}_{a,min}^\delta(\mathcal{O}), \tag{22}$$

$$\overline{Sapr}_a^\delta(\mathcal{O}) = \overline{apr}_{a,min}^\delta(\mathcal{O}), \ \overline{Papr}_a^\delta(\mathcal{O}) = \overline{apr}_{a,max}^\delta(\mathcal{O}), \tag{23}$$

where $\underline{apr}_{a,max}^\delta(\mathcal{O})$ is the lower approximations under the maximum possible covering deriving from the maximum indiscernibility relation and $\overline{apr}_{a,min}^\delta(\mathcal{O})$ is the upper approximations under the minimum possible covering deriving from the minimum indiscernibility relation. These formulae show that we can obtain the lower and upper bounds of approximations without computational complexity, no matter how many possible coverings.

Example 4. We go back to Example 3. Let set $\mathcal{O}$ of objects be $\{o_1, o_3\}$. Using formulae (16) and (17), lower and upper approximations are obtained in each possible covering. For PC_1, $\underline{apr}_{a_1,1}^{0.75}(\mathcal{O}) = \{o_1, o_3\}$, $\overline{apr}_{a_1,1}^{0.75}(\mathcal{O}) = \{o_1, o_3\}$.

For PC_2, $\underline{apr}_{a_1,2}^{0.75}(\mathcal{O}) = \{o_3\}$, $\overline{apr}_{a_1,2}^{0.75}(\mathcal{O}) = \{o_1, o_2, o_3\}$.
For PC_3, $\underline{apr}_{a_1,3}^{0.75}(\mathcal{O}) = \{o_1\}$, $\overline{apr}_{a_1,3}^{0.75}(\mathcal{O}) = \{o_1, o_2, o_3\}$.
For PC_4, $\underline{apr}_{a_1,4}^{0.75}(\mathcal{O}) = \{o_1, o_3\}$, $\overline{apr}_{a_1,4}^{0.75}(\mathcal{O}) = \{o_1, o_3\}$.
For PC_5, $\underline{apr}_{a_1,5}^{0.75}(\mathcal{O}) = \emptyset$, $\overline{apr}_{a_1,5}^{0.75}(\mathcal{O}) = \{o_1, o_2, o_3\}$.
For PC_6, $\underline{apr}_{a_1,6}^{0.75}(\mathcal{O}) = \{o_3\}$, $\overline{apr}_{a_1,6}^{0.75}(\mathcal{O}) = \{o_1, o_2, o_3\}$.
For PC_7, $\underline{apr}_{a_1,7}^{0.75}(\mathcal{O}) = \{o_1\}$, $\overline{apr}_{a_1,7}^{0.75}(\mathcal{O}) = \{o_1, o_2, o_3\}$.
For PC_8, $\underline{apr}_{a_1,8}^{0.75}(\mathcal{O}) = \emptyset$, $\overline{apr}_{a_1,8}^{0.75}(\mathcal{O}) = \{o_1, o_2, o_3\}$.

By using formulae (22) and (23), $\underline{Sapr}_{a_1}^{0.75}(\mathcal{O}) = \emptyset$, $\underline{Papr}_{a_1}^{0.75}(\mathcal{O}) = \{o_1, o_3\}$, $\overline{Sapr}_{a_1}^{0.75}(\mathcal{O}) = \{o_1, o_3\}$, $\overline{Papr}_{a_1}^{0.75}(\mathcal{O}) = \{o_1, o_2, o_3\}$.

Using the lower and upper bounds of approximations denoted by formulae (22) and (23), lower and upper approximations are expressed in interval sets. Certain and possible approximations are the lower and upper bounds of the actual approximation.

Furthermore, the following proposition is valid from formulae (22) and (23).

Proposition 5.

$$\underline{Sapr}_a^\delta(\mathcal{O}) = \{o \mid C(o)_{a,max}^\delta \subseteq \mathcal{O}\}, \ \underline{Papr}_a^\delta(\mathcal{O}) = \{o \mid C(o)_{a,min}^\delta \subseteq \mathcal{O}\},$$
$$\overline{Sapr}_a^\delta(\mathcal{O}) = \{o \mid C(o)_{a,min}^\delta \cap \mathcal{O} \neq \emptyset\}, \ \overline{Papr}_a^\delta(\mathcal{O}) = \{o \mid C(o)_{a,max}^\delta \cap \mathcal{O} \neq \emptyset\},$$

where $C(o)_{a,min}^\delta$ and $C(o)_{a,max}^\delta$ are the minimum and the maximum possibly indiscernible classes of object o on a which are derived from applying formula (14) to minimum and maximum possible indiscernibility relations $PR_{a,min}^\delta$ and $PR_{a,max}^\delta$, respectively.

From this proposition, if the minimum and the maximum possibly indiscernible classes of each object are derived, then the lower and upper bounds of approximations can be obtained. And, $C(o)_{a,min}^\delta$ and $C(o)_{a,max}^\delta$ can be directly derived from the following formula:

$$C(o)_{a,min}^\delta = \{o' \in U \mid (o = o') \vee \forall u \in a(o) \forall v \in a(o') sim(u,v) \geq \delta_a\},$$
$$C(o)_{a,max}^\delta = \{o' \in U \mid \exists u \in a(o) \exists v \in a(o') sim(u,v) \geq \delta_a\}.$$

As a result, this justifies directly using minimum and maximum possibly indiscernible classes from the viewpoint of possible world semantics.[4]

4   Conclusions

We have described the structure of possible coverings under possible world semantics in incomplete information tables with similarity of values. Lots of coverings are derived in an incomplete information table, whereas the covering that is unique is derived in a complete information table. The number of possible coverings may grow exponentially as the number of objects with incomplete information grows. This seems to present some difficulties due to computational complexity of deriving rough sets, but it is not, because the family of possible coverings is a lattice with the minimum and maximum elements. This is also true for the family of maximal descriptions, but is not so for the family of minimal descriptions and the family of sets of close friends.

As Lipski derived the lower and upper bounds of an answer set of a query, we have obtained the lower and upper bounds of approximations. Lower and upper approximations can be derived from only the minimum and maximum coverings by the lattice structure of the family of possible coverings. Therefore, there are no difficulties regarding computational complexity due to the number of incompletely informative objects.

---

[4] This type of justification was first introduced by [22] in extending rough sets to deal with incomplete information.

References

1. Pawlak, Z.: Rough Sets: Theoretical Aspects of Reasoning about Data. Kluwer Academic Publishers, Dordrecht (1991). https://doi.org/10.1007/978-94-011-3534-4
2. Lipski, W.: On semantics issues connected with incomplete information databases. ACM Trans. Database Syst. **4**, 262–296 (1979)
3. Kryszkiewicz, M.: Rules in incomplete information systems. Inf. Sci. **113**, 271–292 (1999)
4. Nakata, M., Sakai, H.: Applying rough sets to information tables containing missing values. In: Proceedings of 39th International Symposium on Multiple-Valued Logic, pp. 286–291. IEEE Press (2009). https://doi.org/10.1109/ISMVL.2009.1
5. Stefanowski, J., Tsoukiàs, A.: Incomplete information tables and rough classification. Comput. Intell. **17**, 545–566 (2001)
6. Yang, T., Li, Q., Zhou, B.: Related family: a new method for attribute reduction of covering information systems. Inf. Sci. **228**, 175–191 (2013)
7. Nakata, M., Sakai, H., Hara, K.: Rough sets based on possibly indiscernible classes in incomplete information tables with continuous values. In: Hassanien, A.E., Shaalan, K., Tolba, M.F. (eds.) AISI 2019. AISC, vol. 1058, pp. 13–23. Springer, Cham (2020). https://doi.org/10.1007/978-3-030-31129-2_2
8. Skowron, A., Stepaniuk, J.: Tolerance approximation spaces. Fund. Inform. **27**, 245–253 (1996)
9. Zakowski, W.: Approximations in the Space (U, Π), Demonstratio Mathematica, **XVI**(3), 761–769 (1983)
10. Bonikowski, Z., Bryniarski, E., Wybraniec-Skardowska, U.: Extensions and intensions in the rough set theory. Inf. Sci. **107**, 149–167 (1998)
11. Zhu, W., Wang, F.: On three types of covering-based rough sets. IEEE Trans. Knowl. Data Eng. **19**(8), 1131–1144 (2007)
12. Yao, Y., Yao, B.: Covering based rough set approximations. Inf. Sci. **200**, 91–107 (2012)
13. Couso, I., Dubois, D.: Rough sets, coverings and incomplete information. Fund. Inform. **108**(3–4), 223–347 (2011)
14. Lin, G., Liang, J., Qian, Y.: Multigranulation rough sets: from partition to covering. Inf. Sci. **241**, 101–118 (2013) http://dx.doi.org/10.1016/j.ins.2013.03.046
15. Chen, D., Wang, C., Hu, Q.: A new approach to attribute reduction of consistent and inconsistent covering decision systems with covering rough sets. Inf. Sci. **177**, 3500–3518 (2007). https://doi.org/10.1016/j.ins.2007.02.041
16. Chen, D., Li, W., Zhang, Z., Kwong, S.: Evidence-theory-based numerical algorithms of attribute reduction with neighborhood-covering rough sets. Int. J. Approximate Reasoning, **55**, 908–923 (2014). http://dx.doi.org/10.1016/j.ijar.2013.10.003
17. Zhang, X., Mei, C.L., Chen, D.G., Li, J.: Multi-confidence rule acquisition oriented attribute reduction of covering decision systems via combinatorial optimization. Knowl.-Based Syst. **50**, 187–197 (2013). https://doi.org/10.1016/j.knosys.2013.06.012
18. Nakata, M., Sakai, H., Hara, K.: Rough sets and rule induction from indiscernibility relations based on possible world semantics in incomplete information systems with continuous domains. In: Hassanien, A.E., Darwish, A. (eds.) Machine Learning and Big Data Analytics Paradigms: Analysis, Applications and Challenges. SBD, vol. 77, pp. 3–23. Springer, Cham (2021). https://doi.org/10.1007/978-3-030-59338-4_1

19. Safari, S., Hooshmandasl, M.R.: On twelve types of covering-based rough sets. Springerplus **5**, 1003 (2016). https://doi.org/10.1186/s40064-016-2670-y
20. Zhu, W.: Relationship among basic concepts in covering-based rough sets. Inf. Sci. **179**, 2478–2486 (2009). https://doi.org/10.1016/j.ins.2009.02.013
21. Liu, Y., Zhu, W.: On three types of covering-based rough sets via definable sets. In: 2014 IEEE International Conference on Fuzzy Systems (FUZZ-IEEE), pp. 1226–1233 (2014)
22. Nakata, M., Sakai, H.: Twofold rough approximations under incomplete information. Int. J. Gen. Syst. **42**, 546–571 (2013). https://doi.org/10.1080/17451000.2013.798898

Group Formation Models Based on Inner Evaluations of Members

Tomoe Entani[✉][iD]

University of Hyogo, Kobe 651-0047, Japan
entani@gsis.u-hyogo.ac.jp

Abstract. Collaborative learning is a useful teaching method in high education and is used in various study areas in the classroom or online. For effective collaboration, group formation is one of the critical factors. It often requires preparation by a teacher. In this study, to help a teacher selection grouping, we propose the models for group formation based on the principles: intra- or inter-group homogeneity or heterogeneity. It depends on the situations what principle is suitable and preferable. We use students' profiles by their inner evaluation to measure the group status, suitable for representing their tendencies. The inner evaluation is denoted as a normalized interval vector of criteria where an interval element indicates a possible evaluation of a criterion. The first and second models are from the intra-group viewpoint by maximizing or minimizing the difference or similarity of members in each group. The students of similar or different tendencies are assigned to the same group. The third model is from the inter-group viewpoint by minimizing differences among groups concerning group tendency. Therefore, balanced groups are obtained.

Keywords: Normalized interval vector · Group formation · Inner evaluation

1 Introduction

In high education, collaborative learning has been increasingly used. Collaborative learning benefits students in terms of higher achievement, greater retention, more positive feelings, and stronger academic self-esteem compared to competitive and individualistic learning [6]. To succeed in the professional world, the students need to acquire teamwork skills and specific technical skills. In various study areas in a classroom or online learning, working as a small group to achieve a common goal is an effective teaching method [9,10]. Group formation is one of the critical factors, and the adequacy of peers is necessary for effective collaboration. The group composition affects the group performance, i.e., poorly formed groups can lead to many possible negative peer group influences. However, few studies have focused on it to improve collaborative learning [2,7].

In practice, there are three kinds of group formation methods used: random selection, student self-selection, and teacher selection [11]. Each has advantages

© Springer Nature Switzerland AG 2022
K. Honda et al. (Eds.): IUKM 2022, LNAI 13199, pp. 29–39, 2022.
https://doi.org/10.1007/978-3-030-98018-4_3

and disadvantages. Random selection is simple, though the students' similarities and differences are not taken into account and can lead to unbalanced groups. Student self-selection positively affects student attitudes, though students who usually prefer to be in their comfort zone learn less from the differences. The formed groups often consist of either above-average students or below-average students so that the groups in a classroom are not balanced. Teacher selection, in other words, criteria-based grouping, is popular in practice, even if it needs preparation. Teachers make the groups balanced and be suitable for the situation or purpose, though some students claim the harmful influence of social loafing [13]. Moreover, the criteria differ in cases, and the availability of students' profiles such as personality and previous marks is not easy. This study focuses on teacher selection and proposes the models to help it.

If we made a selected group representative of a classroom, we would pick up the high-level students according to grades with some communication skills. Such a team possibly performs better and more productive than teams of the other members due to positively influencing each other. Differently from such a selected group, in collaborative learning in a classroom, all the members are assigned to one of the groups. A mix of superior and inferior groups is sometimes not unacceptable so that balanced groups are expected.

Concerning knowledge levels diversity, the simple method is to sort the students from high to low grades and assign one student to each group sequentially. Each group has a mix of students with knowledge levels high and low. Having a high-grade peer in each group is not the only requirement for an adequate group. It is known that group work enhances deep learning through student engagement. Besides study success, student engagement leads to positive outcomes such as persistence and self-esteem.

Although student engagement is essential for group work, there are various definitions. It includes behavior engagement, which refers to participation based on knowledge level, and cognitive engagement, which refers to the mental energy based on communication and human relations [8]. There seem to be two kinds of criteria for a student profile: knowledge level and social interaction. In group formation, teachers consider both of them. The students are sorted from high to low knowledge level by the total grades, and in detail, the specific sub-criteria of rates can be used. On the other hand, it is challenging to sort students from high to low according to their social interaction. The sub-criteria of social interaction include leadership, empathy, and communication skills. Since the sub-criteria stand on personality characteristics, competitive comparison of one's characteristics to the others' is not appropriate. For a student profile, it is reasonable to compare the sub-criteria each other rather than to compare him/her to the others on a criterion. Based on such an inner evaluation on sub-criteria, we find out the student profile, such that s/he is better at communication than leadership.

Exams are common ways for teachers to assess the knowledge level of students. The rates of exams are used to sort the students in a classroom from high to low relatively. Instead of exams, to know the characteristics of a student, a questionnaire is generally used. As a result, for instance, students are scored against criteria such as president, strategist, operative, and finishers [3]. To measure a student's tendency, we prefer a self-evaluation by each student to a teachers' assessment. Similarly to social interaction, the knowledge level or tendency can be found from a student self-evaluation by questionnaire, for instance, asking his/her academic interests. Self-evaluation is not a relative evaluation among the others but an inner evaluation of a student. Therefore, we assume that student profile is denoted with inner-evaluation without regard to knowledge or social interaction.

The ideal formed group status has groups with members who are as similar among themselves as possible, inter-homogeneous, but also empowering the students' individual difference inside such groups, intra-heterogeneous [7]. On the one hand, diversity in a group increases the integration of different ideas from multiple perspectives and enhances behavior engagement. On the other hand, too many challenges to differences in communication styles and feelings of anxiety hinder cognitive engagement. Research showed heterogeneous groups of high-level students are more productive than heterogeneous groups of them, a vice versa in the case of low-level students [13]. It mentions that some count on the others in a heterogeneous group, which may reduce productivity. The interaction relationships of members in a group are worth consideration. In the group formation models in this study, we consider homogeneity or heterogeneity of each group and homogeneity among groups.

This paper outlines as follows. In the next section, we define a normalized interval vector to represent a student's profile of knowledge and social interaction criteria. They are obtained as students' inner evaluations by self-evaluations. In Sect. 3, an individual profile is extended into a group profile. Then, in Sect. 4, based on such the principles of similarity and difference among group members and groups, we propose three models for group formation. In Sect. 5, we illustrate the proposed models with a numerical example and compare the formed groups. Then, we draw the conclusion in Sect. 6.

2 Normalized Interval Vector

This study uses a student profile by self-evaluation of some criteria on knowledge and social interaction. Therefore, it represents each student's inner evaluation among the criteria. The inner evaluation indicates how excellent or lousy the student is on a criterion compared to the other criteria.

We denote member j as a normalized interval vector, $\boldsymbol{X}_j = (X_{j1}, \ldots, X_{jm})^t$, where $X_{jk} = [\underline{x}_{jk}, \overline{x}_{jk}]$ denotes his/her evaluation of criterion k relative to the other criteria $\forall k' \neq k$. An evaluation of criterion is denoted as an interval to reflect its uncertainty or possibility. Denoting a student profile with a crisp value cannot reflect real situations since a student's act usually depends on tasks, periods, the other members, etc. Hence, the evaluation of criterion has some range, and an interval indicates that the evaluation under any specific situation lies from its lower bound to its upper bound.

Let two members, j_1 and j_2, and skills, k_1 and k_2. Even if $X_{j_1 k_1}$ is greater than $X_{j_2 k_1}$, we cannot be sure that member j_1 is better than j_2 by comparing them. It happens that j_1 is, so to speak, a lazy student, and the excellent student, j_2, is better than j_1 under all criteria. However, if $X_{j_1 k_1}$ is greater than $X_{j_1 k_2}$ with regard to member j_1, we are sure that j_1 is better at criterion k_1 than k_2. Since group formation in a classroom is not selecting or making a good group, we use inner evaluations rather than relative evaluations as a student profile in this study. The inner evaluation is denoted as a normalized interval vector.

First, a normalization of interval vector is defined, and then because of n members in a classroom, the properties of more than one normalized interval vector are shown.

Definition 1. *An interval vector* $\boldsymbol{X} = (X_1, \ldots, X_i)$ *each of whose elements is* $X_i = [\underline{x}_i, \overline{x}_i]$ *is normalized, if and only if* $\boldsymbol{X} \in \mathcal{N}$, *such that*

$$\mathcal{N} = \left\{ \boldsymbol{X} \,\middle|\, \sum_{i \neq j} \overline{x}_i + \underline{x}_j \geq 1 \; \forall j \; \sum_{i \neq j} \underline{x}_i + \overline{x}_j \leq 1 \; \forall j, \; \underline{x}_i \leq \overline{x}_i \; \forall i \right\}. \quad (1)$$

The definition is based on interval probability [1, 12]. In the case of real values as $\overline{x}_i = \underline{x}_i = x_i \forall i$, two inequalities are replaced into $\sum_i x_i = 1$ as probability. The redundancy of the intervals to make the sum be one is excluded: the upper and lower bounds cannot be too large and too small, respectively.

When the evaluations of all criteria are no different, this student's profile is denoted as interval vector $([0,1], \ldots, [0,1])$. Moreover, if it is sure that the evaluations are always precisely equal to each other, it is denoted as crisp vector $(1/n, \ldots, 1/n)$. Both vectors are normalized in the sense of interval vector normalization by (1).

The normalized interval vector satisfies the following two propositions [5, 12].

Proposition 1. *Let* $\boldsymbol{X} = (X_1, \ldots, X_n) \in \mathcal{N}$ *be a normalized interval vector. There exist* $x_i \in X_i, \forall i$, *such that* $\sum_i x_i = 1$.

For a crisp value of ith element, $x_i \in X_i$, there are crisp values in the the other elements' intervals, $x_j \in X_j, \forall j \neq i$, which satisfies $x_i + \sum_{j \neq i} x_j = 1$.

Proposition 2. *Let* $\boldsymbol{X}_l = (X_{l1}, \ldots, X_{ln}) \in \mathcal{N}$, $l = \{1, \ldots, n'\}$, *where* $X_{li} = [\underline{x}_{li}, \overline{x}_{li}]$. *Then interval vector* $\boldsymbol{X} = (X_1, \ldots, X_n)$, *where* $X_i = [\underline{x}_i, \overline{x}_i]$, *defined by* $\underline{x}_i = \min_l \underline{x}_{li}$, $\overline{x}_i = \max_l \overline{x}_{li}$, *satisfies* $\boldsymbol{X} \in \mathcal{N}$, *i.e.,* $\boldsymbol{X}$ *is also a normalized interval vector.*

Namely, the compact set of normalized interval vectors is a normalized interval vector.

Proposition 3. *Let* $\boldsymbol{X}_l = (X_{l1}, \ldots, X_{ln}) \in \mathcal{N}$, $l = \{1, \ldots, n'\}$, *where* $X_{li} = [\underline{x}_{li}, \overline{x}_{li}]$, *and* $\boldsymbol{w} = (w_1, \ldots, w_{n'})$, *where* $\sum_l w_l = 1$. *Then interval vector* $\boldsymbol{X} = (X_1, \ldots, X_n)$, *where* $X_i = [\underline{x}_i, \overline{x}_i]$, *defined by* $\underline{x}_i = \sum_l w_l \underline{x}_{li}$, $\overline{x}_i = \sum_l w_l \overline{x}_{li}$, *satisfies* $\boldsymbol{X} \in \mathcal{N}$, *i.e.,* $\boldsymbol{X}$ *is also a normalized interval vector.*

Proposition 3 implies that the weighted sum of normalized interval vectors is normalized.

3   Student Profile and Group Profile

The problem is to divide n members into t groups based on their inner evaluations of m criteria, which we call skills in the following. Namely, there are n students $j = 1, \ldots, n$, who evaluate themselves on m skills, $k = 1, \ldots, m$, and the students in a classroom are divided into t groups, $i = 1, \ldots, t$. We denote a set of students in group i as G_i, which consists of n_i elements, and $\sum_i n_i = n$. As shown in Sect. 2, the profile of student j is explained with a normalized interval vector, $X_j = ([\underline{x}_{jk}, \overline{x}_{jk}], \forall k) \in \mathcal{N}$,

As preparation for interval evaluations of a group, Fig. 1 shows three kinds of relations between two intervals, X_1 and X_2. The similarity of these two intervals is measured using S^1 and S^2 illustrated in the figure. The first S^1 shows the maximum difference of two intervals. Two intervals are more different as the range of integrated interval S^1.

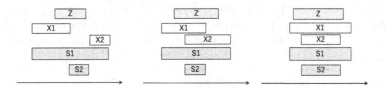

Fig. 1. Similarity and average of two intervals

Then, in general, the difference s^1_{ik} of skill k in group i of n_i members is defined as follows.

$$s^1_{ik} = \min \left(\overline{s^1}_{ik} - \underline{s^1}_{ik} \right)$$
$$s.t. \quad \overline{x}_{jk} \leq \overline{s^1}_{ik}, \forall j \in G_i, \tag{2}$$
$$\underline{s^1}_{ik} \leq \underline{x}_{jk}, \forall j \in G_i.$$

By Proposition 2, the integrated interval vector, $([\underline{s^1}_{i1}, \overline{s^1}_{i1}], \ldots, [\underline{s^1}_{im}, \overline{s^1}_{im}])$ is a normalized interval vector.

The second S^2 corresponds to an intersection of both intervals. Although the left two S^2 in Fig. 1 look the same, there is no common and a common in the left and cent figures. Hence, we denote the similarity of the left figure is negative, and that of the right one is positive [4]. Two intervals are similar as greater the range of intersection S^2. The similarity s^2_{ik} of skill k in group i is defined as follows.

$$s^2_{ik} = \max \left(\overline{s^2}_{ik} - \underline{s^2}_{ik} \right)$$
$$s.t. \quad \overline{s^2}_{ik} \leq \overline{x}_{jk}, \forall j \in G_i, \tag{3}$$
$$\underline{x}_{jk} \leq \underline{s^2}_{ik}, \forall j \in G_i,$$

where s^2_{ik} is less than one if the intervals of all members in group i on skill k have no intersection.

The third Z corresponds to the group profile. The possible evaluation of group performance of skill k of group i is denoted as the sum of all members in the group as follows.

$$Z_{ik} = [\underline{z}_{ik}, \overline{z}_{ki}] = \left[\sum_{j \in G_i} \underline{x}_{jk}, \sum_{j \in G_i} \overline{x}_{jk} \right], \qquad (4)$$

where group lower and upper bounds are obtained from individual members' lower and upper bounds, respectively.

Denote an interval vector of group i, $\boldsymbol{Z}_i = (Z_{i1}, \ldots, Z_{im})^t$, it is normalized by Proposition 3 with the weights $w_l = 1/n, \forall l$. Since the number of members differs in the groups, we use the group performance per member $\boldsymbol{Z}_i' = (Z_{i1}/n_i, \ldots, Z_{im}/n_i)^t$ if we compare the group performances.

4   Group Formation

4.1   Similarity and Difference

This section proposes the problems of dividing n members into t groups based on intra- and inter-group similarities or differences. Therefore, we introduce the variables, $y_{ij}, \forall i, j$, denoting whether member j is assigned to group i or not.

$$y_{ij} = \begin{cases} 1, j \in G_i, \\ 0, \text{else}, \end{cases}$$
$$\sum_j y_{ij} = n_i, \forall i, \qquad (5)$$
$$\sum_i y_{ij} = 1, \forall j,$$

where n_i is the number of members in group i and member j is assigned to one of the groups.

With these variables, the difference of members in a group (2) is rewritten as follows.

$$s_{ik}^1 = \min \, (\overline{s1}_{ik} - \underline{s1}_{ik}),$$
$$s.t. \quad \overline{x}_{jk} y_{ij} \leq \overline{s^1}_{ik}, \forall j, \qquad (6)$$
$$\underline{s^1}_{ik} \leq (M - (M-1)y_{ij})\underline{x}_{jk}, \forall j,$$

where M is a positive large number, and $y_{ij} \forall i, j, \underline{s^1}_{ik}, \overline{s^1}_{ik}$ are variables. In the case of $y_{ij} = 0$, where member j is not in the group i, the first and second constraints are never active.

In the similar way, the similarity of members in a group (3) is rewritten as follows.

$$s_{ik}^2 = \max \, (\overline{s^2}_{ik} - \underline{s^2}_{ik})$$
$$s.t. \quad \overline{s^2}_{ik} \leq (M - (M-1)y_{ij})\overline{x}_{jk}, \forall j, \qquad (7)$$
$$y_{ij}\underline{x}_{jk} \leq \underline{s^2}_{ik}, \forall j.$$

The possible evaluation of group performance (4) is rewritten as follows.

$$Z_{ik} = [\underline{z}_{ik}, \overline{z}_{ki}] = \left[\sum_j \underline{x}_{jk} y_{ij}, \sum_j \overline{x}_{jk} y_{ij} \right]. \qquad (8)$$

4.2 Intra-group and Inter-group

First, we consider intra-group homogeneity or heterogeneity. Excluding social loafing enhances easiness of each student's contribution. Therefore, the members with similar tendencies are assigned to the same groups. The homogeneous groups are formed by maximizing or minimizing similarity or difference of members in each group. By minimizing difference (6), we obtain an in intra-group homogeneity groups as follows.

$$
\begin{aligned}
&\min \ \textstyle\sum_{ik}(\overline{s^1}_{ik} - \underline{s}^1_{ik}), \\
s.t. \ &\underline{s}^1_{ik} \le (M - (M-1)y_{ij})x_{jk}, \forall i,j,k, \\
&\overline{x}_{jk}y_{ij} \le \overline{s^1}_{ik}, \forall i,j,k, \\
&y_{ij} = \{0,1\}, \forall i,j, \ \textstyle\sum_j y_{ij} = n_i, \forall i, \ \textstyle\sum_i y_{ij} = 1, \forall j,
\end{aligned}
\tag{9}
$$

where the variables are $y_{ij}, \forall i,j$ and $\underline{s}^1_{ik}, \overline{s^1}_{ik}, \forall i,k$.

The other model for intra-group homogeneity groups is formulated by maximizing similarity (7).

$$
\begin{aligned}
&\max \ \textstyle\sum_{ik}(\overline{s^2}_{ik} - \underline{s}^2_{ik}), \\
s.t. \ &\overline{s^2}_{ik} \le (M - (M-1)y_{ij})\overline{x}_{jk}, \forall i.j,k, \\
&y_{ij}\underline{x}_{jk} \le \underline{s}^2_{ik}, \forall i,j,k, \\
&y_{ij} = \{0,1\}, \forall i,j, \ \textstyle\sum_j y_{ij} = n_i, \forall i, \ \textstyle\sum_i y_{ij} = 1, \forall j.
\end{aligned}
\tag{10}
$$

In this way, there are two methods for intra-group homogeneity by minimizing difference (9) or maximizing similarity (10). In both models, each criterion and group's difference and similarity are measured independently.

On the contrary, to form the groups of intra-group heterogeneity, the similarity or difference is minimized or maximized. However, the problems by replacing objective functions of (9) and (10) are infeasible. By (10), the similarity of a skill of members in a group is maximized based on the fact that similar members have their intervals of each skill evaluation in common. In other words, the different members do not have their intervals of each skill in common. Each skill's diversity among members leads to the similarity among all skills' evaluations in a group. The following simple example explains this fact.

Assume that two members are denoted with two skills, one of which, named skill 1, is illustrated in the left figure of Fig. 1. From the figures, member 1 $X1$ is less than member 2 $X2$ on skill 1. Since each member's two intervals are normalized by Definition 1, on the other skill 2, member 2 is greater than member 1. Namely, when we compare the first skill of member 1 and the second one of member 2, they are similar. Maximizing the similarity of skills in the group tends to assign the members with different tendencies to the same group.

Hence, we maximize the similarity of skills of all members in a group instead of minimizing the similarity of each skill of members in a group. The problem is formulated as follows.

$$
\begin{aligned}
&\max \ \textstyle\sum_i(\overline{s^2}_i - \underline{s}^2_i), \\
s.t. \ &(M - (M-1)y_{ij})\overline{s^2}_i \le \overline{x}_{jk}, \forall i.j,k, \\
&\underline{x}_{jk} \le y_{ij}\underline{s}^2_i, \forall i.j,k \\
&y_{ij} = \{0,1\}, \forall i,j, \ \textstyle\sum_j y_{ij} = n_i, \forall i, \ \textstyle\sum_i y_{ij} = 1, \forall j,
\end{aligned}
\tag{11}
$$

where the variables are $y_{ij}, \forall i, j$ and $\underline{s^2}_i, \overline{s^2}_i, \forall i$. In other words, $\underline{s^2}_i$ and $\overline{s^2}_i$, such that $\underline{s^2}_{ik} \leq \underline{s^2}_i, \forall k$ and $\overline{s^2}_i \leq \overline{s^2}_{ik}, \forall k$ are introduced into (10).

Next, we consider inter-group homogeneity from the viewpoint of group profile. The group performance is denoted with the sums of the members' evaluations of all skills by (8). In the case of balanced groups, the evaluations of all groups are similar. For such a similarity, all group evaluations of each skill should be in a certain range. The range of group evaluation is minimized as follows.

$$
\begin{aligned}
& \min \sum_k (\overline{z}_k - \underline{z}_k), \\
& s.t. \; n_i \underline{z}_k \leq \underline{z}_{ik}, \forall i, k, \\
& \quad\quad \overline{z}_{ik} \leq n_i \overline{z}_k, \forall i, k, \\
& \quad\quad y_{ij} = \{0, 1\}, \forall i, j, \; \sum_j y_{ij} = n_i, \forall i, \; \sum_i y_{ij} = 1, \forall j,
\end{aligned}
\tag{12}
$$

where in addition to $y_{ij} \forall i, j$, the range of group performance per a member $\underline{z}_k, \overline{z}_k, \forall k$ are variables. Corresponding to group performance per member, when we introduce the maximum $\overline{z}$ and minimum $\underline{z}$, we can replace the objective function with max $(\overline{z} - \underline{z})$ by adding constraints $\underline{z} \leq \underline{z}_k, \overline{z}_k \leq \overline{z}, \forall k$.

The number of members in group i, n_i in (12), is an integer denoted as follows.

$$
n_i = \sum_j x_{ij} \in \left\{ \left\lfloor \frac{n}{t} \right\rfloor, \left\lceil \frac{n}{t} \right\rceil \right\}, \; \forall i,
\tag{13}
$$

where $\left\lfloor \frac{n}{t} \right\rfloor = \left\lceil \frac{n}{t} \right\rceil = \frac{n}{t}$ in the case of divisible n by t.

Thus, the first two non-linear constraints in (12) is relaxed and the problem is reduced into the following linear programming problem.

$$
\begin{aligned}
& \min \sum_k (\overline{z}_k - \underline{z}_k), \\
& s.t. \; \left\lfloor \frac{n}{t} \right\rfloor \underline{z}_k \leq \underline{z}_{ik}, \forall i, k, \\
& \quad\quad \overline{z}_{ik} \leq \left\lceil \frac{n}{t} \right\rceil \overline{z}_k, \forall i, k, \\
& \quad\quad y_{ij} = \{0, 1\}, \forall i, j, \; \sum_j y_{ij} = n_i, \forall i, \; \sum_i y_{ij} = 1, \forall j,
\end{aligned}
\tag{14}
$$

where the group performance of skill k, $Z_{ik} = [\underline{z}_{ik}, \overline{z}_{ik}]$, is in the range from $\left\lfloor \frac{n}{t} \right\rfloor \underline{z}_k$ to $\left\lceil \frac{n}{t} \right\rceil \overline{z}_k$, and those of all skills are in the range from $\underline{z}$ to $\overline{z}$.

Assume a larger group of $\left\lceil \frac{n}{t} \right\rceil$ members. The first constraint is not always active because of $\left\lfloor \frac{n}{t} \right\rfloor \underline{z}_k < \left\lceil \frac{n}{t} \right\rceil \underline{z}_k$ and the second one tend to be active because of $\left\lfloor \frac{n}{t} \right\rfloor \overline{z}_k < \left\lceil \frac{n}{t} \right\rceil \overline{z}_k$. In this sense of controlled skill, we obtain balanced groups.

5   Numerical Example

In the numerical example, there are 11 students in a classroom, and a teacher forms three groups so that each group consists of three or four students. The students' profiles denoted with four criteria, named from skill-1 to skill-4, are given as normalized interval vectors at the left part of Table 1. For instance, student X_1 at the second row is good at skill-4 among four skills but bad at skill-2 with some possibilities. The interval vector in a row represents a student's possible

Table 1. Four kinds of skills of 11 members

Student	Skill-1	Skill-2	Skill-3	Skill-4	Intra-group By (9) or (10)	By (11)	Inter-group similarity By (14)
X_1	[0.10,0.14]	[0.02,0.1]	[0.10,0.14]	0.70	1	1	1
X_2	[0.02,0.11]	0.77	[0.02,0.11]	[0.09,0.11]	1	3	1
X_3	[0.19,0.23]	0.58	[0.08,0.19]	[0.04,0.12]	1	2	2
X_4	0.7	0.14	[0.02,0.13]	[0.03,0.14]	1	2	3
X_5	[0.02,0.12]	[0.1,0.12]	0.73	[0.02,0.15]	2	1	3
X_6	[0.03,0.21]	[0.13,0.21]	[0.03,0.13]	0.63	2	3	2
X_7	0.11	[0.02,0.11]	0.77	0.11	2	2	3
X_8	0.65	[0.04,0.11]	0.22	[0.02,0.09]	3	1	3
X_9	0.11	[0.02,0.11]	0.76	[0.02,0.11]	3	1	2
X_{10}	[0.03,0.14]	0.14	0.69	[0.03,0.14]	3	3	1
X_{11}	0.8	0.09	[0.02,0.09]	[0.02,0.09]	3	3	1

inner-evaluation of criteria, different from the exam scores for students' competitive comparison. Comparing the intervals in a row is valid, though comparing those in a column is meaningless.

The groups by the proposed models are shown at the right part of Table 1 for a comparison. In real situations, a teacher decides a principle for grouping, which is suitable for the case, beforehand. Then, one of the intra-group homogeneous, intra-group heterogeneous, and inter-group homogeneous models is applied. To compare the groups by four models, we obtained adjusted rand index (ARI), used in k-means clustering to measure the similarity between two clusterings. If two clusterings are the same, ARI is 1, decreasing as they become less similar. In this example, those of three pairs of three formed groups are less than 0.1, so each is unique.

First, we compare the groups by intra-group homogeneity model (9) or (10) and intra-group heterogeneity model (11). The students assigned to groups 1 and 2 by the former model are assigned to three different groups by the latter model. For intra-group similarity, the tendencies of three students in each group are similar, though they are not in the same group for intra-group diversity. Even though the principles of these two models are opposite, student pairs of (X_3, X_4) and (X_{10}, X_{11}) are in the same group by both models. One of the reasons is the group size constraint.

Next, we compare the groups by intra- and inter-group homogeneity models (9) and (14). Similar to the former comparison, the students in groups 1 and 3 by the intra-group model are in three different groups by the inter-group model. It is reasonable to distribute the students with similar tendencies for group equilibrium. The intra-group heterogeneous model or inter-group homogeneity model maximizes the difference in each group or minimizes differences among groups. At the same time, both models make groups to reduce the similarity of members' profiles in each group. It is because their principles are opposite from that of the intra-homogeneity group. This fact is supported by Table 2, where we calculated the objective function values of (9), (10), (11), and (14), which are the four models of intra-group difference, intra-group similarity, and inter-group

38   T. Entani

similarity, with their optimal solutions. The second and third rows are more
similar to each other than the first row. According to three measurements, the
intra-homogeneity model at the first row deviates from the intra-group heterogeneity and inter-group homogeneity models at the bottom two rows.

Table 2. Four measurements in objective functions

	Intra-group			Inter-group
	Difference	Similarity	Skill-similarity	Similarity
Min. diff. (9) or max. sim. (10)	5.78	0.54	−1.77	0.73
Min. sim. by max. skill-sim. (11)	7.90	−0.46	−0.56	0.66
Balanced group by max. group-sim. (14)	8.62	−1.35	−1.35	0.54

6 Conclusion

We have proposed group formation models from intra-group homogeneity and
heterogeneity and inter-group homogeneity. A student profile is denoted as a
normalized interval vector of criteria. A teacher can choose one of them suitable
for the situation and principle. From the intra-group viewpoint, we measure the
similarity of members by a range of integrated intervals or an intersection of
criteria. For members' homogeneity, the integrated interval of all members is in
a small range, or the intersection is maximized. On the other hand, for members' heterogeneity, the interval evaluations of all criteria are similar by having
more intersections. As a result, the members with similar or different tendencies
are assigned to the same groups. From the inter-group viewpoint, the similarity
of groups is measured as the range of group performance interval by summing
up the members' intervals, and the range is minimized. As a result, the groups
are balanced concerning group evaluations of criteria. In this study, a member's
profile is denoted as an interval vector of criteria. The interval indicates a student's possible evaluation, and in any specific case, it lies between its lower and
upper bounds. Hence, in the proposed models, the similarity, the difference, and
the sum of intervals are measured based on the possibilistic viewpoint focusing
on the lower and upper bounds. Depending on the lower or upper bound of an
interval is a limitation of the proposed method. The interval evaluation can be
more precise if a unique case is assumed. Since the members affected each other,
we will specify a crisp value in an interval or reduce the possibility in future
work.

Acknowledgments. This work was partially supported by JSPS KAKENHI Grant
Number JP19K04885.

References

1. de Campos, L.M., Huete, J.F., Moral, S.: Probability intervals: a tool for uncertain reasoning. Internat. J. Uncertain. Fuzziness Knowl.-Based Syst. **2**(2), 167–196 (1994)
2. Chen, C.M., Kuo, C.H.: An optimized group formation scheme to promote collaborative problem-based learning. Comput. Educ. **133**, 94–115 (2019)
3. Dias, T.G., Borges, J.: A new algorithm to create balanced teams promoting more diversity. Eur. J. Eng. Educ. **42**, 1365–1377 (2017)
4. Entani, T.: Group interval weights based on conjunction approximation of individual interval weights. Int. J. Anal. Hierarchy Process **7**(3), 1–5 (2015)
5. Entani, T., Inuiguchi, M.: Pairwise comparison based interval analysis for group decision aiding with multiple criteria. Fuzzy Sets Syst. **274**(1), 79–96 (2015)
6. Johnson, D.W., Johnson, R.T., Smith, K.: The state of cooperative learning in postsecondary and professional settings. Educ. Psychol. Rev. **19**, 15–29 (2007). https://doi.org/10.1007/s10648-006-9038-8
7. Moreno, J., Ovalle, D.A., Vicari, R.M.: A genetic algorithm approach for group formation in collaborative learning considering multiple student characteristics. Comput. Educ. **58**(1), 560–569 (2012)
8. Poort, I., Jansen, E., Hofman, A.: Does the group matter? effects of trust, cultural diversity, and group formation on engagement in group work in higher education. Higher Educ. Res. Dev. 1–16 (2020)
9. Prince, M.: Does active learning work a review of the research. J. Eng. Educ. **93**, 223–231 (2007)
10. Romanow, D., Napier, N.P., Cline, M.K.: Using active learning, group formation, and discussion to increase student learning: a business intelligence skills analysis. J. Inf. Syst. Educ. **31**, 218–230 (2020)
11. Sanz-Martínez, L., Martínez-Monés, A., Bote-Lorenzo, M.L., Muñoz-Cristóbal, J.A., Dimitriadis, Y.: Automatic group formation in a MOOC based on students' activity criteria. In: Lavoué, É., Drachsler, H., Verbert, K., Broisin, J., Pérez-Sanagustín, M. (eds.) EC-TEL 2017. LNCS, vol. 10474, pp. 179–193. Springer, Cham (2017). https://doi.org/10.1007/978-3-319-66610-5_14
12. Sugihara, K., Ishii, H., Tanaka, H.: Interval priorities in AHP by interval regression analysis. Eur. J. Oper. Res. **158**(3), 745–754 (2004)
13. Wichmann, A., Hecking, T., Elson, M., Christmann, N., Herrmann, T., Hoppe, H.U.: Group formation for small-group learning: are heterogeneous groups more productive? OpenSym (2016)

Preference-Based Assessment of Organisational Risk in Complex Environments

Silvia Carpitella[1]([⊠]) [iD] and Joaquín Izquierdo[2] [iD]

[1] Department of Decision-Making Theory, Institute of Information Theory and Automation, Czech Academy of Sciences, Prague, Czech Republic
carpitella@utia.cas.cz

[2] Fluing-Institute for Multidisciplinary Mathematics, Universitat Politècnica de València, Valencia, Spain
jizquier@upv.es

Abstract. This paper proposes a preference-based approach for optimising the process of organisational risk assessment in complex and uncertain environments, where significant decision-making factors may be interconnected. Organisational risks are herein treated from the perspective of the work-related stress risk involving psycho-physical factors crucial for the safety and well-being of human resources. The traditional Health and Safety Executive (HSE) model commonly used for stress evaluation in working environments is herein improved by first applying the Analytic Network Process (ANP) to weight management standards (MS). This technique has been chosen to avoid neglecting potential relations bounding MS with each other. Finally, Fuzzy Cognitive Maps (FCMs) are used to study dependence among significant stress factors. In such a direction, the support offered by the fuzzy set theory is relevant to deal with subjective evaluations of preference. The case of an Italian airport is analysed to demonstrate the applicability of the approach, and managerial insights are discussed.

Keywords: Organisational risk · Decision-making · Analytic Network Process · Fuzzy Cognitive Maps · Complexity management

1 Motivation and State of the Art

The occurrence of organisational risks may have a strong impact on human resources' safety. This category of risks is related to organisational shortcomings and includes work-related stress as well as psychological factors as fundamental elements of analysis. Specifically, not only are psychological factors crucial for human well-being and professional achievement but also hugely influence operational performance by contributing to generating company results on the whole [7]. Increasing attention is devoted nowadays to research on psychological factors within entrepreneurial realities. In this context, proper assessments

© Springer Nature Switzerland AG 2022
K. Honda et al. (Eds.): IUKM 2022, LNAI 13199, pp. 40–52, 2022.
https://doi.org/10.1007/978-3-030-98018-4_4

of the risk of work-related stress greatly contribute to approaching and better understanding how to manage these types of factors. The existence of particular indicators is investigated, as well as work conditions that may cause discomfort and stress for workers, leading to their poor performance and dissatisfaction.

The risk of work-related stress is commonly assessed by preliminary analysing such conditions as professional environment, working hours, monotony or fragmentation of tasks, uncertainty, excessive or insufficient workload, relationships among colleagues and superiors, and so on [9]. These factors could potentially harm the psycho-physical health of workers, especially if they have to act in synergy with each other. In any case, evidence demonstrates that they coexist in almost any work environment, reducing organisational effectiveness. This is the reason why work-related stress risk assessment has to be implemented by companies, as established by the existing international standards.

Among the various methodologies used to purse such a type of evaluation [10,15], we here discuss the integrated management approach developed by the British agency Health and Safety Executive (HSE). The evaluation model applies the perspective of Research & Development activities, aiming at scientifically demonstrating the entity of repercussions of work-related stress on general health conditions of individuals. The HSE model analyses six main areas or management standards (MS), by proposing a structured interview to workers in the form of an inquiring questionnaire tool called the MS indicator tool [8]. Each item of the questionnaire refers to a specific MS. The goal consists in investigating critical organisational aspects to be improved by contributing to the creation of a research network system in the field of occupational health and safety. Specifically, the HSE indicator tool is specially focused on physical and psychical consequences as well as progressive alterations of lifestyle and behavior of workers. Given the huge complexity and the uncertainty characterising this field, the present paper proposes a methodological framework by combining the Analytic Network Process (ANP), a well-known decision-making technique, with the *ah hoc* generation of Fuzzy Cognitive Maps (FCMs), the latter being particularly suitable for managing uncertainty when subjective preference evaluations are required [3]. As an artificial intelligence technique capable of effectively supporting decision-making [1], FCM integrates characteristics of fuzzy sets and neural networks. As reported by López and Ishizaka [6], FCMs have been successfully hybridized with several multi-criteria decision-making techniques so far. In particular, by mentioning a work of research specifically integrating ANP and FCM [14], the authors underline as FCMs ,ay support in the calculation of local and/or global weights of a set of decision-making elements. Considering this evidence, we aim to exploit the strengths derived from such a methodological integration. To the best of the authors' knowledge, it is the first time that ANP and FCMs are combined for improving the process of organisational risk management in terms of work-related stress assessment.

With these preliminaries, the six MS considered by the HSE model will be first analysed and their mutual importance will be established by means of the ANP. Second, a suitable FCM will be built to study relations of dependence bounding the main aspects investigated by the HSE indicator tool. This integration can positively contribute to the topic of research by effectively highlighting critical issues so that possibilities of improvement of working conditions in complex environment can be real.

The research is organised as follows. Traditional HSE methodology is discussed in Sect. 2, where the items of the indicator tool are associated to the corresponding MS. Section 3 provides methodological details about the preference-based approach. An Italian airport has been analysed for the real application of Sect. 4, airports being extremely complex organisations where many stressful factors may likely impact on employees conditions. Conclusions of Sect. 5 close the paper by discussing potential future developments.

2   HSE Management Standards for Organisational Risks

MS may be classified according to three organisational dimensions: 1. **content** (cnt), referring to general pressures workers may feel because of work characteristics, 2. **context** (cxt), referring to work environment, human relations and cooperation, 3. **awareness** (aws), referring to the personal perception of workers about their own contribution and involvement. Within these three main dimension groups, six MS are identified as key areas that, when not properly managed, are associated with health problems and lower productivity as well as increasing probability of injuries and rates of sickness absence.

- MS_1, demand: it includes such aspects as workload, tasks and environment;
- MS_2, control: it refers to autonomy of people in the way they lead their job;
- MS_3, support, it includes encouragement and resources from the company;
- MS_4, relationship, it refers to managing conflict and unacceptable behaviour;
- MS_5, role, it considers the clear understanding about specific working roles;
- MS_6, change, it refers to change management and communication processes.

Specifically, MS_1 and MS_2 belong to the content dimension, MS_3 and MS_4 refer to the context dimension, while MS_5 and MS_6 are related to the awareness dimension. The interesting idea behind MS-based approach is that companies have the possibility of benchmarking their current practices of organisational risk evaluation by designing related measures to enhance stress management performance. The HSE indicator tool aims to support this process. Thirty-five items are randomly proposed to workers and the related answers can be provided according to a linguistic scale. Analysing the questionnaire from a structural point of view, that is to say, by connecting specific items to MS, is useful to further elaborate employees' responses. This classification will help to easily understand if the standards are achieved or not. In such a direction, Table 1 organises the items of the questionnaire by associating them to the corresponding MS to ease the evaluation of the most critical area(s).

The HSE model based on the six described MS can be hence considered as an integrated approach to design and optimise the simultaneous management of stressful factors, usually interacting with each other in real contexts. Such an interaction would lead to the amplification of the effects that these factors

Table 1. Decision-making elements under analysis

ID	MS	Items of management standard indicator tool
MS_1	Demand	DE_1 I clearly understand the expectations about my work
		DE_2 I do not experience difficulties when I have to combine job requests coming from diverse people and/or operational units
		DE_3 I know how to perform my job and all the related tasks
		DE_4 I usually have deadlines not extremely difficult to meet
		DE_5 I do not have to perform particularly hard activities
		DE_6 I do not neglect private issues because of my work
		DE_7 I do not feel high pressure due to overtime work
		DE_8 I do not have to be very quick when leading operations
		DE_9 I never fail in satisfactorily meeting my deadlines
MS_2	Control	CO_1 I can autonomously decide when to have a break
		CO_2 I can decide the rhythm at which my tasks are performed
		CO_3 I can make decisions about the organisation of my work
		CO_4 I am free to take enough breaks
		CO_5 I have freedom of choice about the content of my tasks
		CO_6 I can express my opinions about how to perform my tasks
		CO_7 I have flexible working hours
MS_3	Support	SU_1 I am supported by my colleagues for difficult work
		SU_2 I receive effective information that is helpful for my activity
		SU_3 I can rely on my boss should I experience any problem
		SU_4 I receive the help and support I need from colleagues
		SU_5 I can openly discuss with my boss if I am annoyed
		SU_6 I use to dialogue with my colleagues about my problems
		SU_7 I am supported in emotionally demanding tasks
		SU_8 I often receive encouragement by my boss
MS_4	Relationship	RE_1 I do not experience personal harassment in the form of rude words or bad behavior from other colleagues and/or superiors
		RE_2 Frictions or conflicts among colleagues are rare
		RE_3 I am not bullied nor subjected to any restriction
		RE_4 I have the respect that I deserve from my colleagues
		RE_5 Relationships in my workplace are not strained
MS_5	Role	RO_1 I have clear my duties and responsibilities
		RO_2 I have clear the objectives and goals of my department
		RO_3 I have a clear understanding about the importance of my work in pursuing the overall goals of the organization
MS_6	Change	CH_1 I have sufficient opportunities to ask managers for explanations about any change related to my work
		CH_2 Staff is always consulted about potential changes
		CH_3 I clearly understand the practical effects of those changes happening in my work environment

would have if they were isolated. Getting a comprehensive knowledge about MS is essential to lead the risk assessment process according to the particular characteristics of the organisation under analysis.

3   Methodological Approach

In this section we provide methodological details of the techniques we are going to integrate for supporting procedures of organisational risk assessment. The purpose consists in providing a scientifically sound support for dealing with complex environments, where elements of evaluation are typically highly interconnected, and quantitative assessments of variables may be difficult.

The ANP will attribute degrees of importance to MS by taking into account the existence of complex relations of mutual dependence. FCMs will help to understand which items - among those belonging to the mainly critical WS - are the most significant to promote proper management actions.

3.1   ANP to Weight Management Standards

The ANP, first implemented by Thomas Saaty [13] as a development of the Analytic Hierarchy Process (AHP) [12], is a decision-making tool widely applied to assess the main elements of a problem (also called nodes). The goal consists in calculating a vector of weights by considering the possible interdependence among the nodes. In the present paper, the ANP application is conducted to evaluate the set of six MS discussed in the previous section. The practical application will be led by collecting preference judgments with the help of an expert in the field. The ANP technique is implemented as described next [4].

- Representing the decision-making problem by means of a hierarchical structure, clearly characterising nodes. Once the structure has been fixed, relations of interdependence among the nodes have to be formalised. At this stage, opinions provided by the expert will be important to highlight and characterise any possible relation.
- Building the influence matrix, in which relations identified during the previous stage are formalised. The influence matrix is a squared block matrix, whose size equals the total number of nodes and whose entries a_{ij} are equal to 1 if a relation of dependence between element j over element i exists, 0 otherwise. The influence matrix acts as a template for the non-zero elements of the unweighted supermatrix described next.
- Building the unweighted supermatrix (following the non-zero-entry structure of the influence matrix) by pairwise comparing those nodes for which a relation of dependence has been identified ($a_{ij} = 1$), and by calculating weights for the corresponding elements, for example by making use of the AHP, as we will propose in our application. The calculated weights will be the entries of the unweighted supermatrix.
- Producing the weighted supermatrix by means of a normalisation procedure. The sums of the columns of the weighted supermatrix will be equal to one and, in such a way, the matrix gets stochastic.
- Obtaining the limit matrix by raising to powers the weighted supermatrix. All the columns of the limit matrix are equal, and each one of them represents the global priorities, which will have to be eventually normalised to produce the sought information.

- Formalising the final vectors of weights, which embody the interdependence accumulated throughout the successive powering of the weighted supermatrix. Broadly speaking, elements with associated higher values should have major prominence in leading the decision-making process.

3.2   FCM for Analysing Dependence Relations

FCMs [5] enables to analyse complex decision-making problems by modelling and understanding relationships of dependence coexisting within a set of elements [2]. Relations are represented by means of linguistic variables treated as fuzzy numbers. Indirect effects and total effects (namely IE and TE) from element C_i to element C_j are described by using such linguistic evaluations e_{ij} as *much*, *some* and *a lot*, to be translated to fuzzy numbers. Figure 1 shows as an example the FCM proposed in [5], whose network is used to formalise the next equations.

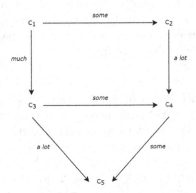

Fig. 1. Example of FCM developed by Kosko [5]

C_1 and C_5 can be connected with each other by means of three possible casual paths, which we herein indicate as $P_1(1 - 2 - 4 - 5)$, $P_2(1 - 3 - 5)$ and $P_3(1 - 3 - 4 - 5)$. Tree indirect effects between C_1 and C_5 are associated to these paths (IE_1, IE_2 and IE_3):

$$IE_1(C_1, C_5) = \min\{e_{12}, e_{24}, e_{45}\} = \min\{some, a\,lot, some\} = some; \qquad (1)$$

$$IE_2(C_1, C_5) = \min\{e_{13}, e_{15}\} = \min\{much, a\,lot\} = much; \qquad (2)$$

$$IE_3(C_1, C_5) = \min\{e_{13}, e_{34}, e_{45}\} = \min\{much, some, some\} = some. \qquad (3)$$

Furthermore, apart from evaluating indirect effects IE, the total effect TE of element C_1 over element C_5 has to be taken into account. The total effect corresponds to the maximum evaluation associated to the three indirect effects, which in our case is:

$$TE(C_1, C_5) = \max\{IE_1(C_1, C_5), IE_2(C_1, C_5), IE_3(C_1, C_5)\} = much. \qquad (4)$$

This result means that, on the whole, element C_1 imparts *much* causality to element C_5. Linguistic evaluations are translated to fuzzy numbers, i.e. triangular or trapezoidal fuzzy numbers, collected into input matrices and represented by a map graphically showing the entity of the relations among elements.

4 Real Case Study of an Italian Airport

4.1 Context Description

The civil airport sector has gone through deep structural modifications and developments over the last few decades. International airports no longer operate as mere providers of infrastructure services, but they can be considered as actual complex business organisations, offering a wide plurality of services with the consequent need of designing and implementing suitable cost management strategies. To such an aim, airport managers dedicate plenty of efforts to the diversification of income sources with the purpose of generating revenues from many diverse activities. This aspect also refers to the aggressive competition among international airports caused by such processes as liberalisation and privatisation, with consequent management of increasing passengers' flows as well as portfolio of routes and affiliated airlines.

Some preliminaries are herein reported to complement the context description according to definitions provided by regulatory sources, is necessary. An *aerodrome* is an area with well-defined boundaries, dedicated to the such activities as landing, take-off and ground movement operations from both civil and aviation military aircrafts, used for commercial, entertainment or training purposes. An *airport* is an aerodrome provided with additional infrastructures that are aimed at offering services for management of aircraft, passengers and goods. An airport is hence a highly complex environment, where the organic organisation of multiple and varied activities is required from several companies that have to simultaneously coexist and operate in the same physical area. The capability for promptly responding to precise standards and practices aimed at minimising risks is clearly crucial, something that has to be verified by proper airport certification processes.

We are herein analysing an Italian airport classified within the *small* category, which registered a yearly flow of around 500.000 passengers in the period antecedent to the outbreak of the COVID-19 pandemic. The company in charge of the airport management has been operating for several years by integrating as much as possible the administration of areas, infrastructures and plants, and by taking special care of maintenance activities. Furthermore, business processes are periodically reviewed in order to improve the quality of services, to optimise costs, operational times and profits. The organisation has the characteristics of a multi-business company, pushing towards continuous strategic consolidation by means of two main criteria. First, the clear attribution of responsibility to the various professional roles promotes a flexible structure, and second, staff activities have been centralised according to human resources management, maintenance and development, administration, finance and control, environment, safety and security, and so on. A total of seventy-eight employees are distributed to the related areas of competence. In this context, organisational aspects are clearly fundamental and proper actions of organisational risk assessment need to be implemented and continuously updated.

4.2 Results and Discussion

The HSE indicator tool previously presented in Sect. 2 has been analysed for the airport of reference with the support of a safety specialist. The responsible of the safety and security system in charge at the airport under consideration has been involved in view of his wide experience on organisational issues. As already illustrated, the present application implements an in-depth analysis of the HSE indicator tool making use of the combination between ANP and FCM, preliminary to the stage of employees' interviews. We specify that interviews will be led and recorded for producing the journal extension of the present work, where we also plan to carry out comparisons with other methodological approaches for organisational risk assessment. For example, methods proposed by Italian regulation authorities such as the national institute for occupational accident insurance (Italian acronym: INAIL) and/or the health and safety prevention service of Verona Province may be object of future evaluation.

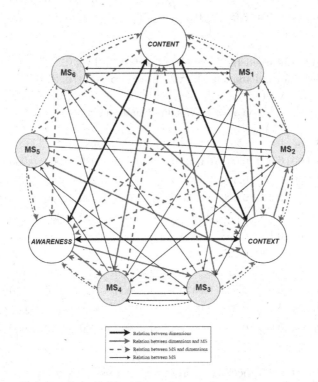

Fig. 2. Relationships linking dimensions and management standards

We now proceed by first applying the ANP technique to calculate the importance weights of the six MS. This will be done by first building the structure of interactions (shown in Fig. 2) formalising relations of dependence among MS with respect to the three main dimensions of reference discussed in Sect. 2.

The unweighted supermatrix (Table 2) has been built by means of the influence relations and preferences established by the involved expert, who was asked to pairwise compare the elements with identified relations of dependence.

Table 2. Unweighted supermatrix

	Goal	Cnt	Cxt	Aws	MS_1	MS_2	MS_3	MS_4	MS_5	MS_6
Goal	0.000	0.000	0.000	0.000	0.000	0.000	0.000	0.000	0.000	0.000
Cnt	0.500	0.000	0.700	0.250	0.000	0.000	0.400	0.400	0.000	0.000
Cxt	0.250	0.500	0.000	0.750	1.000	1.000	0.000	0.000	1.000	1.000
Aws	0.250	0.500	0.300	0.000	0.000	0.000	0.600	0.600	0.000	0.000
MS_1	0.199	0.232	0.185	0.174	0.000	0.000	0.000	0.200	0.000	0.200
MS_2	0.199	0.191	0.250	0.200	0.500	0.000	0.300	0.200	0.400	0.200
MS_3	0.199	0.114	0.225	0.159	0.200	0.000	0.000	0.600	0.300	0.300
MS_4	0.124	0.120	0.177	0.093	0.000	0.000	0.700	0.000	0.300	0.300
MS_5	0.148	0.239	0.088	0.185	0.000	1.000	0.000	0.000	0.000	0.000
MS_6	0.131	0.104	0.075	0.189	0.300	0.000	0.000	0.000	0.000	0.000

The weights obtained by AHP are reported in Table 2, whose columns have been normalised for calculating the weighted supermatrix Table 3.

Table 3. Weighted supermatrix

	Goal	Cnt	Cxt	Aws	MS_1	MS_2	MS_3	MS_4	MS_5	MS_6
Goal	0.000	0.000	0.000	0.000	0.000	0.000	0.000	0.000	0.000	0.000
Cnt	0.250	0.000	0.350	0.125	0.000	0.000	0.200	0.200	0.000	0.000
Cxt	0.125	0.250	0.000	0.375	0.500	0.500	0.000	0.000	0.500	0.500
Aws	0.125	0.250	0.150	0.000	0.000	0.000	0.300	0.300	0.000	0.000
MS_1	0.100	0.116	0.093	0.087	0.000	0.000	0.000	0.100	0.000	0.100
MS_2	0.100	0.096	0.125	0.100	0.250	0.000	0.150	0.100	0.200	0.100
MS_3	0.100	0.057	0.113	0.080	0.100	0.000	0.000	0.300	0.150	0.150
MS_4	0.062	0.060	0.089	0.047	0.000	0.000	0.350	0.000	0.150	0.150
MS_5	0.074	0.120	0.044	0.093	0.000	0.500	0.000	0.000	0.000	0.000
MS_6	0.066	0.052	0.038	0.095	0.150	0.000	0.000	0.000	0.000	0.000

The limit matrix has been then processed by raising the weighted supermatrix to successive powers until convergence. Table 4 reports the values of any of the columns of the limit matrix as well as the weights of MS in percentage. We can observe that the context dimension (cxt) is, on the whole, the most critical in terms of organisational risk management, having associated a weight of 47.56%.

Table 4. Dimensions and MS weights

Dim.	Limit matrix value	% weight	WS	Limit matrix value	% weight
Cnt	2.55E+15	27.27%	MS_1	1.15E+15	12.28%
			MS_2	2.19E+15	23.48%
Cxt	4.44E+15	47.56%	MS_3	1.82E+15	19.52%
			MS_4	1.67E+15	17.87%
Aws	2.35E+15	25.17%	MS_5	1.81E+15	19.42%
			MS_6	6.93E+15	7.43%

However, when we look at the single standards, higher weights correspond to MS_2 (control) and MS_3 (support), respectively referring to the content (cnt) and context (cxt) dimensions. These results indicate that, instead of focusing just on the most critical dimension and on the related MS of support and relationship, it would be preferable to dedicate special attention to the control MS (together with the support MS) for better managing stressful conditions of employees.

The last stage of the application consists in building the FCM for obtaining the total effects associated to relevant items of evaluation (items from Table 1). In such a way, specific aspects that can play a key role for promoting the efficient management of the work-related stress risk can be formally highlighted. Such a type of analysis offers opportunities for pursuing overall organisational optimisation. This is herein achieved by collecting fuzzy preference relations translating evaluations of mutual influence between pairs of elements, again expressed by the responsible of the safety and security system in charge as follows: VL (Very Low), L (Low), M (Medium), H (High), VH (Very High). We are herein reporting the FCM related to the MS of control and support, that are the standards with major significance resulting from the previous ANP application and in need of being managed with priority. The procedure has been initialised by collecting linguistic preferences from our expert, reported in Table 5.

These evaluations have been translated into trapezoidal fuzzy numbers and successively defuzzified by following the procedure implemented in [11]. The last column of Table 5 indicates the total effect of each item, obtained as the maximum between the two values of indirect effects.

The corresponding defuzzified matrix is not herein reported because of the limited space allowed. However, defuzzified values constitute the numerical values of input for building the FCM of Fig. 3, reproduced by iterating the Mental Modeler software. The map shows 106 connections, identified for 15 items, an average of 7.07 connections per item. Items CO_2 and CO_3 have associated evaluations of *medium* total effects for the control MS, while items SU_7 and SU_8 have associated evaluations of *high* total effects for the support MS.

Table 5. Connection matrix

ID	CO$_1$	CO$_2$	CO$_3$	CO$_4$	CO$_5$	CO$_6$	CO$_7$	SU$_1$	SU$_2$	SU$_3$	SU$_4$	SU$_5$	SU$_6$	SU$_7$	SU$_8$	IE	TE
CO$_1$	0	VH	H	VH	VL	VL	L	0	0	0	0	0	0	0	0	VL	L
CO$_2$	VH	0	VH	VH	L	L	M	0	0	0	0	0	0	0	0	L	M
CO$_3$	VH	VH	0	H	M	H	M	0	M	0	0	H	0	0	0	M	M
CO$_4$	VH	H	M	0	L	L	M	0	0	0	0	0	0	0	0	L	L
CO$_5$	L	M	M	L	0	VH	L	0	0	0	0	VL	0	0	0	VL	VL
CO$_6$	L	M	M	L	VH	0	L	0	VH	VH	0	0	0	0	0	L	L
CO$_7$	H	H	H	H	L	L	0	0	0	0	0	0	0	0	0	L	L
SU$_1$	0	0	0	0	0	0	0	0	M	H	VH	H	H	H	H	M	M
SU$_2$	0	0	L	0	0	VH	0	M	0	L	L	M	L	L	M	L	M
SU$_3$	0	0	0	0	0	VH	0	H	M	0	H	VH	M	H	VH	M	M
SU$_4$	0	0	0	0	0	0	0	VH	M	M	0	M	VH	VH	H	M	M
SU$_5$	0	0	H	0	0	VL	0	H	0	VH	H	0	H	M	VH	VL	VL
SU$_6$	0	0	0	0	0	0	0	VH	0	M	VH	H	0	VH	M	M	M
SU$_7$	0	0	0	0	0	0	0	H	VH	H	VH	H	H	0	H	H	H
SU$_8$	0	0	0	0	0	0	0	H	H	VH	H	VH	H	VH	0	H	H
IE	L	M	L	L	VL	VL	L	M	M	L	L	VL	L	L	M	–	–

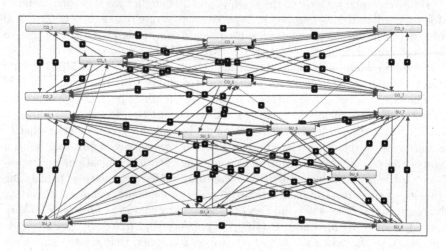

Fig. 3. FCM showing relationships among items of control and support MS

From a practical point of view, these results indicate that stressful conditions concerning standards of control would be realistically reduced if workers received less pressures concerning the rhythm and the organisation of their tasks. Furthermore, support standards would benefit if more attention was paid to such aspects as moral support and encouragement. FCM demonstrates as the discussed factors are mainly related with all the others, so that their priority management would imply general enhancement of working conditions. By implementing the procedure based on ANP and FCM is then clear that specific aspects can be identified and improved for managing work-related stress and for broadly reducing organisational risk at the airport herein presented.

5   Conclusions and Future Research

This research proposes a methodological integration between ANP and FCM as a novel application to the field of organisational risk management in complex business environments. First, ANP is helpful to establish priority organisational standards by analysing relations of dependence among MS. And second, FCM can highlight specific factors that influence global stressful conditions by effectively managing uncertainty. We designed an improved version of the HSE model for work-related stress risk evaluation. Our framework is capable of unveiling those items of the indicator tool that are in need of prominent attention on the basis of mutual relations of influence. Also, our model is less generic than the HSE tool, since it can be personalised according to the specific context of reference by involving inner expert preferences. Our approach was applied to an Italian airport company with meaningful outcome.

Future lines of work will aim to customise even more the tool of work-related analysis by referring to specific homogeneous groups of workers who share similar tasks, being then subjected to risks of similar nature. A decision support system elaborating answers provided by workers may be implemented to support in analysing personal perceptions of workers about significant stressful conditions.

Acknowledgements. The research is financially supported by the Czech Science Foundation, Grant No. 19-06569S.

References

1. Axelrod, R.: Structure of Decision: The Cognitive Maps of Political Elites. Princeton University Press, Princeton (2015)
2. Bakhtavar, E., Valipour, M., Yousefi, S., Sadiq, R., Hewage, K.: Fuzzy cognitive maps in systems risk analysis: a comprehensive review. Complex Intell. Syst. **7**(2), 621–637 (2020). https://doi.org/10.1007/s40747-020-00228-2
3. Carpitella, S., Mzougui, I., Izquierdo, J.: Fuzzy cognitive maps for knowledge-oriented human risk management in industry. In: 26th ISSAT International Conference on Reliability and Quality in Design, RQD 2021, pp. 134–138. International Society of Science and Applied Technologies (2021)
4. Carpitella, S., Mzougui, I., Benítez, J., Carpitella, F., Certa, A., Izquierdo, J., La Cascia, M.: A risk evaluation framework for the best maintenance strategy: the case of a marine salt manufacture firm. Reliab. Eng. Syst. Saf. **205**, 107265 (2021)
5. Kosko, B.: Fuzzy cognitive maps. Int. J. Man Mach. Stud. **24**(1), 65–75 (1986)
6. López, C., Ishizaka, A.: A hybrid FCM-AHP approach to predict impacts of offshore outsourcing location decisions on supply chain resilience. J. Bus. Res. **103**, 495–507 (2019)
7. Marcatto, F., et al.: Work-related stress risk factors and health outcomes in public sector employees. Saf. Sci. **89**, 274–278 (2016)
8. Marcatto, F., Colautti, L., Larese Filon, F., Luis, O., Ferrante, D.: The HSE management standards indicator tool: concurrent and construct validity. Occup. Med. **64**(5), 365–371 (2014)

9. Peeters, L.J., Holland, K.L., Huddlestone-Holmes, C., Boulton, A.J.: A spatial causal network approach for multi-stressor risk analysis and mapping for environmental impact assessments. Sci. Total Environ. **802**, 149845 (2022)

10. Persechino, B., et al.: Work-related stress risk assessment in Italy: a methodological proposal adapted to regulatory guidelines. Saf. Health Work **4**(2), 95–99 (2013)

11. Poomagal, S., Sujatha, R., Kumar, P.S., Vo, D.V.N.: A fuzzy cognitive map approach to predict the hazardous effects of malathion to environment (air, water and soil). Chemosphere **263**, 127926 (2020)

12. Saaty, T.L.: A scaling method for priorities in hierarchical structures. J. Math. Psychol. **15**(3), 234–281 (1977)

13. Saaty, T.L.: Fundamentals of the analytic network process. In: Proceedings of the 5th International Symposium on the Analytic Hierarchy Process, pp. 12–14 (1999)

14. Yu, R., Tzeng, G.H.: A soft computing method for multi-criteria decision making with dependence and feedback. Appl. Math. Comput. **180**(1), 63–75 (2006)

15. Zoni, S., Lucchini, R.G.: European approaches to work-related stress: a critical review on risk evaluation. Saf. Health Work **3**(1), 43–49 (2012)

A Data-Driven Weighting Method Based on DEA Model for Evaluating Innovation Capability in Banking

Nu Dieu Khue Ngo[✉] and Van-Nam Huynh

Japan Advanced Institute of Science and Technology, Nomi, Japan
{khuengo,huynh}@jaist.ac.jp

Abstract. The innovation capability evaluation is in fact a multi-criteria decision-making problem that requires aggregating multiple innovation management practices into a composite innovation capability index. In such multi-criteria decision-making, assigning appropriate weights to criteria is a critical and difficult task. However, the literature related to innovation capability evaluation mainly used the weighting methods based on subjective expert opinions. These conventional methods have problems when dealing with complex multi-criteria data. This study aims to develop a method for automatically determining the weights of multiple innovation management practices for evaluating innovation capability in banking based on data envelopment analysis (DEA) model without input. The results will show the typical importance weights of innovation management practices for each bank which are then used to derive an aggregated index objectively representing the innovation capability level of each bank. A case study of three banks in Vietnam was adopted from the prior study to show the applicability of the proposed method.

Keywords: Data-driven weighting · Data envelopment analysis (DEA) · Innovation capability · Banking

1 Introduction

The fourth industrial revolution with digitization and the explosion of many new technologies such as artificial intelligence, big data, and cloud computing brings great opportunities for the development in the production and business processes. Organizations across sectors have been putting many efforts into exploiting new technologies to innovate their products/services in order to survive in the digital economy. The pivotal role of innovation in the competitive advantage and success of a company is firmly confirmed in the literature [1,2]. According to [3], a company can only effectively innovate if it has innovation capability (IC). IC is a significant determinant of continuous innovations to respond to the dynamic market environment and also firm performance [4,5]. Therefore, the IC evaluation is a serious problem that organizations must consider to comprehend their IC levels and find out important areas in the innovation management process that should

© Springer Nature Switzerland AG 2022
K. Honda et al. (Eds.): IUKM 2022, LNAI 13199, pp. 53–66, 2022.
https://doi.org/10.1007/978-3-030-98018-4_5

be focused on to improve the IC level for achieving better innovative performance as well as higher business performance.

Because IC is a multidimensional process [6,7], the IC evaluation can be considered as a multi-criteria decision-making (MCDM) problem which requires taking into account multiple criteria (in this study, multiple innovation management practices (IMPs)). Some of the IMPs used for measuring IC in the prior studies are strategic planning [8,9], organization [10,11], resource management [12,13], technology management [14,15], research and development (R&D) [16,17], knowledge management [13,18], network and collaboration [8,19]. In MCDM, weighting and aggregating of criteria are major tasks in developing composite indicators [20]. Especially, different sets of weights lead to different ranking outcomes, so the weighting method should be fair. To derive an overall evaluation on the IC of a company, we first need to determine the weights of different IMPs for each company that are then used for computing the composite innovation capability index ($CICI$) of that company.

In the literature on the IC evaluation, the widely used weighting methods have been relied on subjective opinions from experts such as the analytic hierarchy process (AHP) [10], fuzzy measures [17,21]. However, it is difficult, time-consuming, and even costly to get such information from experts, especially in case there are complex and changing multiple criteria. One of the common subjective weighting methods is the AHP which requires subjective judgments of experts to make pairwise comparisons among criteria from which the weights of criteria are obtained. When the number of criteria is high, the experts may face difficulties to deal with many comparisons, sometime they may be confused. It is the reason why the weighting methods that require external or prior information was criticized by [22]. Moreover, the prior studies only applied the same set of weights for different companies. This may cause disagreement among the companies because each company may have its own business strategies that lead to different preferences in developing particular IMPs. To overcome the shortcomings of subjective weighting methods, further consideration can be placed on developing objective weighting methods that can endogenously drive the weights of criteria based on data without referring to any prior or external information. Up to now, far too little attention has been paid to applying data-driven weighting methods in the IC evaluation. This indicates a need to develop a weighting method based on the collected data of IMPs to be applied in evaluating the IC of companies. Several data-driven weighting methods such as DEA, or Genetic Algorithm (GA) can be considered.

The purpose of this paper is to develop a data-driven weighting method based on DEA to determine the typical set of weights of IMPs for each bank or the IMPs focused/ignored by each bank and thereby compute an overall IC evaluation ($CICI$) for each bank based on aggregating multiple IMPs and sub-IMPs. DEA is one of the popular methods for developing composite indices in MCDM, it can select the best possible weights of IMPs for each bank by giving higher weights for better IMPs and therefore give objective evaluations on the IC of banks. To illustrate the applicability and validity of the proposed method, the data of IMPs and sub-IMPs on a case study of three banks in Vietnam was taken from the literature [23]. The data on sub-IMPs were first averaged to obtain the scores of IMPs. The data-driven weighting method developed based on DEA model was then employed to determine the weights of IMPs for each bank that were finally used to aggregate IMPs into a composite index $(CICI)$. The research findings could be used as a basis for benchmarking the most innovative banks and potentially support bank managers in proposing effective strategies for properly allocating innovation resources in order to upgrade their IC and achieve better innovative performance.

This study makes two contributions to the innovation literature as well as the practices of innovation management. First, this study can be considered as one of the first attempts that apply a data-driven weighting method (DEA without input) for evaluating IC. Second, this study will contribute to a deeper understanding of the important IMPs that each bank is focusing on and the corresponding IC levels of banks, based on which some useful lessons can be drawn for innovation management in banking.

The remaining part of this paper proceeds as follows: Sect. 2 reviews theories of IC evaluation and DEA models. Section 3 is concerned with the proposed method by this study. The empirical results of using the proposed IC evaluation method in the case study of three banks in Vietnam are displayed in Sect. 4. Section 5 presents the conclusions of this study.

2   Literature Review

2.1   IC Evaluation

Innovation can be defined as beneficial changes in organizations to create new or improved products/services and thereby to improve business performance [24–27]. Successful innovations require a wide combination of many different assets, resources, and capabilities that facilitate the development of new or improved products/services to better satisfy market needs (also known as IC) [16,28–31]. According to [32], IC refers to the capability of utilizing innovation strategies, technological processes, and innovative behaviors. Lawson and Samson proposed seven constructs in developing IC including strategy, competence, creative idea, intelligence, culture, organization, and technology [14]. As IC is a complex concept that is multi-dimensional and impossible to be measured by a single dimension [33], multiple IMPs must be considered to evaluate the IC of a company.

On account of the role of improving IC for successful innovation, IC evaluation has become one of the dominant streams in the innovation research literature. The common approach for evaluating IC in the previous works was based on multiple IMPs to comprehensively apprehend all necessary capabilities for organizations to effectively innovate. However, particular authors may adapt different IMPs according to the research contexts and also used different techniques to aggregate all IMPs into a single index showing the IC level of a firm. Wang et al. [17] applied a non-additive measure and fuzzy integral method to evaluate the overall performance of technological IC in Taiwanese hi-tech companies. Five factors including innovation-decision, manufacturing, capital, R&D, and marketing capabilities with various qualitative and quantitative criteria were considered in their research. Cheng and Lin [21] proposed a fuzzy expansion of the Technique for Order of Preference by Similarity to Ideal Solution (TOPSIS) to measure the technological IC of Taiwanese printed circuit board firms taking into account seven criteria comprising planning and commitment of the management, knowledge and skills, R&D, marketing, information and communication, operation, and external environment. Wang and Chang [10] presented a hierarchical system to diagnose the innovation value of hi-tech innovation projects considering five main dimensions (strategy innovation, organization innovation, resource innovation, product innovation, and process innovation) and their fifteen secondary dimensions. By adopting the AHP, the main dimensions are found in the descending order of importance to the firm's innovation performance: process innovation, resources innovation, product innovation, strategic innovation, and organizational innovation. Boly et al. [9] adopted a multi-criteria approach and value test method to measure the IC of French small and medium-sized manufacturing companies based on 15 IMPs: strategies management, organization, moral support, process improvement, knowledge management, competence management, creativity, interactive learning, design, project management, project portfolio management, R&D, technology management, customer relationship management, and network management. The evaluated companies were then categorized into four innovative groups (proactive, preactive, reactive, passive) based on their IC levels.

The literature review reveals that many attempts have been made to evaluate IC in manufacturing sectors [9,17,21]. However, there are limited numbers of studies that focus on IC evaluation in the service sector, particularly in the banking sector. In fact, banks are also keenly focusing on innovating their services by adopting new technologies to promptly deliver their services, improve banking experiences for customers, and thereby stay competitive in the market [34]. Innovation becomes a core business value of banks nowadays, it helps banks to explore new opportunities for stable development, and long-term success [35,36]. It is widely approved that innovation in each sector has different unique characteristics [37]; therefore, banks cannot apply the same innovation management policies as manufacturing sectors when developing their new services. Thus, there is an emerging need for a study dedicated to evaluating the IC of banks. In an effort to fill this gap, this study will contribute a method for IC evaluation in banking by investigating the importance weights of IMPs in the banking context as well as determining the overall IC level of banks to be evaluated.

2.2 DEA Models

DEA, proposed by [38], is used to measure the efficiency of decision-making units (DMUs) that is obtained as the maximum of a ratio of a weighted sum of outputs to a weighted sum of inputs. For each particular DMU, the weights are chosen to maximize its efficiency. For example, to calculate the efficiency of a DMU k in a set of all DMUs to be measured K:

$$\text{Maximize: } e_k = \frac{\sum_{i=1}^{n} w_i y_{ik}}{\sum_{j=1}^{m} u_j x_{jk}} \tag{1}$$

subject to

$$e_{k'} = \frac{\sum_{i=1}^{n} w_i y_{ik'}}{\sum_{j=1}^{m} u_j x_{jk'}} \leq 1; \quad \forall k' \in K$$

$$w_i, u_j \geq 0; \quad i = 1, ..., n; \quad j = 1, ..., m$$

where e_k and $e_{k'}$ are the efficiency of DMU k and DMU k', k and $k' \in K$; n and m are the number of outputs and the number of inputs, respectively; w_i and u_j are the weight of the i-th output $(i = 1, ..., n)$ and the weight of the j-th input $(j = 1, ..., m)$, respectively; $y_{ik'}$ is the value of the i-th output of DMU k'; $x_{jk'}$ is the value of the j-th output of DMU k'. The maximization (Eq. (1)) selects the most favorable set of weights for the DMU k whose score is being optimized while the constraints allow. To compute the efficiency of the other DMUs, it just needs to change what to maximize in Eq. (1). The advantage of the DEA model is that it can endogenously derive the different preference profiles for each DMU and thus provide a more objective evaluation for DMUs than the approaches that determine weights based on subjective information from experts.

DEA has become one of the commonly used techniques that can resolve the subjectivity problem in developing composite indicators. Although the original DEA requires outputs and inputs to be specified, several authors have proposed DEA-like models to solve the problems that there is no input. For instance, Zhou et al. [39] presented the best practice model in which a DEA-like model without input was used to obtain the different weights for each DMU. Their approach allows each DMU to pick its own most favorable weights to maximize its aggregated score. However, extreme weighting of sub-indicators may occur, so this approach becomes unrealistic and comes with low discriminating power. To alleviate this shortcoming, Hatefi and Torabi [40] proposed a common weights approach, the same weights are applied to compute scores for all DMUs, to improve discriminating power. The authors used an optimization model to select the weights that minimize the largest deviation among the scores' deviations from 1. This means the selected weights will maximize the lowest score. Thus, this approach has a drawback as the worst performing DMU controls the final weights.

3   Data-Driven Weighting Method Based on DEA Model

In this study, a data-driven weighting method based on DEA model is proposed to compute composite indices representing IC levels of banks $(CICI)$. However, in our formulation, the proposed DEA model has no input and several revisions in constraint conditions compared with the original DEA model.

The IC evaluation in banking follows the two-level hierarchy: the upper level contains IMPs and the lower level comprises the sub-IMPs related to each IMP in the upper level. The sub-IMPs are assessed using a five-point Likert scale (from 1-very bad to 5-very good) to show how efficiently those practices are achieved at the evaluated banks. Accordingly, there are two levels of aggregation to calculate the $CICI$ of these banks. The first level of aggregation (lower level aggregation) is to aggregate sub-IMPs of an IMP to determine the development degree of this

IMP at each bank. The second level of aggregation (upper level aggregation) aims to aggregate IMPs to derive the overall IC of each bank ($CICI$).

3.1 Lower Level Aggregation

Let B be the set of all banks to be evaluated. Considering a bank $b \in B$, the development degree of IMP i at bank b is determined as follows:

$$\bar{x}_i^{(b)} = \frac{1}{N_i} \sum_{j=1}^{N_i} x_{ij}^{(b)}, \quad i \in \{1, ..., N\} \tag{2}$$

where $\bar{x}_i^{(b)}$ is the development degree of IMP i at bank b, $\bar{x}_i^{(b)} \in [1, 5]$; $x_{ij}^{(b)}$ is the score of the j-th sub-IMP of the i-th IMP of bank b, $x_{ij}^{(b)} \in [1, 5]$; N_i is the number of sub-IMPs associated with IMP i; N is the number of IMPs.

According to Eq. (2), the development degree of an IMP is obtained by averaging the scores of all sub-IMPs related to this IMP, in other words, the weights of sub-IMPs are equal. Equal weighting is applied because the relation between IMPs and their measurement items (sub-IMPs) is not causal [9]. Moreover, we prioritize to determine the different weights of IMPs in the upper level of aggregation to specify critical IMPs that much decide the IC of banks.

3.2 Upper Level Aggregation

For the upper level aggregation, we first determine the optimal set of weights of IMPs for each bank so that it will maximize the $CICI$ of the bank being evaluated. The optimal weights for each bank is calculated based on the data of IMPs obtained in the lower level aggregation.

Considering a bank $b \in B$ (B is the set of all banks to be evaluated), let $W^{(b)} = \{w_1^{(b)}, ..., w_N^{(b)}\}$ be the optimal set of weights for maximizing the $CICI$ of bank b, $CICI^{(b)} \in [1, 5]$. The optimal set of weights for bank b is determined by solving the optimization problem below:

$$\text{Maximize} \quad CICI^{(b)} = \sum_{i=1}^{N} \bar{x}_i^{(b)} \times w_i^{(b)} \tag{3}$$

subject to

$$0 \leq w_i^{(b)} \leq 1 \quad \text{and} \quad \sum_{i=1}^{N} w_i^{(b)} = 1, \quad i \in \{1, ..., N\} \tag{4}$$

where $\bar{x}_i^{(b)}$ is the development degree of IMP i at bank b, $\bar{x}_i^{(b)} \in [1, 5]$; $w_i^{(b)}$ is the weight of IMP i in the optimal set of weights $W^{(b)}$ for bank b; N is the number of IMPs. The above optimization problem is converted into a linear programming problem that can be solved by a linear programming solver (such as Scipy package for Python).

It is worth noting that the most ideal $CICI$ value that a bank can reach is 5, but in practice, the $CICI$ values are usually lower than 5. Therefore, we set the threshold of $CICI$ as $5 - \epsilon, \epsilon \in [0, 4]$. One more constraint condition is added to solve the above optimization problem: The $CICI$ values of all banks in the set B must be equal or lower than $5 - \epsilon$ when applying the optimal weights for bank b being optimized.

$$\sum_{i=1}^{N} \bar{x}_i^{(b')} \times w_i^{(b)} \leq 5 - \epsilon; \quad \epsilon \in [0, 4]; \quad \forall b' \in B \tag{5}$$

It is clear that, if the value of ϵ is low, extreme weighting may occur with higher weights for better IMPs, which leads to a high standard deviation of weight values. When ϵ is increased, the standard deviation of weight values will be reduced. At the standard deviation of weight values equals 0, equal weighting happens. The selection of ϵ is optional, depending on the evaluator's preference. ϵ can be chosen so that the corresponding standard deviation of weight values is in the range between its highest value and its lowest value. If the evaluator prefers the weights toward extreme weighting to clearly show the best practices of each bank, ϵ is selected at the corresponding standard deviation of weight values near its highest value. In contrast, in case the evaluator prefers the weights toward equal weighting, ϵ is chosen so that the corresponding standard deviation of weight values is close to its lowest value. In this study, we tend to choose the standard deviation of weight values in the middle area of its possible range to balance extreme weighting and equal weighting.

The optimal set of weights for a bank can disclose which IMPs that this bank is focusing on. By comparing with other banks, we can explore the strengths and weaknesses of each bank on different IMPs.

4   An Illustrated Example

This example is adopted from the research of [23] on evaluating the IC of three banks in Vietnam. The concept of IC in their research was defined based on the Pareto analysis - a statistical technique to select the major tasks which the management should put more effort into. As a result, 11 IMPs were chosen as critical practices in innovation management process: managing strategy (MS), managing resource (MR), organizing (OR), managing idea (MI), improving process (IP), marketing (MA), R&D (RD), interactive learning (IL), managing portfolio (MP), managing knowledge (MK), and managing technology (MT). The 44 measurement items/sub-IMPs measuring the 11 IMPs were adapted from [8–13, 15, 16, 19, 41–47], which ensures the reliability and validity of the measurement scale as they were verified through peer-reviewed previous research (see Table 1). In their data collection [23], five experts in banking fields individually responded to the questionnaire to rate the development degrees of sub-IMPs in the three evaluated banks, enormously called Bank a, Bank b, and Bank c, using a five-point Likert-scale ranging from 1 (very bad) to 5 (very good). The final scores of 44 sub-IMPs for the three banks (shown in Table 2) were obtained by averaging the assessment scores of the five experts.

Table 1. IMPs and sub-IMPs

No	IMPs	Sub-IMPs
1	MS	MS1: Set clear innovation goals in business strategies MS2: Widely disseminate innovation strategies throughout the bank MS3: Managers dedicatedly encourage innovation practices MS4: Effective use methods supporting decision making to create business strategies
2	MR	MR1: Provide proper resources for innovation MR2: Manage adaptive and diverse capital sources MR3: Concentrate on employing talented employees MR4: Regularly schedule training programs for providing necessary knowledge to develop new services
3	OR	OR1: Organizational culture and atmosphere assist innovative activities OR2: Reward employees for their innovation achievements OR3: Tolerate failures in doing something new OR4: Develop interactive communication systems among employees in the bank
4	MI	MI1: Have a validated process to gather ideas from various divisions in the bank MI2: Collaborate with outside organizations for idea development MI3: Have a quick procedure to evaluate new ideas MI4: Use a test markets before launching new services
5	IP	IP1: Structure innovation processes IP2: Assign facilitators supporting innovation activities IP3: Schedule regular meetings to inspect innovation activities IP4: Managers usually examine the development of innovation projects
6	MA	MA1: Keep great associations with clients MA2: Have capable sales employees MA3: Evaluate the levels of customer satisfaction after sales MA4: Create a positive brand image in clients' minds
7	RD	RD1: Structure R&D programs RD2: Upgrade funds for R&D activities RD3: Enhance cooperation across different functional departments RD4: Hold regular meetings to discuss R&D subjects
8	IL	IL1: Boost interactive learning activities IL2: Assign managers who are responsible for interactive learning activities IL3: Hole meetings to evaluate the completed innovation projects IL4: Disseminate experiences obtained from past projects all through the bank
9	MP	MP1: Business strategies fit with investment portfolios MP2: Analyze all proceeding projects based on multiple criteria MP3: Have periodic reports on the allocation of resources to projects MP4: Assure the balance between long-term and short-term, and high-risk and low-risk projects
10	MK	MK1: Identify and update employees' knowledge to satisfy job requirements MK2: Encourage knowledge sharing at work MK3: Classify and store knowledge for employees to easily access MK4: Adapt knowledge dissemination methods
11	MT	MT1: Increase the integration of new technologies into banking products as a key success factor MT2: Plan scenarios to predict the trend of new technologies MT3: Capture the technologies competitors are using MT4: Technologies acquired from the external fit the infrastructures and operations of the bank

Table 2. Scores of 44 sub-IMPs for three banks in Vietnam

Bank	MS1	MS2	MS3	MS4	MR1	MR2	MR3	MR4	OR1	OR2	OR3	OR4	MI1	MI2	MI3	MI4	IP1	vIP2	IP3	IP4	MA1	MA2
a	4.4	4.2	3.8	4.0	3.4	4.2	4.0	3.4	3.8	4.0	3.6	3.4	3.2	3.4	3.2	3.4	3.4	3.8	3.8	3.6	4.0	3.8
b	4.6	4.4	4.4	4.4	4.0	4.0	4.6	4.2	4.4	4.0	3.6	4.2	3.8	4.0	3.8	4.2	4.2	4.2	4.2	4.2	4.6	4.2
c	4.0	4.0	4.8	4.4	4.0	4.8	4.4	4.2	3.4	4.2	3.4	4.0	4.2	3.6	3.8	3.6	4.2	4.0	4.0	3.8	4.4	3.6

Bank	MA3	MA4	RD1	RD2	RD3	RD4	IL1	IL2	IL3	IL4	MP1	MP2	MP3	MP4	MK1	MK2	MK3	MK4	MT1	MT2	MT3	MT4
a	3.6	3.8	3.6	3.8	3.6	3.6	4.0	3.4	3.6	3.2	4.2	3.4	3.8	3.8	4.6	3.8	3.8	4.6	4.4	3.6	3.8	3.4
b	4.2	4.4	4.4	4.0	4.4	4.2	4.0	4.0	4.2	4.0	4.4	4.2	4.2	4.0	4.0	4.2	4.0	4.0	4.2	4.2	4.4	4.2
c	3.8	4.2	4.0	3.8	3.6	3.6	4.2	4.2	4.2	4.0	4.4	4.4	4.0	4.2	4.2	4.2	4.0	3.8	3.8	3.8	3.8	4.2

The IC evaluation for the three banks is composed of two levels of aggregations as shown in Fig. 1. In the lower level aggregation, the 4 sub-IMPs associated with each IMP at each bank are aggregated. Eq. (2) with the values of Table 3 gives the average scores of the 11 IMPs for the three banks in the sample.

To aggregate the 11 IMPs in the upper level, we first need to determine the optimal weights of the 11 IMPs for each of the three banks by solving model (3) under the constraints (4) and (5). ϵ in the constraints (5) was run with the initial value of 0 and the increased step size of 0.05. Figure 2 shows different values of ϵ and corresponding standard deviations of weight values. In this study, we chose $\epsilon = 0.85$ for Bank a, $\epsilon = 0.65$ for Bank b, and $\epsilon = 0.70$ for Bank c so that the corresponding standard deviations of weight values are in the middle area of its possible range. Table 4 displays the optimal set of weights for each bank at the chosen ϵ. As a final result, the $CICI$ values of Bank a, Bank b, and Bank c were determined to be 4.15, 4.35, and 4.30, respectively using each bank's optimal sets of weights. According to that, Bank b is the most innovative bank among the three evaluated bank. This results were verified by comparing with the ranking of the same three banks based on subjective models in [23].

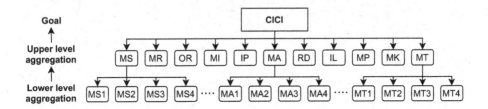

Fig. 1. Hierarchical structure of IMPs and sub-IMPs for evaluating IC in banking

Table 3. Average scores of 11 IMPs for three banks in Vietnam

	MS	MR	OR	MI	IP	MA	RD	IL	MP	MK	MT
Bank a	4.10	3.75	3.70	3.30	3.65	3.80	3.65	3.55	3.80	4.20	3.80
Bank b	4.45	4.20	4.05	3.95	4.20	4.35	4.25	4.05	4.20	4.05	4.25
Bank c	4.30	4.35	3.75	3.80	4.00	4.00	3.75	4.15	4.25	4.05	3.90

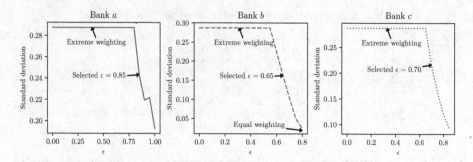

Fig. 2. Different ϵ values and corresponding standard deviations of weight values

Table 4. Optimal weights and $CICI$ for each bank

	w_{MS}	w_{MR}	w_{OR}	w_{MI}	w_{IP}	w_{MA}	w_{RD}	w_{IL}	w_{MP}	w_{MK}	w_{MT}	$CICI$
$W^{(a)}(\epsilon=0.85)$	0.051	0.009	0.011	0.006	0.008	0.013	0.009	0.007	0.013	0.859	0.014	$CICI^{(a)}=4.15$
$W^{(b)}(\epsilon=0.65)$	0.603	0.036	0.029	0.022	0.043	0.073	0.055	0.024	0.038	0.024	0.052	$CICI^{(b)}=4.35$
$W^{(c)}(\epsilon=0.70)$	0.072	0.771	0.008	0.011	0.016	0.014	0.007	0.028	0.042	0.02	0.011	$CICI^{(c)}=4.30$

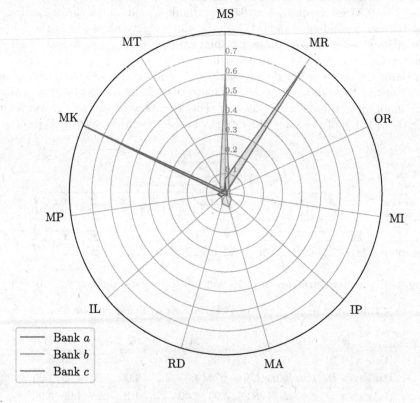

Fig. 3. Weights of IMPs in three banks in Vietnam

5   Conclusion

This study proposes a data-driven weighting method based on DEA to solve a multi-criteria problem that is then applied in evaluating the IC of the three banks in Vietnam. The proposed method can determine the optimal set of weights for maximizing each bank's IC. This way contributes an objective evaluation or ranking approach on IC without bias toward any banks. Based on the optimal set of weights of each bank, we can point out which IMPs each bank is focusing on (strengths) or ignoring (weaknesses). Particularly, by applying the proposed method in the case of the three banks in Vietnam, we found distinctive IMPs of each bank as follows:

- Bank a: This bank was found to pay attention to only two IMPs (MK "managing knowledge" and MS "managing strategies") while almost neglecting the rest of IMPs. It must be noted that most IMPs in Bank a have the least implemented levels among the three banks, except for MK. Generally, the IC level of Bank a is lower than the other two banks.
- Bank b: Except for MI "managing ideas" where its score is a bit lower than other IMPs, Bank b widely develops other IMPs, especially focuses on managing strategies, marketing, R&D, managing technologies, and improving processes. Most IMPs have the implemented levels generally higher than the other banks. Globally, this bank may be considered as being most seriously pursuing innovation activities.
- Bank c: This bank puts more efforts into managing resources, managing strategies, and managing portfolio while keeping good levels on improving processes, marketing, interactive learning, and managing knowledge. It is at low levels in organizing, managing ideas, R&D, and managing technologies.
- It can also be noticed that all of the three banks, specially the most innovative bank (Bank b) give prominence to managing strategies in innovation management, which proves that strategies management is an important practice in innovation management in Vietnamese banks. The above-mentioned points are graphically described in Fig. 3.

The research results also reveal the ranking of the three banks based on their IC. In details, Bank b is the most innovative bank among the three banks, the next is Bank c, and Bank a was ranked last. The findings provide a basis for bank managers to improve their innovation management policies to upgrade their IC. Specifically, to increase the IC level, a bank can strengthen its IC by prioritizing to allocate more resources into the most important IMPs that have the strongest weights such as strategies management, marketing, and R&D as the most innovative bank (Bank b) does.

This study is limited by the a small sample size with only three banks in Vietnam. The future study should use a bigger sample size to establish a greater degree of applicability and validity of the proposed method. In addition, the discriminating power among the evaluated banks is still low (in case of comparing the IC levels between Bank b and Bank c). Considerably more work will need to be done to develop other methods that can create a more distinguishable ranking, for example using multi-objective approach.

References

1. Crossan, M.M., Apaydin, M.: A multi-dimensional framework of organizational innovation: a systematic review of the literature. J. Manage. Stud. **47**(6), 1154–1191 (2010)
2. Soliman, F.: Does innovation drive sustainable competitive advantages? J. Mod. Account. Audit. **9**(1), 130–143 (2013)
3. Laforet, S.: A framework of organisational innovation and outcomes in SMEs. Int. J. Entrepreneurial Behavior Res. **17**(4), 380–408 (2011). https://doi.org/10.1108/13552551111139638
4. Slater, S.F., Hult, G.T.M., Olson, E.M.: Factors influencing the relative importance of marketing strategy creativity and marketing strategy implementation effectiveness. Ind. Mark. Manage. **39**(4), 551–559 (2010)
5. Mone, M.A., McKinley, W., Barker, V.L., III.: Organizational decline and innovation: a contingency framework. Acad. Manage. Rev. **23**(1), 115–132 (1998)
6. Raghuvanshi, J., Garg, C.P.: Time to get into the action: unveiling the unknown of innovation capability in Indian MSMEs. Asia Pac. J. Innov. Entrep. **12**(3), 279–299 (2018). https://doi.org/10.1108/APJIE-06-2018-0041
7. Saunila, M., Ukko, J.: A conceptual framework for the measurement of innovation capability and its effects. Balt. J. Manage. **7**(4), 355–375 (2012)
8. Rejeb, H.B., Morel-Guimarães, L., Boly, V., et al.: Measuring innovation best practices: improvement of an innovation index integrating threshold and synergy effects. Technovation **28**(12), 838–854 (2008)
9. Boly, V., Morel, L., Camargo, M., et al.: Evaluating innovative processes in French firms: methodological proposition for firm innovation capacity evaluation. Res. Policy **43**(3), 608–622 (2014)
10. Wang, T., Chang, L.: The development of the enterprise innovation value diagnosis system with the use of systems engineering. In: Proceedings 2011 International Conference on System Science and Engineering, pp. 373–378. IEEE (2011)
11. Yam, R.C., Lo, W., Tang, E.P., Lau, A.K.: Analysis of sources of innovation, technological innovation capabilities, and performance: an empirical study of Hong Kong manufacturing industries. Res. Policy **40**(3), 391–402 (2011)
12. Tidd, J., Thuriaux-Alemán, B.: Innovation management practices: cross-sectorial adoption, variation, and effectiveness. R&D Manag. **46**(S3), 1024–1043 (2016)
13. Liu, L., Jiang, Z.: Influence of technological innovation capabilities on product competitiveness. Ind. Manag. Data Syst. **116**(5), 883–902 (2016)
14. Lawson, B., Samson, D.: Developing innovation capability in organisations: a dynamic capabilities approach. Int. J. Innov. Manag. **5**(03), 377–400 (2001)
15. Koc, T., Ceylan, C.: Factors impacting the innovative capacity in large-scale companies. Technovation **27**(3), 105–114 (2007)
16. Guan, J.C., Yam, R.C., Mok, C.K., Ma, N.: A study of the relationship between competitiveness and technological innovation capability based on DEA models. Eur. J. Oper. Res. **170**(3), 971–986 (2006)
17. Wang, C.-H., Lu, I.-Y., Chen, C.-B.: Evaluating firm technological innovation capability under uncertainty. Technovation **28**(6), 349–363 (2008)

18. Yang, C., Zhang, Q., Ding, S.: An evaluation method for innovation capability based on uncertain linguistic variables. Appl. Math. Comput. **256**, 160–174 (2015)

19. Sumrit, D., Anuntavoranich, P.: Using DEMATEL method to analyze the causal relations on technological innovation capability evaluation factors in Thai technology-based firms. Int. Trans. J. Eng. Manag. Appl. Sci. Technol. **4**(2), 81–103 (2013)

20. Saisana, M., Tarantola, S.: State-of-the-art report on current methodologies and practices for composite indicator development, vol. 214. Citeseer (2002)

21. Cheng, Y.-L., Lin, Y.-H.: Performance evaluation of technological innovation capabilities in uncertainty. Proc. Soc. Behav. Sci. **40**, 287–314 (2012)

22. Chung, W.: Using DEA model without input and with negative input to develop composite indicators. In: 2017 IEEE International Conference on Industrial Engineering and Engineering Management (IEEM), pp. 2010–2013. IEEE (2017)

23. Ngo, N.D.K., Le, T.Q., Tansuchat, R., Nguyen-Mau, T., Huynh, V.-N.: Evaluating innovation capability in banking under uncertainty. IEEE Trans. Eng. Manag. 1–18 (2022). https://doi.org/10.1109/TEM.2021.3135556

24. Damanpour, F.: Organizational complexity and innovation: developing and testing multiple contingency models. Manag. Sci. **42**(5), 693–716 (1996)

25. Rogers, M.: The definition and measurement of innovation. Melbourne Institute of Applied Economic and Social Research, University of Melbourne, Melbourne (1998)

26. Du Plessis, M.: The role of knowledge management in innovation. J. Knowl. Manag. **11**(4), 20–29 (2007)

27. Baregheh, A., Rowley, J., Sambrook, S.: Towards a multidisciplinary definition of innovation. Manag. Decis. **47**, 1323–1339 (2009)

28. Sen, F.K., Egelhoff, W.G.: Innovative capabilities of a firm and the use of technical alliances. IEEE Trans. Eng. Manag. **47**(2), 174–183 (2000)

29. Christensen, J.F.: Asset profiles for technological innovation. Res. Policy **24**(5), 727–745 (1995)

30. Szeto, E.: Innovation capacity: working towards a mechanism for improving innovation within an inter-organizational network. TQM Mag. **12**(2), 149–158 (2000)

31. Burgelman, R., Christensen, C., Wheelwright, S.: Strategic Management of Technology and Innovation. Mc-Graw-Hill, New York (2009)

32. Wang, C.L., Ahmed, P.K.: The development and validation of the organisational innovativeness construct using confirmatory factor analysis. Eur. J. Innov. Manag. **7**(4), 303–313 (2004)

33. Guan, J., Ma, N.: Innovative capability and export performance of Chinese firms. Technovation **23**(9), 737–747 (2003)

34. Parameswar, N., Dhir, S., Dhir, S.: Banking on innovation, innovation in banking at ICICI bank. Glob. Bus. Organ. Excell. **36**(2), 6–16 (2017)

35. Ikeda, K., Marshall, A.: How successful organizations drive innovation. Strategy Leadersh. **44**(3), 9–19 (2016). https://doi.org/10.1108/SL-04-2016-0029

36. Tajeddini, K., Trueman, M., Larsen, G.: Examining the effect of market orientation on innovativeness. J. Mark. Manag. **22**(5–6), 529–551 (2006)

37. Drejer, I.: Identifying innovation in surveys of services: a Schumpeterian perspective. Res. Policy **33**(3), 551–562 (2004)

38. Charnes, A., Cooper, W.W., Rhodes, E.: Measuring the efficiency of decision making units. Eur. J. Oper. Res. **2**(6), 429–444 (1978)

39. Zhou, P., Ang, B., Poh, K.: A mathematical programming approach to constructing composite indicators. Ecol. Econ. **62**(2), 291–297 (2007)

40. Hatefi, S., Torabi, S.: A common weight MCDA-DEA approach to construct composite indicators. Ecol. Econ. **70**(1), 114–120 (2010)
41. Cooper, R.G., Edgett, S.J., Kleinschmidt, E.J.: Benchmarking best NPD practices-II. Res. Technol. Manag. **47**(3), 50–59 (2004)
42. Easingwood, C.J.: New product development for service companies. J. Prod. Innov. Manag. **3**(4), 264–275 (1986)
43. Oke, A.: Innovation types and innovation management practices in service companies. Int. J. Oper. Prod. Manag. **27**(6), 564–587 (2007)
44. Kuczmarski, T.: Winning new product and service practices for the 1990s. Kuczmarski & Associates, Chicago (1994)
45. Cooper, R.G., Edgett, S.J., Kleinschmidt, E.J.: Benchmarking best NPD practices-I. Res. Technol. Manag. **47**(1), 31–43 (2004)
46. Griffin, A.: PDMA research on new product development practices: updating trends and benchmarking best practices. J. Prod. Innov. Manag. Int. Publ. Prod. Dev. Manag. Assoc. **14**(6), 429–458 (1997)
47. Yam, R.C., Guan, J.C., Pun, K.F., Tang, E.P.: An audit of technological innovation capabilities in Chinese firms: some empirical findings in Beijing, China. Res. Policy **33**(8), 1123–1140 (2004)

Decision Analysis with the Set of Normalized Triangular Fuzzy Weight Vectors in Fuzzy AHP

Shigeaki Innan[✉] and Masahiro Inuiguchi

Graduate School of Engineering Science, Osaka University Toyonaka, Osaka, Japan
innan@inulab.sys.es.osaka-u.ac.jp

Abstract. Intervals and fuzzy numbers have been introduced to the analytic hierarchy process (AHP) reflecting the vagueness of the decision maker. In this paper, we propose a fuzzy AHP approach to multiple criteria decision analysis. First we investigate the normalized fuzzy weight vector estimation problem under a given fuzzy pairwise comparison matrix (PCM). After reviewing a previous approach to the estimation problem, we show the non-uniqueness of the normalized fuzzy weight vector associated with a consistent fuzzy PCM. Those normalized fuzzy weight vectors associated with the same consistent fuzzy PCM are at the same distance from the given PCM. Therefore, we require that all such fuzzy weight vectors should be the solutions to the estimation problem. As the previous estimation method does not satisfy this requirement, the estimation method is modified so that all such normalized fuzzy weight vectors are estimated. Then a decision analysis with all such normalized fuzzy weight vectors is proposed. The stability of the best alternative can be scrutable as the range of alternative orderings is analyzed by the proposed approach.

Keywords: AHP · Triangular fuzzy number · Non-uniqueness · Maximin rule

1 Introduction

The analytic hierarchy process (AHP) [1] is one of the most widely used method in multiple criteria decision analysis. AHP has also been studied for various applications. In AHP, the decision maker makes pairwise comparisons between alternatives/criteria. In a pairwise comparison, the relative importance is given as the evaluation value. The conventional AHP requires decision-makers to make precise judgments. However, it is difficult to obtain precise judgments from decision makers because their judgments often contain vagueness. In order to deal with this problem, a method of representing relative importance by interval values [2,8] and fuzzy numbers [3,7] has been considered. In those methods, we obtain pairwise comparison matrices with intervals and fuzzy numbers.

© Springer Nature Switzerland AG 2022
K. Honda et al. (Eds.): IUKM 2022, LNAI 13199, pp. 67–78, 2022.
https://doi.org/10.1007/978-3-030-98018-4_6

In this paper, we focus on the case when the components of pairwise comparison matrices are fuzzy numbers, i.e., a fuzzy AHP. In the fuzzy AHP, the method for estimating priority weights from fuzzy pairwise comparisons are proposed. An estimation method of crisp weights from a fuzzy pairwise comparison matrix based on the fuzzy preference programming method has been proposed in [7]. On the other hand, a method for estimating the normalized fuzzy weight vector from a fuzzy pairwise comparison matrix by solving the linear goal programming model has been proposed in [3].

In this paper, we investigate the problem estimating the normalized fuzzy weight vector from a fuzzy pairwise comparison matrix. We consider the case where all fuzzy components of both the pairwise comparison matrix and the weight vector are given by triangular fuzzy numbers. Namely, we assume that the decision maker expresses her/his evaluation of each relative importance of the i-th criterion/alternative to the j-th one by a triangular fuzzy number. Therefore, we obtain a triangular fuzzy pairwise comparison matrix (TFPCM). As the relative importance is given by a triangular fuzzy number, the fuzzy weights are assumed to be triangular fuzzy numbers so that their ratios can approximate well the triangular fuzzy number components of TFPCM.

Under this situation, an estimation method of a normalized triangular fuzzy weight vector has already been proposed. However, we consider that it is not a unique solution. From the solution, we obtain a consistent TFPCM by calculating ratios between the obtained fuzzy weights. We show that the solution associated with the consistent TFPCM is not unique. From this fact, we consider that all normalized triangular fuzzy weight vectors associated with a consistent TFPCM should be solutions to the estimated problem, because the appropriateness of any of those solutions is same because it can be defined by the distance of the consistent TFPCM from the given one. Then we propose a method obtaining all solutions to the estimation problem. It is shown that those solutions are usually represented by a parameter and a normalized fuzzy weight vector. Then the decision analysis based on those solutions of the normalized fuzzy weight vector estimation problem is proposed. By this decision analysis, the range of alternative orderings is analyzed. Through this analysis, the stability of the decision maker's preference as well as conceivable best solutions can be scrutable.

This paper is organized as follows. In Sect. 2, we formally define a TFPCM and explain briefly a conventional method for estimating a normalized triangular fuzzy weight vector from a given TFPCM. In Sect. 3, we describe the non-uniqueness of the solution of the estimation problem and propose a method for obtaining all solutions from the normalized triangular fuzzy weight vector obtained by the conventional method. In Sect. 4, we calculate the total utility value of each alternative as a fuzzy number. In Sect. 5, a numerical example is given to demonstrate the usefulness of the proposed method. In Sect. 6, the concluding remarks are given.

2   The Previous Fuzzy AHP

In the AHP, multiple criteria decision-making problem is structured in a hierarchy of criteria and alternatives. Then, the criteria and alternatives are evaluated in each level of hierarchy. In this paper, we assume that the evaluation values of the alternatives for each criterion are given by the decision maker. The evaluation value for each criterion is given by the decision maker through pairwise comparisons between the criteria. From these pairwise comparison, we can obtain pairwise comparison matrices (PCMs). In the AHP, a weight vector $w = (w_1, w_2, \ldots, w_n)^{\mathrm{T}}$ for criteria is estimated from a PCM $A = (a_{ij})_{n \times n}$. In the conventional AHP, (i, j)-th component a_{ij} of PCM A shows the relative importance of the i-th criterion over the j-th criterion. If human judgments are precise, a_{ij} is equal to $w_i/w_j, i, j \in N = \{1, 2, \ldots, n\}$. However, due to the vagueness of decision maker's judgements, we may assume $a_{ij} \approx w_i/w_j, i, j \in N$. Then, weights $w_i, i \in N$ are estimated by minimizing the sum of deviation between a_{ij} and $w_i/w_j, i, j \in N$.

On the other hand, the method that reflects the vagueness of the decision maker's evaluation is to change the evaluation value of each element of the PCM to a fuzzy number, and to obtain weights from these evaluation values. This approach is called fuzzy AHP. In the fuzzy AHP, we estimate the fuzzy weight vector $\tilde{W} = (\tilde{w}_1, \tilde{w}_2, \ldots \tilde{w}_n)^{\mathrm{T}}$ from the fuzzy PCM $\tilde{A}$. In this paper, we mainly treat the case of triangular fuzzy numbers in fuzzy AHP.

Firstly, we consider a triangular fuzzy pairwise comparison matrix (TFPCM):

$$\tilde{A} = (\tilde{a}_{ij})_{n \times n} = \begin{bmatrix} 1 & \cdots & (a_{1n}^{\mathrm{L}}, a_{1n}^{\mathrm{M}}, a_{1n}^{\mathrm{R}}) \\ \vdots & (a_{ij}^{\mathrm{L}}, a_{ij}^{\mathrm{M}}, a_{ij}^{\mathrm{R}}) & \vdots \\ (a_{n1}^{\mathrm{L}}, a_{n1}^{\mathrm{M}}, a_{n1}^{\mathrm{R}}) & \cdots & 1 \end{bmatrix}, \quad (1)$$

where $a_{ij}^{\mathrm{L}} = 1/a_{ji}^{\mathrm{R}}, a_{ij}^{\mathrm{M}} = 1/a_{ji}^{\mathrm{M}}, a_{ij}^{\mathrm{R}} = 1/a_{ji}^{\mathrm{L}}, i, j \in N(i \neq j)$. The TFPCM $\tilde{A}$ can be split into three crisp matrices:

$$A_{\mathrm{L}} = \begin{bmatrix} 1 & \cdots & a_{1n}^{\mathrm{L}} \\ \vdots & a_{ij}^{\mathrm{L}} & \vdots \\ a_{n1}^{\mathrm{L}} & \cdots & 1 \end{bmatrix}, \ A_{\mathrm{M}} = \begin{bmatrix} 1 & \cdots & a_{1n}^{\mathrm{M}} \\ \vdots & a_{ij}^{\mathrm{M}} & \vdots \\ a_{n1}^{\mathrm{M}} & \cdots & 1 \end{bmatrix}, \ A_{\mathrm{R}} = \begin{bmatrix} 1 & \cdots & a_{1n}^{\mathrm{R}} \\ \vdots & a_{ij}^{\mathrm{R}} & \vdots \\ a_{n1}^{\mathrm{R}} & \cdots & 1 \end{bmatrix}. \quad (2)$$

As with the crisp PCM, each element of the TFPCM indicates the relative importance between criteria. If the TFPCM is precise, then the evaluation value of the paired comparison and the ratio of the fuzzy weights are equivalent, in short, $\tilde{a}_{ij} = (a_{ij}^{\mathrm{L}}, a_{ij}^{\mathrm{M}}, a_{ij}^{\mathrm{R}}) = \tilde{w}_i/\tilde{w}_j = (w_i^{\mathrm{L}}/w_j^{\mathrm{R}}, w_i^{\mathrm{M}}/w_j^{\mathrm{M}}, w_i^{\mathrm{R}}/w_j^{\mathrm{L}}), i, j \in N$ but $j \neq i$, where triangular fuzzy weights $\tilde{w}_i = (w_i^{\mathrm{L}}, w_i^{\mathrm{M}}, w_i^{\mathrm{R}}), i \in N$. From these equations, we can obtain the following equations:

$$A_{\mathrm{L}} W_{\mathrm{R}} = W_{\mathrm{R}} + (n - 1) W_{\mathrm{L}}, \quad (3)$$

$$A_{\mathrm{R}} W_{\mathrm{L}} = W_{\mathrm{L}} + (n - 1) W_{\mathrm{R}}, \quad (4)$$

$$A_{\mathrm{M}} W_{\mathrm{M}} = n W_{\mathrm{M}}, \quad (5)$$

where $W_{\mathrm{L}} = (w_1^{\mathrm{L}}, \ldots, w_n^{\mathrm{L}})^{\mathrm{T}}$, $W_{\mathrm{M}} = (w_1^{\mathrm{M}}, \ldots, w_n^{\mathrm{M}})^{\mathrm{T}}$, $W_{\mathrm{R}} = (w_1^{\mathrm{R}}, \ldots, w_n^{\mathrm{R}})^{\mathrm{T}}$. However, even with fuzzy AHP, it is difficult to obtain a precise fuzzy PCM due to the vagueness of the decision maker's evaluation. In other words, Eqs. (3)–(5) are often not hold. Therefore, the deviation vectors in Eqs. (3)–(5) are defined:

$$E^+ - E^- = (A_{\mathrm{L}} - I)W_{\mathrm{R}} - (n-1)W_{\mathrm{L}}, \tag{6}$$

$$\varGamma^+ - \varGamma^- = (A_{\mathrm{R}} - I)W_{\mathrm{L}} - (n-1)W_{\mathrm{R}}, \tag{7}$$

$$\varDelta = (A_{\mathrm{M}} - nI)W_{\mathrm{M}}, \tag{8}$$

where deviation vectors $E^+ = (\varepsilon_1^+, \ldots, \varepsilon_n^+)^{\mathrm{T}}$, $E^- = (\varepsilon_1^-, \ldots, \varepsilon_n^-)^{\mathrm{T}}$, $\varGamma^+ = (\gamma_1^+, \ldots, \gamma_n^+)^{\mathrm{T}}$, $\varGamma^- = (\gamma_1^-, \ldots, \gamma_n^-)^{\mathrm{T}}$, $\varDelta = (\delta_1, \ldots, \delta_n)^{\mathrm{T}}$, I is $n \times n$ unit matrix, $\varepsilon_i^+, \varepsilon_i^-, \gamma_i^+, \gamma_i^-, \delta_i, i \in N$ are deviation variables. These deviation variables are non-negative. The triangular fuzzy weight vector $\tilde{W}$ is estimated from the TFPCM $\tilde{A}$ by minimizing the sum of the deviation variables in Eqs. (6)–(8).

In addition, the normalization condition is considered as a property that the fuzzy weight vector should satisfy. The normalization condition [6] of the triangular fuzzy weight vector $\tilde{W}$ are written as

$$\sum_{i \in N \backslash j} w_i^{\mathrm{R}} + w_j^{\mathrm{L}} \geq 1, \quad \sum_{i \in N \backslash j} w_i^{\mathrm{L}} + w_j^{\mathrm{R}} \leq 1, \ j \in N, \quad \sum_{i \in N} w_i^{\mathrm{M}} = 1. \tag{9}$$

Under the constraints of Eqs. (6)–(9), the fuzzy weight vector can be estimated by minimizing the sum of deviation variables. Therefore, we can obtain the following linear goal programming (LGP) model for estimating the fuzzy weight vector. This model was proposed by Y. -M. Wang et al. [3]

$$
\begin{aligned}
\text{minimize} \quad & e^{\mathrm{T}}(E^+ + E^- + \varGamma^+ + \varGamma^- + \varDelta) \\
\text{subject to} \quad & (A_{\mathrm{L}} - I)W_{\mathrm{R}} - (n-1)W_{\mathrm{L}} - E^+ + E^- = 0, \\
& (A_{\mathrm{R}} - I)W_{\mathrm{L}} - (n-1)W_{\mathrm{R}} - \varGamma^+ + \varGamma^- = 0, \\
& (A_{\mathrm{M}} - nI)W_{\mathrm{M}} - \varDelta = 0, \\
& \sum_{i \in N \backslash j} w_i^{\mathrm{R}} + w_j^{\mathrm{L}} \geq 1, \ j \in N, \\
& \sum_{i \in N \backslash j} w_i^{\mathrm{L}} + w_j^{\mathrm{R}} \leq 1, \ j \in N, \\
& \sum_{i \in N} w_i^{\mathrm{M}} = 1, \\
& w_i^{\mathrm{R}} \geq w_i^{\mathrm{M}} \geq w_i^{\mathrm{L}} \geq \epsilon, \ i \in N, \\
& E^+, E^-, \varGamma^+, \varGamma^-, \varDelta \geq 0,
\end{aligned}
\tag{10}
$$

where $e^{\mathrm{T}} = (1, 1, \ldots, 1)$. This is the LGP model for obtaining the triangular fuzzy weight vector $\tilde{W} = (W_{\mathrm{L}}, W_{\mathrm{M}}, W_{\mathrm{R}})$ from TFPCM $\tilde{A} = (A_{\mathrm{L}}, A_{\mathrm{M}}, A_{\mathrm{R}})$.

3 The Proposed Fuzzy AHP

In the previous section, triangular fuzzy weight vector estimation method from TFPCM is introduced. Before describing our proposed approach, we show the non-uniqueness of normalized fuzzy weight vector associated with a consistent fuzzy PCM. We consider the following example:

$$\tilde{A} = \begin{bmatrix} (1,1,1) & (7/8,8/7,3/2) & (7/6,8/5,9/4) \\ (2/3,7/8,8/7) & (1,1,1) & (1,7/5,2) \\ (4/9,5/8,6/7) & (1/2,5/7,1) & (1,1,1) \end{bmatrix}.$$

$\tilde{A}$ is a consistent fuzzy PCM. Then, we can obtain following normalized fuzzy weight vectors associated with $\tilde{A}$:

$$\tilde{W} = \begin{bmatrix} (0.35,0.40,0.45) \\ (0.30,0.35,0.40) \\ (0.20,0.25,0.30) \end{bmatrix}, \begin{bmatrix} (0.3333,0.40,0.4286) \\ (0.2857,0.35,0.3810) \\ (0.1905,0.25,0.2857) \end{bmatrix}, \begin{bmatrix} (0.3684,0.40,0.4737) \\ (0.3158,0.35,0.4210) \\ (0.2105,0.25,0.3158) \end{bmatrix}.$$

This example shows that the solution associated with the consistent TFPCM is not unique. From this result, we consider that all normalized triangular fuzzy weight vectors associated with a consistent TFPCM should be solutions to the estimated problem, because the appropriateness of any of those solutions is same because it can be defined by the distance of the consistent TFPCM from the given one. Therefore, in this section, we propose a method obtaining all solutions to estimation problem. In order to obtain all solutions of normalized fuzzy weight vector estimation problem, we propose the following fractional programming model, which is modified from the fuzzy weight estimation model proposed by Wang et al. [3]:

$$\text{minimize} \quad \frac{2e^{\mathrm{T}}(E^+ + E^- + \Gamma^+ + \Gamma^-)}{e^{\mathrm{T}}(W_{\mathrm{L}} + W_{\mathrm{R}})} + e^{\mathrm{T}}\Delta \tag{11}$$
$$\text{subject to } \textit{constraints of } (10).$$

By solving the model (11) with $e^{\mathrm{T}}(W_{\mathrm{L}} + W_{\mathrm{R}}) = 2$, one of all solutions to the problem of estimating the normalized fuzzy weight vector $\tilde{W}$ is estimated. Then, those solutions are represented by a parameter t and estimated normalized fuzzy weight vector $\tilde{W}$. In other words, when the normalized triangular fuzzy weight $w_i^{\mathrm{L}}, w_i^{\mathrm{M}}, w_i^{\mathrm{R}}, i \in N$, is estimated by (11) with $e^{\mathrm{T}}(W_{\mathrm{L}} + W_{\mathrm{R}}) = 2$, the all solutions are represented by $(tw_i^{\mathrm{L}}, w_i^{\mathrm{M}}, tw_i^{\mathrm{R}}), i \in N, t \in [t^{\mathrm{L}}, t^{\mathrm{R}}]$, where $t^{\mathrm{L}}, t^{\mathrm{R}} \in \mathbb{R}$. In the all solutions described above, $w_i^{\mathrm{M}}, i \in N$ are not multiplied by t and do not change because the normalized fuzzy weights must satisfy the normalization condition (9) and the sum of $w_i^{\mathrm{M}}, i \in N$ must be equal to 1.

Next, we consider the range of t. The normalization condition (9) must be satisfied for $tw_i^{\mathrm{L}}, tw_i^{\mathrm{R}}, i \in N$ where $w_i^{\mathrm{L}}, w_i^{\mathrm{R}}, i \in N$ are multiplied by t. Hence, the following inequalities need to hold for $tw_i^{\mathrm{L}}, tw_i^{\mathrm{R}}, i \in N$:

$$t \cdot \min_{j \in N} \left(\sum_{i \in N \backslash j} w_i^{\mathrm{R}} + w_j^{\mathrm{L}} \right) \geq 1, \quad t \cdot \max_{j \in N} \left(\sum_{i \in N \backslash j} w_i^{\mathrm{L}} + w_j^{\mathrm{R}} \right) \leq 1. \tag{12}$$

Also, since $w_i^{\mathrm{L}}, w_i^{\mathrm{R}}, i \in N$ are multiplied by t, but $w_i^{\mathrm{M}}, i \in N$ remain unchanged, the following inequalities must hold from the properties of triangular fuzzy weights, $w_i^{\mathrm{R}} \geq w_i^{\mathrm{M}} \geq w_i^{\mathrm{L}}, i \in N$:

$$t \cdot \min_{k \in N} \left(\frac{w_k^{\mathrm{R}}}{w_k^{\mathrm{M}}} \right) \geq 1, \; t \cdot \max_{k \in N} \left(\frac{w_k^{\mathrm{L}}}{w_k^{\mathrm{M}}} \right) \leq 1. \tag{13}$$

In the range of t satisfying inequalities (12) and (13), $(tw_i^{\mathrm{L}}, w_i^{\mathrm{M}}, tw_i^{\mathrm{R}}), i \in N$ are the solutions of normalized triangular fuzzy weights estimation problem. Therefore, the lower bound values t^{L} and upper bound values t^{R} for the range of t satisfying all inequalities (12) and (13) are defined by

$$t^{\mathrm{L}} = \max \left\{ \max_{k \in N} \frac{w_k^{\mathrm{M}}}{w_k^{\mathrm{R}}}, \; \frac{1}{\displaystyle\min_{i \in N} \left(w_i^{\mathrm{L}} + \sum_{j \in N \setminus i} w_j^{\mathrm{R}} \right)} \right\}, \tag{14}$$

$$t^{\mathrm{R}} = \min \left\{ \min_{k \in N} \frac{w_k^{\mathrm{M}}}{w_k^{\mathrm{L}}}, \; \frac{1}{\displaystyle\max_{i \in N} \left(w_i^{\mathrm{R}} + \sum_{j \in N \setminus i} w_j^{\mathrm{L}} \right)} \right\}. \tag{15}$$

In summary, $(tw_i^{\mathrm{L}}, w_i^{\mathrm{M}}, tw_i^{\mathrm{R}})$, $i \in N$, $t \in [t^{\mathrm{L}}, t^{\mathrm{R}}]$ are all solutions of normalized triangular fuzzy weights estimation problem. In Sect. 5, the numerical example shows the estimation of the triangular fuzzy weights, the calculation of the range of t, and the variation of the weights with t for the TFPCM given by the decision maker.

4   The Calculation of Total Utility of Alternatives

In this section, we introduce the method to calculate the total utility values of alternatives. In this paper, we assume that the utility values of alternatives in each criterion are given. Then, we estimate the weights of criteria from a given PCM. In the conventional AHP, the total utility value of an alternative is determined by the weighted sum of utility values. Then alternatives are ranked by the total utility values. However, we treat the fuzzy weight of criteria in this paper. Therefore, the total utility value cannot be determined by a weighted sum of utility values as in the conventional method. On the other hand, in interval AHP, the method of ranking alternatives when the weights of the criteria are interval values has been studied [4]. In that study, the alternatives are ranked by the minimum total utility values of the alternatives. In other words, the alternatives are ranked according to the maximin rule. In addition, ranking alternatives based on the maximum total utility value can be considered (i.e., maximax rule). In the case where the weights of criteria are fuzzy numbers, we calculate the minimum and maximum total utility values for the α-cuts of the fuzzy weights, at each $\alpha \in [0,1]$.

Let $u_i(o), i \in N$ be the utility value of alternative o in view of i-th criterion. Let $\tilde{w}_i, i \in N$ be the estimated fuzzy weight of i-th criterion. In these setting, we consider the minimum total utility values $U_{\min}(o)$ of an alternative o. The minimum total utility value $[U_{\min}(o)]_\alpha$ for the α-cut of weights $[\tilde{w}_i]_\alpha$ is defined by the minimum weighted sum of utility values, Accordingly, $[U_{\min}(o)]_\alpha$ is defined by

$$[U_{\min}(o)]_\alpha = \min \left\{ \sum_{i \in N} w_i u_i(o) \; \middle| \; w_i \in [\tilde{w}_i]_\alpha, \; i \in N, \sum_{i \in N} w_i = 1 \right\}. \quad (16)$$

The larger $[U_{\min}(o)]_\alpha$ is, the more preferable alternative o is.

On the other hand, we consider the maximum total utility values $U_{\max}(o)$ of an alternative o. The maximum total utility value $[U_{\max}(o)]_\alpha$ for the α-cut of weights $[\tilde{w}_i]_\alpha$ is defined by the maximum weighted sum of utility values, Accordingly, $[U_{\max}(o)]_\alpha$ is defined by

$$[U_{\max}(o)]_\alpha = \max \left\{ \sum_{i \in N} w_i u_i(o) \; \middle| \; w_i \in [\tilde{w}_i]_\alpha, \; i \in N, \sum_{i \in N} w_i = 1 \right\}. \quad (17)$$

The larger $[U_{\max}(o)]_\alpha$ is, the more preferable alternative o is.

When the estimated fuzzy weight $\tilde{w}_i, i \in N$ of the i-th criterion is a triangular fuzzy weight, the total utility value of the alternative o can be expressed as a triangular fuzzy number $\tilde{U}(o) = (U(o)^L, U(o)^M, U(o)^R)$. $U(o)^M$ is obtained by summing the utility values weighted by the weights $w_i^M, i \in N$, because $[\tilde{w}_i]_1 = w_i^M, i \in N$ and the sum of $w_i^M, i \in N$ is 1. $U(o)^L$ and $U(o)^R$ are obtained from $U_{\min}(o)$ and $U_{\max}(o)$. In other words, $U(o)^L = [U_{\min}(o)]_0$ and $U(o)^R = [U_{\max}(o)]_0$. In order to rank alternatives, following value are used [3]:

$$U(o)^C = \frac{1}{3}(U(o)^L + U(o)^M + U(o)^R), \quad (18)$$

where $U(o)^C$ is centroid of fuzzy utility value $\tilde{U}(o)$. The larger $U(o)^C$ is, the more preferable alternative o is.

5   A Numerical Example

In this section, we estimate the triangular fuzzy weight vector from the TFPCM given by the decision maker and derive the solutions of fuzzy weight vector estimation problem. It also shows the transitions in the ranking alternatives within the all solutions.

Consider a multiple criteria decision problem with five criteria $c_1, c_2, \ldots, c_5$. Then, 5×5 triangular fuzzy comparison matrix given by decision maker below:

$$
\tilde{A} = \begin{bmatrix}
(1,1,1) & (3/2,2,5/2) & (5/2,3,7/2) & (5/2,3,7/2) & (7/2,4,9/2) \\
(2/5,1/2,2/3) & (1,1,1) & (3/2,2,5/2) & (3/2,2,5/2) & (5/2,3,7/2) \\
(2/7,1/3,2/5) & (2/3,1/2,2/5) & (1,1,1) & (2/3,1,3/2) & (3/2,2,5/2) \\
(2/7,1/3,2/5) & (2/3,1/2,2/5) & (2/3,1,3/2) & (1,1,1) & (3/2,2,5/2) \\
(2/9,1/4,2/7) & (2/5,1/3,2/7) & (2/3,1/2,2/5) & (2/3,1/2,2/5) & (1,1,1)
\end{bmatrix}.
$$

Each element of A represents relative importance between c_i and c_j. By solving triangular fuzzy weight estimation model (11) with $e^T(W_L + W_R) = 2$ for the above triangular fuzzy comparison matrix, we obtain the following normalized triangular fuzzy weight vector:

$$
\tilde{W} = \begin{bmatrix}
(0.3799, 0.4045, 0.4186) \\
(0.2179, 0.2450, 0.2700) \\
(0.1167, 0.1369, 0.1615) \\
(0.1167, 0.1369, 0.1615) \\
(0.0694, 0.0767, 0.0878)
\end{bmatrix}.
$$

From estimated triangular fuzzy weight vector $\tilde{W}$, we estimate the all solutions $(tw_i^L, w_i^M, tw_i^R), i \in N, t \in [t^L, t^R]$. t^L and t^R are calculated by (14) and (15). Then, $t^L = 0.9665$ and $t^R = 1.0495$. As an example of the solutions of fuzzy weight vectors estimation problem, the fuzzy weight vectors $\tilde{W}_{(t^L)}$ at $t = t^L$ and $\tilde{W}_{(t^R)}$ at $t = t^R$, respectively, are obtained as follows:

$$
\tilde{W}_{(t^L)} = \begin{bmatrix}
(0.3672, 0.4045, 0.4045) \\
(0.2106, 0.2450, 0.2609) \\
(0.1128, 0.1369, 0.1561) \\
(0.1128, 0.1369, 0.1561) \\
(0.0671, 0.0767, 0.0849)
\end{bmatrix}, \quad
\tilde{W}_{(t^R)} = \begin{bmatrix}
(0.3987, 0.4045, 0.4393) \\
(0.2286, 0.2450, 0.2834) \\
(0.1225, 0.1369, 0.1695) \\
(0.1225, 0.1369, 0.1695) \\
(0.0728, 0.0767, 0.0922)
\end{bmatrix}.
$$

The fuzzy weight vector $\tilde{W}_{(t^L)}$ is also estimated by solving the LGP model (10).

In addition, we assume three alternatives o_1, o_2, o_3 in this multiple criteria decision problem. In this example, the utility values of those alternatives under the five criteria are given in Table 1.

Table 1. The utility values of alternatives in each criterion.

	c_1	c_2	c_3	c_4	c_5
o_1	0.24	0.23	0.08	0.23	0.22
o_2	0.12	0.46	0.21	0.10	0.11
o_3	0.22	0.19	0.45	0.06	0.08

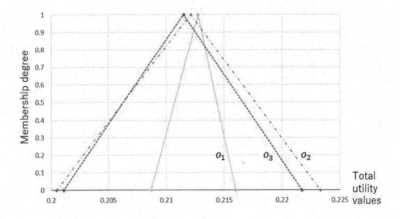

Fig. 1. Total utility values of three alternatives ($t = 1$)

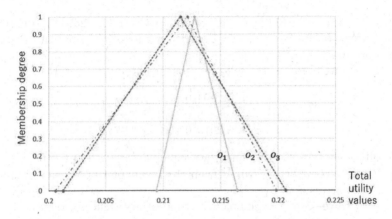

Fig. 2. Total utility values of three alternatives ($t = t^{\mathrm{L}}$)

Under these settings, the total utility values of the three alternatives $\tilde{U} = (\tilde{U}(o_1), \tilde{U}(o_2), \tilde{U}(o_3))^{\mathrm{T}}$ are calculated by using (16) and (17). As an example of the total utility value vector $\tilde{U}$ under the all solutions of fuzzy weight vector estimation problem, the total utility value vectors $\tilde{U}_{(t=1)}$ at $t = 1$, $\tilde{U}_{(t=t^{\mathrm{L}})}$ at $t = t^{\mathrm{L}}$ and $\tilde{W}_{(t=t^{\mathrm{R}})}$ at $t = t^{\mathrm{R}}$, respectively, are obtained. In addition, the vector of centroid of total utility value $U^{\mathrm{C}} = [U(o_1)^{\mathrm{C}}, U(o_2)^{\mathrm{C}}, U(o_3)^{\mathrm{C}}]^{\mathrm{T}}$ is obtained by (18). From U^{C}, we can obtain ranking alternatives.

For $t = 1$, we obtained the following total utility value vector $\tilde{U}_{(t=1)}$:

$$\tilde{U}_{(t=1)} = \begin{bmatrix} (0.2087, 0.2127, 0.2160) \\ (0.2005, 0.2121, 0.2233) \\ (0.2011, 0.2115, 0.2217) \end{bmatrix}.$$

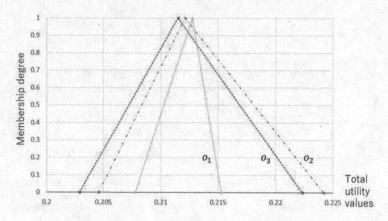

Fig. 3. Total utility values of three alternatives $(t = t^{\mathrm{R}})$

This result is shown in Fig. 1. Then, we obtain $U^{\mathrm{C}} = [0.2125, 0.2120, 0.2114]^{\mathrm{T}}$. Since $U(o_1)^{\mathrm{C}} > U(o_2)^{\mathrm{C}} > U(o_3)^{\mathrm{C}}$ from the result of U^{C}, we can obtain the ranking alternatives $o_1 \succ o_2 \succ o_3$.

For $t = t^{\mathrm{L}}$, we obtained the following total utility value vector $\tilde{U}_{(t=t^{\mathrm{L}})}$:

$$\tilde{U}_{(t=t^{\mathrm{L}})} = \begin{bmatrix} (0.2094, 0.2127, 0.2164) \\ (0.2006, 0.2121, 0.2198) \\ (0.2013, 0.2115, 0.2207) \end{bmatrix}.$$

This result is shown in Fig. 2. Then, we obtain $U^{\mathrm{C}} = [0.2129, 0.2108, 0.2111]^{\mathrm{T}}$. Since $U(o_1)^{\mathrm{C}} > U(o_3)^{\mathrm{C}} > U(o_2)^{\mathrm{C}}$ from the result of U^{C}, we can obtain the ranking alternatives $o_1 \succ o_3 \succ o_2$.

For $t = t^{\mathrm{R}}$, we obtained the following total utility value vector $\tilde{U}_{(t=t^{\mathrm{R}})}$:

$$\tilde{U}_{(t=t^{\mathrm{R}})} = \begin{bmatrix} (0.2078, 0.2127, 0.2153) \\ (0.2046, 0.2121, 0.2242) \\ (0.2029, 0.2115, 0.2223) \end{bmatrix}.$$

This result is shown in Fig. 3. Then, we obtain $U^{\mathrm{C}} = [0.2119, 0.2136, 0.2122]^{\mathrm{T}}$. Since $U(o_2)^{\mathrm{C}} > U(o_3)^{\mathrm{C}} > U(o_1)^{\mathrm{C}}$ from the result of U^{C}, we can obtain the ranking alternatives $o_2 \succ o_3 \succ o_1$.

From these results, it is confirmed that the ranking of alternatives transitions among the all solutions from fuzzy weight vector estimation problem. Therefore, we derive all the values of t when the ranking alternatives changes in this numerical example. The result of the transition of the centroids of total utility values alternatives and the ranking alternatives in $[t^{\mathrm{L}}, t^{\mathrm{R}}]$ are shown in Figs. 4 and 5.

These result shows the ranking alternatives changes with $t = 0.9776, 1.0102$ and 1.0430. At $t = 0.9776$, the ranking of o_2 and o_3, at $t = 1.0102$, the ranking of o_1 and o_2 and at $t = 1.0430$, the ranking of o_1 and o_3 are reversed. Since the ranking reversal occurred three times in $[t^{\mathrm{L}}, t^{\mathrm{R}}]$, the ranking alternatives cannot be uniquely determined in this example.

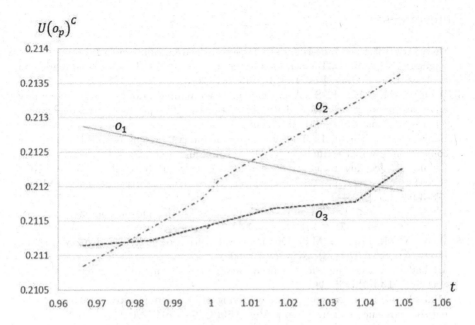

Fig. 4. Transition of ranking score of three alternatives

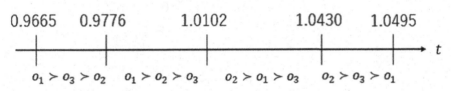

Fig. 5. Transition of ranking alternatives by t

6   Concluding Remarks

In this paper, the non-uniqueness of the solution to the estimation problem of the normalized triangular fuzzy weight vector is suggested under a given triangular pairwise comparison matrix. It is shown that all solutions are obtained easily by a parameter and a solution obtained by the conventional estimation method with an additional constraint. As all solutions are ordered by a parameter, the transition of alternative orderings with respect to the parameter change is shown. This analysis enables us to see the stability of the decision maker's preference and the best alternatives. The extension of the proposed approach to the case of trapezoidal fuzzy pairwise comparison matrices is one of the future topics.

Acknowledgements. This work is supported by JSPS KAKENHI Grant Number JP18H01658.

References

1. Satty, T.L.: The Analytic Hierarchy Process. McGraw-Hill, New York (1980)
2. Sugihara, K., Ishii, H., Tanaka, H.: Interval priorities in AHP by interval regression analysis. Eur. J. Oper. Res. **158**(3), 745–754 (2004)
3. Wang, Y.-M., Chin, K.-S.: A linear goal programming priority method for fuzzy analytic hierarchy process and its applications in new product screening. Int. J. Approx. Reason. **49**(2), 451–465 (2008)
4. Inuiguchi, M., Torisu, I.: The advantage of interval weight estimation over the conventional weight estimation in AHP in ranking alternatives. In: Huynh, V.-N., Entani, T., Jeenanunta, C., Inuiguchi, M., Yenradee, P. (eds.) IUKM 2020. LNCS (LNAI), vol. 12482, pp. 38–49. Springer, Cham (2020). https://doi.org/10.1007/978-3-030-62509-2_4
5. Wang, Y.-M., Yang, J.-B., Xu, D.-L., Chin, K.-S.: On the centroids of fuzzy numbers. Fuzzy Sets Syst. **157**(7), 919–926 (2006)
6. Wang, Y.-M., Elhag, T.M.S.: On the normalization of interval and fuzzy weights. Fuzzy Sets Syst. **157**(18), 2456–2471 (2006)
7. Mikhailov, L.: Deriving priorities from fuzzy pairwise comparison judgements. Fuzzy Sets Syst. **134**(3), 365–385 (2003)
8. Mikhailov, L.: A fuzzy approach to deriving priorities from interval pairwise comparison judgements. Eur. J. Oper. Res. **159**(3), 687–704 (2004)

Optimization and Statistical Methods

Coyote Optimization Algorithm with Linear Convergence for Global Numerical Optimization

Hsin-Jui Lin and Sheng-Ta Hsieh[✉]

Asia Eastern University of Science and Technology, New Taipei City, Taiwan
fo013@mail.aeust.edu.tw

Abstract. In particular, the popularity of computational intelligence has accelerated the study of optimization. Coyote Optimization Algorithm (COA) is a new meta heuristic optimization. It is pays attention to the social structure and experience exchange of coyotes. In this paper, the coyote optimization algorithm with linear convergence (COALC) is proposed. In order to explore a huge search space in the pre-optimization stage and to avoid premature convergence, the convergence factor is also involved. Thus, the COALC will explore a huge search space in the early optimization stage to avoid premature convergence. Also, the small area is adopted in the later optimization stage to effectively refine the final solution, while simulating a coyote killed by a hunter in the environment. It can avoid the influence of bad solutions. In experiments, ten IEEE CEC2019 test functions is adopted. The results show that the proposed method has rapid convergence, and a better solution can be obtained in a limited time, so it has advantages compared with other related methods.

Keywords: Functional optimization · Swarm intelligence · Global optimization problems · Coyote Optimization Algorithm · Coyote optimization algorithm with linear convergence

1 Introduction

Optimization can now solve many problems in the real world, including civil engineering, construction, electromechanical, control, financial, health management, etc. There are significant results [1–5], whether it is applied to image recognition, feature extraction, machine learning, and deep learning model training, The optimization algorithm can be used to adjust [6–8]. The optimization method can make the traditional researcher spend a lot of time to establish the expert system to adjust and optimize, and greatly reduce the time required for exploration. With the exploration of intelligence, the complexity of the problem gradually increases.

In the past three decades, meta-heuristic algorithms that simulate the behavior of nature have received a lot of attention, for example, Particle Swarm Optimization (PSO) [9], Differential Evolution (DE) [10], Crow Search algorithm (CSA) [11], Grey Wolf Optimization (GWO) [12], Coyote Optimization Algorithm (COA) [13], Whale Optimization Algorithm (WOA) [14], Honey Badger Algorithm (HBA) [15] and Red fox

© Springer Nature Switzerland AG 2022
K. Honda et al. (Eds.): IUKM 2022, LNAI 13199, pp. 81–91, 2022.
https://doi.org/10.1007/978-3-030-98018-4_7

optimization algorithm (RFO) [16]. These meta-heuristic algorithms inspired by natural behavior are highly efficient in optimization problems. Performance and ease of application have been improved and applied to various problems.

GWO simulates the class system in the predation process of wolves in nature, and divides gray wolves into four levels. Through the domination and leadership between the levels, the gray wolves are driven to find the best solution. This method prevents GWO from falling into the local optimum solution, and use the convergence factor to make the algorithm use a longer moving distance to perform a global search in the early stage, and gradually tends to a local search as time changes. Arora *et al.* [17] mixed GWO and CSA use the flight control parameters and modified linear control parameters in CSA to achieve the balance in the exploration and exploration process, and use it to solve the problem of function optimization.

The idea of COA comes from the coyotes living in North America. Unlike most meta heuristic optimization, which focuses on the predator relationship between predators and prey, COA focuses on the social structure and experience exchange of coyotes. It has a special algorithm structure. Compared with GWO, although alpha wolf (best value) is still used to guide, it does not pay attention to the ruling rules of beta wolf (second best value) and delta wolf (third best value), and balance the global search in the optimization process and local search. Li *et al.* [18] changed the COA differential mutation strategy, designed differential dynamic mutation disturbance strategy and adaptive differential scaling factor, and used it in the fuzzy multi-level image threshold in order to change the COA iteration to the local optimum, which is prone to premature convergence. Get better image segmentation quality.

RFO is a recently proposed meta-heuristic algorithm that imitates the life and hunting methods of red foxes. It simulates red foxes traveling through the forest to find and prey on prey. These two methods correspond to global search and local search, respectively. The hunting relationship between hunters makes RFO converge to an average in the search process.

In this paper, the combination of convergence factors allows COA to implement better exploration, and adds the risk of coyotes being killed by hunters in nature and the ability to produce young coyote to increase local exploration.

In summary, Sect. 2 is a brief review of COA methods and Sect. 3 describes the project scope and objectives of the proposed method. Section 4 shows the experimental results of proposed algorithms and other algorithms on test functions. Finally, the conclusions are in Sect. 5.

2 Standard Coyote Optimization Algorithm

Coyote Optimization Algorithm (COA) was proposed by Pierezan *et al.* [13] in 2018, it has been widely used in many fields due to its unique algorithm structure [19–21]. In COA, the coyote population is divided into N_p groups, and each group contains N_i coyotes, so the total number of coyotes is $N_p * N_i$, at the start of the COA, the number of coyotes in each group has the same population. Each coyote represents the solution of the optimization problem, and is updated and eliminated in the iteration.

COA effectively simulates the social structure of the coyote, that is, the decision variable $\vec{x}$ of the global optimization problem. Therefore, the social condition *soc* (decision variable set) formed at the time when the i_{th} coyote of the p_{th} ethnic group is an integer t is written as:

$$soc_i^{p,t} = \vec{x} = (x_1, x_2, \ldots, x_d) \tag{1}$$

where d is the search space dimension of the optimization problem, the first initialize the coyote race group. Each coyote is randomly generated in the search space. The i_{th} coyote in the race group p is expressed in the j_{th} dimension as:

$$soc_{(i,j)}^{(p,t)} = LB_j + r_j \times (UB_j - LB_j), \quad j = 1, 2, \ldots d \tag{2}$$

where LB_j and UB_j represent the lower and upper bounds of the search space, and r_j is a random number generated in the range of 0 to 1. The current social conditions of the i_{th} coyote, as can be shown in (3):

$$fit_i^{p,t} = f\left(soc_i^{p,t}\right) \tag{3}$$

In nature, the size of a coyote group does not remain the same, and individual coyotes sometimes leave or be expelled from the group alone, become a single one or join another group. COA defines the individual coyote outlier probability P_e as:

$$P_e = 0.005 \times N_i^2 \tag{4}$$

When the random number is less than P_e, the wolf will leave one group and enter another group. COA limits the number of coyotes per group to 14. And COA adopts the optimal individual (*alpha*) guidance mechanism:

$$alpha^{p,t} = \left\{ soc_i^{p,t} | arg_{i=1,2,\ldots,d} \max fit\left(soc_i^{p,t}\right) \right\} \tag{5}$$

In order to communicate with each other among coyotes, the cultural tendency of coyote is defined as the link of all coyotes' social information:

$$cult_j^{p,t} = \begin{cases} O_{\frac{N_i+1}{2},j}^{p,t}, & N_i \text{ is odd} \\ \dfrac{O_{\frac{N_i+1}{2},j}^{p,t} + O_{\frac{N_i+1}{2},j}^{p,t}}{2}, & otherwise \end{cases} \tag{6}$$

The cultural tendency of the wolf pack is defined as the median of the social status of all coyotes in the specific wolf pack, $O^{p,t}$ is the median of all individuals in the population p in the jth dimension at the tth iteration.

The birth of the *pup* is a combination of two parents (coyote selected at random) and environmental influences as:

$$pup_j^{p,t} = \begin{cases} soc_{r_1,j}^{p,t}, & rand_j < P_s \text{ or } j = j_1 \\ soc_{r_2,j}^{p,t}, & rand_j \geq P_s + P_a \text{ or } j = j_2 \\ R_j, & otherwise \end{cases} \tag{7}$$

Among them, r_1 and r_2 are the random coyotes of two randomly initialized packages, j_1 and j_2 are two random dimensions in the space. Therefore, the newborn coyotes can be inherited through random selection by parents, or new social condition can be produced by random, while P_s and P_a influenced by search space dimension are the scatter probability and the associated probability respectively, as shown in (8) and (9). R_j is a random number in the search space of the j_{th} dimension, and $rand_j$ is a random number between 0 and 1.

$$P_s = 1/d \tag{8}$$

$$P_a = (1 - P_s)/2 \tag{9}$$

In order to maintain the same population size, COA uses coyote group that do not have environmental adaptability ω and the number of coyotes in the same population φ, when $\varphi = 1$ the only coyote in ω dies and $\varphi > 1$ the oldest coyote in ω dies, and pup survives, and when $\varphi < 1$, pup alone cannot satisfy the survival condition. At the same time, in order to show the cultural exchange in the population, set influence led by the alpha wolf (δ_1) and the influence by the group (δ_2), the δ_1 guided by the optimal individual makes the coyote close to the optimal value, and the δ_2 guided by the coyote population reduces the probability of falling into the local optimal value, where cr_1 and cr_2 are Two random coyotes, δ_1 and δ_2 are written as:

$$\delta_1 = alpha^{p,t} - soc_{cr_1}^{p,t} \tag{10}$$

$$\delta_2 = cult^{p,t} - soc_{cr_2}^{p,t} \tag{11}$$

After calculating the two influencing factors δ_1 and δ_2, using the pack influence and the alpha, the new social condition (12) of the coyote is initialized by two random numbers between 0 and 1, and the new social condition (13) (the position of the coyote) is evaluated.

$$new\_soc_i^{p,t} = soc_i^{p,t} + r_1 \times \delta_1 + r_2 \times \delta_2 \tag{12}$$

$$new\_fit_i^{p,t} = f\left(new\_soc_i^{p,t}\right) \tag{13}$$

Finally, according to the greedy algorithm, update the new social condition (the position of the coyote) as (14), and the optimized solution of the problem is the coyote's can best adapt to the social conditions of the environment.

$$soc_i^{p,t+1} = \begin{cases} new\_soc_i^{p,t}, & new\_fit_i^{p,t} < fit_i^{p,t} \\ soc_i^{p,t}, & otherwise \end{cases} \tag{14}$$

3 Coyote Optimization Algorithm with Linear Convergence

It is important for optimization algorithms to strike a balance between exploration and exploration. In the classic COA, the position update distance (12) of the coyote is calculated by multiplying two random numbers between 0 and 1 and the influencing factors δ_1 and δ_2. This method makes the position of the coyote tend to an average. As a result, the global search capability in the early stage of the algorithm is insufficient, and the local search cannot be performed in depth in the later stage. At the same time, when calculating the social culture of coyotes (6), they will be dragged down by the poorly adapted coyotes, resulting in poor final convergence.

In order to overcome the limitations of the above conventional COA, the linear convergence strategy of GWO is adopted, and the linear control parameter (a) is calculated by follows.

$$a = 2 - (2 \times t/Max_{iter}) \tag{15}$$

And calculate two random moving vectors A to replace two random numbers, so that the algorithm can move significantly in the early stage to obtain a better global exploration, and in the later stage can perform a deep local search, so that the algorithm can converge in a limited time give better results. The value of a is 2 from the beginning of the iteration, and decreases linearly to 0 with the iteration. Therefore, the movement vector A is calculated by follows.

$$A = 2 * a * r_1 * a \tag{16}$$

Among them, r_1 is a random number between 0 and 1. Therefore, the social conditions of the new coyote will be generated by follows (17). The pseudo code of proposed method is presented (see Fig. 1).

$$new\_soc_i^{p,t} = soc_i^{p,t} + A_1 \times \delta_1 + A_2 \times \delta_2 \tag{17}$$

COA uses the average of coyote information to form social culture, but it is easily affected by the coyote with the lowest adaptability, making iterative early-stage algorithms unable to quickly converge to a better range. Therefore, referring to the hunting relationship between the red fox and the hunter in the RFO, and applying it in the COA to simulate the situation where the coyote strays into the range of human activities and is hunted, the probability of the coyote being killed by the hunter is H (18), by The linear control parameter is calculated and rounded. With time, H will gradually decrease to 0. In the later stage of the algorithm, this mechanism is not used to avoid falling into the local optimal solution.

$$H = [N_i * (a * 0.1)] \tag{18}$$

```
1: Initialize N_p packs with N_i coyotes each (Eq.2).

2: Verify the coyote's adaptation (Eq.3).

3: while stopping criterion is not achieved do

4:      for each p pack do

5:          Define the alpha coyote of the pack (Eq.5).

6:          Compute the social tendency of the pack (Eq.6).

7:          for each c coyotes of the p pack do

8:              Calculate a and A(Eq.15 and Eq.16).

9:              Update the social condition (Eq. 17).

10:             Evaluate the new social condition (Eq.13).

11:             Adaptation (Eq.14).

12:         end for

13:         Calculate the H(Eq.18).

14:         if H >= 1 then

15:             Replace Coyote(Eq.19).

16:             Birth and death (Eq.7).

17:     end for

18:     Transition between packs (Eq.4).

19:     Update the coyotes' ages.

20: end while

21: Select the best adapted coyote.
```

Fig. 1. Pseudo code of proposed method.

In order to maintain the total population, new coyotes will be produced. Therefore, new coyotes will be born from combining information of the best coyote ($best_1$) and second-best coyote ($best_2$) in the group. The location of the newborn coyotes (19), k is a random vector in the range [0, 1]. Therefore, the pseudocode of COALC is showed below after the formula is replaced.

$$new\_soc_i^{p,t} = k * \frac{soc_{best_1}^{p,t} + soc_{best_2}^{p,t}}{2} \tag{19}$$

4   Experimental Settings and Experimental Results

4.1   Benchmarks Functions and Algorithms Setup

Table 1. Global optimization, dimensions and search range of ten CEC 2019 test functions

No.	Function name	$F_i^* = F_i(x^*)$	D	Range
f_1	Storn's Chebyshev Polynomial Fitting Problem	1	9	$[-8192, 8192]$
f_2	Inverse Hilbert Matrix	1	16	$[-16384, 16384]$
f_3	Lennard-Jones Minimum Energy Cluster	1	18	$[-4, 4]$
f_4	Rastrigin's Function	1	10	$[-100, 100]$
f_5	Griewank's Function	1	10	$[-100, 100]$
f_6	Weierstrass Function	1	10	$[-100, 100]$
f_7	Modified Schwefel's Function	1	10	$[-100, 100]$
f_8	Expanded Schaffer's F6 Function	1	10	$[-100, 100]$
f_9	Happy Cat Function	1	10	$[-100, 100]$
f_{10}	Ackley Function	1	10	$[-100, 100]$

The proposed COALC method uses the 10 benchmark functions shown in Table 1 in the IEEE CEC2019 test function (CEC2019) [22] to extend our benchmark test, where F_i^* is the global optimum and D is the dimension of the optimization problem. These benchmarks vary according to the number, dimensionality, and search space of the local optimal classifications. In the CEC2019 function, the functions f_1, f_2, and f_3 are completely dependent on the parameters and do not rotate (or shift). Among them, f_1 and f_2 are error functions that need to rely on highly conditional solutions, and f_3 is a way to simulate atomic interaction. It is difficult to find the best solution directly in the function f_9, and the optimization algorithm must perform a deep search in the circular groove. F_4, f_5, f_6, f_7, f_8 and f_{10} are classic optimization problems.

In the benchmark test, this paper compared the proposed COALC, standard COA, ICOA, GWO, and RFO. In the experiment, COALC, standard COA and ICOA defined the coyote population number parameter N_p as 6, and the coyote N_i in each group was set as 5. The number of gray wolves and foxes for GWO and RFO is set to 30, and the above optimization algorithms are based on their original settings, only need to mention the parameters of the method. Therefore, all running comparison heuristics have 30 number of species. The experiment was run on Python 3.8.8.

4.2 Comparison and Analysis of Experimental Results

Table 2. Experiment results of five optimizers

No.	COALC	COA [13]	ICOA [18]	GWO [12]	RFO [16]
f_1	0.00E+00	6.42E+05	3.97E+06	8.93E+04	**0.00E+00**
	±4.81E+00	±4.77E+05	±2.84E+06	±1.75E+05	**±0.00E+00**
f_2	**3.28E+00**	1.12E+03	2.63E+03	3.99E+02	4.00E+00
	±5.66E−02	±3.62E+02	±9.09E+02	±2.83E+02	±0.00E+00
f_3	**1.49E+00**	2.27E+00	5.72E+00	1.86E+00	5.47E+00
	±8.50E−01	±1.46E+00	±1.64E+00	±1.90E+00	±1.90E+00
f_4	**1.26E+01**	1.34E+01	1.78E+01	2.00E+01	9.64E+01
	±4.26E+00	±5.32E+00	±5.89E+00	±1.07E+01	±1.64E+01
f_5	0.27E+00	**0.18E+00**	0.70E+00	0.86E+00	3.89E+01
	±9.33E−02	**±8.31E−02**	±1.63E−01	±5.45E−01	±1.46E+01
f_6	**1.17E+00**	1.92E+00	3.52E+00	1.63E+00	9.03E+00
	±8.34E−01	±1.26E+00	±1.37E+00	±1.26E+00	±8.84E−01
f_7	5.86E+02	**4.93E+02**	6.39E+02	8.20E+02	2.00E+03
	±2.07E+02	**±2.20E+02**	±2.30E+02	±3.34E+02	±2.18E+02
f_8	**2.71E+00**	2.82E+00	3.06E+00	2.82E+00	3.98E+00
	±4.14E−01	±2.85E−01	±3.21E−01	±3.87E−01	±1.87E−01
f_9	**0.18E+00**	0.22E+00	0.22E+00	0.20E+00	1.37E+00
	±5.77E−02	±7.12E−02	±8.73E−02	±7.12E−02	±6.52E−01
f_{10}	**1.83E+01**	2.01E+01	2.01E+01	2.05E+01	2.07E+01
	±5.13E+00	±5.10E−02	±5.54E−02	±1.20E−01	±1.33E−01

Each optimizer is performed 25 independent runs on the CEC2019, and the stopping criterion is equal to the number of ethnic groups * 500 iterations. Thus, the maximum fitness evaluation (FEs) is set as 15,000. The average error obtained from the global optimum and standard deviation is shown in Table 2, and the best performance is shown in bold. In Fig. 2 can be seen that for most of the benchmark functions, COALC can find the best solution compared to other methods. COALC acquires better exploration capabilities by inheriting the relationship between hunter and prey of RFO, and has more outstanding capabilities than other methods in f_1 and f_2 functions.

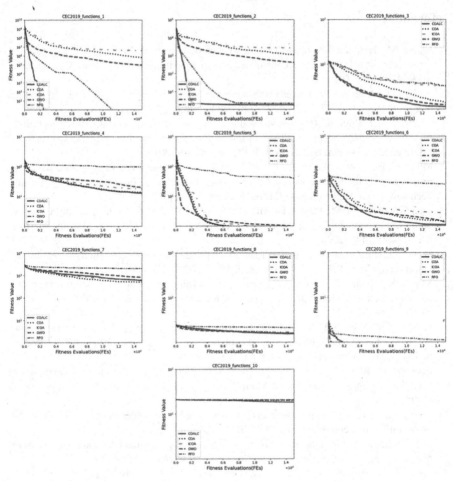

Fig. 2. Median convergence characteristics of five optimizers.

5   Conclusion and Future Research

In this paper, the main contribution is to propose an improved COA algorithm with the convergence factor of the GWO algorithm and the elimination of the worst coyote mechanism, and named it COALC. This method allows COA to acquire better exploration and exploration capabilities in a limited time through the convergence factor, and at the same time eliminates poor coyotes to improve the convergence speed of COA. Finally, Results of experimental benchmark tests have shown that the proposed COALC and recent metaheuristic algorithms such as COA, ICOA, GWO, and RFO, etc., and are evaluated in the CEC2019 test function. In most cases, a better global solution can be obtained than other algorithms.

References

1. Cheng, M., Tran, D.: Two-phase differential evolution for the multiobjective optimization of time–cost tradeoffs in resource-constrained construction projects. IEEE Trans. Eng. Manage. **61**(3), 450–461 (2014)
2. Al-Timimy, A., et al.: Design and losses analysis of a high power density machine for flooded pump applications. IEEE Trans. Ind. Appl. **54**(4), 3260–3270 (2018)
3. Chabane, Y., Ladjici, A.: Differential evolution for optimal tuning of power system stabilizers to improve power systems small signal stability. In: Proceedings of 2016 5th International Conference on Systems and Control (ICSC), pp. 84–89 (2016)
4. Münsing, E., Mather, J., Moura, S.: Blockchains for decentralized optimization of energy resources in microgrid networks. In: Proceedings of 2017 IEEE Conference on Control Technology and Applications (CCTA), pp. 2164–2171 (2017)
5. Lucidi, S., Maurici, M., Paulon, L., Rinaldi, F., Roma, M.: A simulation-based multiobjective optimization approach for health care service management. IEEE Trans. Autom. Sci. Eng. **13**(4), 1480–1491 (2016)
6. Huang, C., He, Z., Cao, G., Cao, W.: Task-driven progressive part localization for fine-grained object recognition. IEEE Trans. Multimed. **18**(12), 2372–2383 (2016)
7. Mistry, K., Zhang, L., Neoh, S.C., Lim, C.P., Fielding, B.: A micro-GA embedded PSO feature selection approach to intelligent facial emotion recognition. IEEE Trans. Cybern. **47**(6), 1496–1509 (2017)
8. Haque, M.N., Noman, M.N., Berretta, R., Moscato, P.: Optimising weights for heterogeneous ensemble of classifiers with differential evolution. In: Proceedings of 2016 IEEE Congress on Evolutionary Computation (CEC), pp. 233–240 (2016)
9. Eberhart, R., Kennedy, J.: A new optimizer using particle swarm theory. In: Proceedings of Sixth International Symposium on Micro Machine and Human Science (MHS), pp. 39–43 (1995)
10. Storn, R., Price, K.: Differential evolution–a simple and efficient heuristic for global optimization over continuous spaces. J. Glob. Optim. **11**(4), 341–359 (1997)
11. Askarzadeh, A.: A novel metaheuristic method for solving constrained engineering optimization problems: Crow search algorithm. Comput. Struct. **169**, 1–12 (2016)
12. Mirjalili, S., Mirjalili, S.M., Lewis, A.: Grey wolf optimizer. Adv. Eng. Softw. **69**, 46–61 (2014)
13. Pierezan, J., Coelho, L.D.S.: Coyote optimization algorithm: a new metaheuristic for global optimization problems. In: Proceedings of 2018 IEEE Congress on Evolutionary Computation (CEC), pp. 1–8 (2018)
14. Mirjalili, S., Lewis, A.: The whale optimization algorithm. Adv. Eng. Softw. **95**, 51–67 (2016)
15. Hashim, F.A., Houssein, E.H., Hussain, K., Mabrouk, M.S., Al-Atabany, W.: Honey Badger Algorithm: new metaheuristic algorithm for solving optimization problems. Math. Comput. Simul. **192**, 84–110 (2021)
16. Połap, D., Woźniak, M.: Red fox optimization algorithm. Expert Syst. Appl. **166**, 114107 (2021)
17. Arora, S., Singh, H., Sharma, M., Sharma, S., Anand, P.: A new hybrid algorithm based on grey wolf optimization and crow search algorithm for unconstrained function optimization and feature selection. IEEE Access **7**, 26343–26361 (2019)
18. Li, L., Sun, L., Xue, Y., Li, S., Huang, X., Mansour, R.F.: Fuzzy multilevel image thresholding based on improved coyote optimization algorithm. IEEE Access **9**, 33595–33607 (2021)
19. Abdelwanis, M.I., Abaza, A., El-Sehiemy, R.A., Ibrahim, M.N., Rezk, H.: Parameter estimation of electric power transformers using coyote optimization algorithm with experimental verification. IEEE Access **8**, 50036–50044 (2020)

20. Diab, A.A.Z., Sultan, H.M., Do, T.D., Kamel, O.M., Mossa, M.A.: Coyote optimization algorithm for parameters estimation of various models of solar cells and PV modules. IEEE Access **8**, 111102–111140 (2020)
21. Boursianis, A.D., et al.: Multiband patch antenna design using nature-inspired optimization method. IEEE Open J. Antennas Propag. **2**, 151–162 (2021)
22. Price, K.V., Awad, N.H., Ali, M.Z., Suganthan, P.N.: Problem definitions and evaluation criteria for the 100-digit challenge special session and competition on single objective numerical optimization. Technical Report. Nanyang Technological University, Singapore (2018)

Accounting for Gaussian Process Imprecision in Bayesian Optimization

Julian Rodemann[✉] and Thomas Augustin

Department of Statistics, Ludwig-Maximilians-Universität München (LMU),
Munich, Germany
rodemann@stat.uni-muenchen.de

Abstract. Bayesian optimization (BO) with Gaussian processes (GP) as surrogate models is widely used to optimize analytically unknown and expensive-to-evaluate functions. In this paper, we propose Prior-mean-RObust Bayesian Optimization (PROBO) that outperforms classical BO on specific problems. First, we study the effect of the Gaussian processes' prior specifications on classical BO's convergence. We find the prior's mean parameters to have the highest influence on convergence among all prior components. In response to this result, we introduce PROBO as a generalization of BO that aims at rendering the method more robust towards prior mean parameter misspecification. This is achieved by explicitly accounting for GP imprecision via a prior near-ignorance model. At the heart of this is a novel acquisition function, the generalized lower confidence bound (GLCB). We test our approach against classical BO on a real-world problem from material science and observe PROBO to converge faster. Further experiments on multimodal and wiggly target functions confirm the superiority of our method.

Keywords: Bayesian optimization · Imprecise Gaussian process · Imprecise probabilities · Prior near-ignorance · Model imprecision · Robust optimization

1 Introduction

Bayesian optimization (BO) is a popular method for optimizing functions that are expensive to evaluate and do not have any analytical description ("black-box-functions"). Its applications range from engineering [8] to drug discovery [16] and COVID-19 detection [2]. BO's main popularity, however, stems from machine learning, where it has become one of the predominant hyperparameter optimizers [15] after the seminal work of [22].

Julian Rodemann would like to thank the scholarship program of Evangelisches Studienwerk Villigst for the support of his studies and Lars Kotthoff for providing data as well as Christoph Jansen and Georg Schollmeyer for valuable remarks.
Open Science: Code to reproduce all findings and figures presented in this paper is available on a public repository: github.com/rodemann/gp-imprecision-in-bo.

© Springer Nature Switzerland AG 2022
K. Honda et al. (Eds.): IUKM 2022, LNAI 13199, pp. 92–104, 2022.
https://doi.org/10.1007/978-3-030-98018-4_8

BO approximates the target function through a surrogate model. In the case of all covariates being real-valued, Gaussian Process (GP) regression is the most popular model, while random forests are usually preferred for categorical and mixed covariate spaces. BO scalarizes the surrogate model's mean and standard error estimates through a so-called acquisition function[1], that incorporates the trade-off between exploration (uncertainty reduction) and exploitation (mean optimization). The arguments of the acquisition function's minima are eventually proposed to be evaluated. Algorithm 1 describes the basic procedure of Bayesian optimization applied on a problem of the sort: $\min_{x \in \mathcal{X}} \Psi(x)$, where $\Psi : \mathcal{X}^p \rightarrow \mathbb{R}$, $\mathcal{X}^p$ a p-dimensional covariate space. Here and henceforth, minimization is considered without loss of generality.

Algorithm 1. Bayesian Optimization

1: create an initial design $D = \{(x^{(i)}, \Psi^{(i)})\}_{i=1,...,n_{init}}$ of size n_{init}
2: **while** termination criterion is not fulfilled **do**
3:     **train** a surrogate model (SM) on data D
4:     **propose** x^{new} that optimizes the acquisition function $AF(SM(x))$
5:     **evaluate** Ψ on x^{new} and **update** $D \leftarrow D \cup (x^{new}, \Psi(x^{new}))$
6: **end while**
7: **return** $\arg\min_{x \in D} \Psi(x)$ and respective $\Psi(\arg\min_{x \in D} \Psi(x))$

Notably, line 4 imposes a new optimization problem, sometimes referred to as "auxiliary optimization". Compared to $\Psi(x)$, however, $AF(SM(x))$ is analytically traceable. It is a deterministic transformation of the surrogate model's mean and standard error predictions, which are given by line 3. Thus, evaluations are cheap and optima can be retrieved through naive algorithms, such as grid search, random search or the slightly more advanced focus search[2], all of which simply evaluate a huge number of points that lie dense in $\mathcal{X}$. Various termination criteria are conceivable with a pre-specified number of iterations being one of the most popular choices.[3]

As stated above, GP regressions are the most common surrogate models in Bayesian optimization for continuous covariates. The main idea of functional regression based on GPs is to specify a Gaussian process a priori (a GP prior distribution), then observe data and eventually receive a posterior distribution over functions, from which inference is drawn, usually by mean and variance prediction. In more general terms, a GP is a stochastic process, i.e. a set of random variables, any finite collection of which has a joint normal distribution.

---

[1] Also referred to as infill criterion.
[2] Focus search shrinks the search space and applies random search, see [4, p. 7].
[3] BO's computational complexity depends on the SM. In case of GPs, it is $\mathcal{O}(n^3)$ due to the required inversion of the covariance matrix, where n is total number of target function evaluations.

Definition 1 (Gaussian Process Regression). *A function $f(\boldsymbol{x})$ is said to be generated by a Gaussian process $\mathcal{GP}\left(m(\boldsymbol{x}), k(\boldsymbol{x}, \boldsymbol{x}')\right)$ if for any finite vector of data points $(x_1, ..., x_n)$, the associated vector of function values $\boldsymbol{f} = (f(x_1), ..., f(x_n))$ has a multivariate Gaussian distribution: $\boldsymbol{f} \sim \mathcal{N}\left(\boldsymbol{\mu}, \boldsymbol{\Sigma}\right)$, where $\boldsymbol{\mu}$ is a mean vector and $\boldsymbol{\Sigma}$ a covariance matrix.*

Hence, Gaussian processes are fully specified by a mean function $m(\boldsymbol{x}) = \mathbb{E}[f(\boldsymbol{x})]$ and a kernel[4] $k_\theta(\boldsymbol{x}, \boldsymbol{x}') = \mathbb{E}\left[\left(f(\boldsymbol{x}) - \mathbb{E}[f(\boldsymbol{x})]\right)\left(f(\boldsymbol{x}') - \mathbb{E}[f(\boldsymbol{x}')]\right)\right]$ such that $f(\boldsymbol{x}) \sim \mathcal{GP}\left(m(\boldsymbol{x}), k_\theta(\boldsymbol{x}, \boldsymbol{x}')\right)$, see e.g. [18, p. 13]. The mean function gives the trend of the functions drawn from the GP and can be regarded as the best (constant, linear, quadratic, cubic etc.) approximation of the GP functions. The kernel gives the covariance between any two function values and thus, broadly speaking, determines the function's smoothness and periodicity.

The paper at hand is structured as follows. Section 2 conducts a sensitivity analysis of classical Bayesian optimization with Gaussian processes. As we find the prior's mean parameters to be the most influential prior component, Sect. 3 introduces PROBO, a method that is robust towards prior mean misspecification. Section 4 describes detailed experimental results from benchmarking PROBO to classical BO on a problem in material science. We conclude by a brief discussion of our method in Sect. 5.

2 Sensitivity Analysis

2.1 Experiments

The question arises quite naturally how sensitive Bayesian optimization is towards the prior specification of the Gaussian process. It is a well-known fact that classical inference from GPs is sensitive with regard to prior specification in the case of small n. The less data, the more the inference relies on the prior information. What is more, there exist detailed empirical studies such as [21] that analyze the impact of prior mean function and kernel on the posterior GP for a variety of real-world data sets. We systematically investigate to what extent this translates to BO's returned optima and convergence rates.[5] Analyzing the effect on optima and convergence rates is closely related, yet different. Both viewpoints have weaknesses: Focusing on the returned optima means conditioning the analysis on the termination criterion; considering convergence rates requires the optimizer to converge in computationally feasible time. To avoid these downsides, we analyze the mean optimizations paths.

Definition 2 (Mean Optimization Path). *Given R repetitions of Bayesian optimization applied on a test function $\Psi(\boldsymbol{x})$ with T iterations each, let $\Psi(\boldsymbol{x}^*)_{r,t}$*

---

[4] Also called covariance function or kernel function.

[5] To the best of our knowledge, this is the very first systematic assessment of GP prior's influence on BO.

Accounting for Gaussian Process Imprecision in Bayesian Optimization

be the best incumbent target value at iteration $t \in \{1, ..., T\}$ from repetition $r \in \{1, ..., R\}$. The elements

$$MOP_t = \frac{1}{R} \sum_{r=1}^{R} \Psi(x^*)_{r,t}$$

shall then constitute the T-dimensional vector MOP, which we call mean optimization path (MOP) henceforth.

As follows from Definition 1, specifying a GP prior comes down to choosing a mean function and a kernel. Both kernel and mean function are in turn determined by a functional form (e.g. linear trend and Gaussian kernel) and its parameters (e.g. intercept and slope for the linear trend and a smoothness parameter for the Gaussian kernel). Hence, we vary the GP prior with regard to the mean functional form $m(\cdot)$, the mean function parameters, the kernel functional form $k(\cdot, \cdot)$ and the kernel parameters (see Definition 1). We run the analysis on 50 well-established synthetic test functions from the R package smoof [5]. The functions are selected at random, stratified across the covariate space dimensions $1, 2, 3, 4$ and 7. For each of them, a sensitivity analysis is conducted with regard to each of the four prior components. The initial design (line 1 in Algorithm 1) of size $n_{init} = 10$ is randomly sampled anew for each of the $R = 40$ BO repetitions with $T = 20$ iterations each. This way, we make sure the results do not depend on a specific initial sample. For each test function we obtain an accumulated difference (AD) of mean optimization paths.

Definition 3 (Accumulated Difference of Mean Optimization Paths).
Consider an experiment comparing S different prior specifications on a test function with R repetitions per specification and T iterations per repetition. Let the results be stored in a $T \times S$-matrix of mean optimization paths for iterations $t \in \{1, ..., T\}$ and prior specification $s \in \{1, ..., S\}$ (e.g. constant, linear, quadratic etc. trend as mean functional form) with entries $MOP_{t,s} = \frac{1}{R} \sum_{r=1}^{R} \Psi(x^*)_{r,t,s}$. The accumulated difference (AD) for this experiment shall then be:

$$AD = \sum_{t=1}^{T} \left(\max_{s} MOP_{t,s} - \min_{s} MOP_{t,s} \right).$$

2.2 Results of Sensitivity Analysis

The AD values vary strongly across functions. This can be explained by varying levels of difficulty of the optimization problem, mainly influenced by modality and smoothness. Since we are interested in an overall, systematic assessment of the prior's influence on Bayesian optimization, we sum the AD values over the stratified sample of 50 functions. This absolute sum, however, is likely driven by some hard-to-optimize functions with generally higher AD values or by the scale of the functions' target values.[6] Thus, we divide each AD value by the mean AD

---

[6] Note that neither accumulated differences (Definition 3) nor mean optimization paths (Definition 2) are scale-invariant.

of the respective function. Table 1 shows the sums of these relative AD values. It becomes evident that the optimization is affected the most by the functional form of the kernel and the mean parameters, while kernel parameters and the mean functional form play a minor role.

Table 1. Sum of relative ADs of all 50 MOPs per prior specification. Comparisons between mean and kernel are more valid than between functional form and parameters.

Mean functional form	Kernel functional form	Mean parameters	Kernel parameters
42.49	68.20	77.91	11.40

2.3 Discussion of Sensitivity Analysis

Bayesian optimization typically deals with expensive-to-evaluate functions. As such functions imply the availability of few data, it comes at no surprise that the GP's predictions in BO heavily depend on the prior. Our results suggest this translates to BO's convergence. It is more sensitive towards the functional form of the kernel than towards those of the mean function and more sensitive towards the mean function's parameters than towards those of the kernel, which appear to play a negligible role in BO's convergence.

The kernel functional form determines the flexibility of the GP and thus has a strong effect on its capacity to model the functional relationship. What is more interesting, the mean parameters' effect may not only stem from the modeling capacity but also from the optimizational nature of the algorithm. While unintended in statistical modeling, a systematic under- or overestimation may be beneficial when facing an optimization problem. Further research on interpreting the effect of the GP prior's components on BO's performance is recommended.

2.4 Limitations of Sensitivity Analysis

Albeit the random sample of 50 test functions was drawn from a wide range of established benchmark functions, the analysis does by far not comprise all types of possible target functions, not to mention real-world optimization problems. Additionally, the presented findings regarding kernel and mean function parameters are influenced by the degree of variation, the latter being a subjective choice. Statements comparing the influence of the functional form with the parameters are thus to be treated with caution. Yet, the comparison between kernel and mean function parameters is found valid, as both have been altered by the same factors.

What weighs more, interaction effects between the four prior components were partly left to further research. The reported AD values for mean parameters and mean functional forms were computed using a Gaussian kernel. Since other kernels may interact differently with the mean function, the analysis was revisited using a power exponential kernel as well as a Matérn kernel. As we observe only small changes in AD values, the sensitivity analysis can be seen as relatively robust in this regard, at least with respect to these three widely-used kernels.

3 Prior-Mean-Robust Bayesian Optimization

While a highly popular hyperparameter optimizer in machine learning [15], Bayesian optimization itself – not without a dash of irony – heavily depends on its hyperparameters, namely the Gaussian process prior specification. The sensitivity analysis in Sect. 2 has shown that the algorithm's convergence is especially sensitive towards the mean function's parameters.

In light of this result, it appears desirable to mitigate BO's dependence on the prior by choosing a prior mean function that expresses a state of ignorance. Recall that Bayesian optimization is typically used for "black-box-functions", where very little, if any, prior knowledge exists. The classical approach would be to specify a so-called non-informative prior over the mean parameters. However, such a prior is not unique [3] and choosing different priors among the set of all non-informative priors would lead to different posterior inferences [12]. Thus, such priors cannot be regarded as fully uninformative and represent indifference rather than ignorance. Principled approaches would argue that this dilemma cannot be solved within the framework of classical precise probabilities. Methods working with sets of priors have thus attracted increasing attention, see e.g. [1,19]. Truly uninformative priors, however, would entail sets of all possible probability distributions and thus lead to vacuous posterior inference. That is, prior beliefs would not change with data, which would make learning impossible. [3] thus propose prior *near*-ignorance models as a compromise that conciliates learning and *almost* non-informative priors. In the case of Gaussian processes, so-called imprecise Gaussian processes (IGP) are introduced by [11] as prior near-ignorance models for GP regression. The general idea of an IGP is to incorporate the model's imprecision regarding the choice of the prior's mean function parameter, given a constant mean function and a fully specified kernel. In the case of univariate regression, given a base kernel $k_\theta(x,x')$ and a degree of imprecision $c > 0$, [11, Definition 2] defines a constant mean imprecise Gaussian process as a set of GP priors:

$$\mathcal{G}_c = \left\{ GP\left(Mh, k_\theta(x,x') + \frac{1+M}{c}\right) : h = \pm 1, M \geq 0 \right\} \qquad (1)$$

It can be shown that $\mathcal{G}_c$ verifies prior near-ignorance [11, p. 194] and that $c \to 0$ yields the precise model [11, p. 189]. Note that the mean functional form (constant) as well as both kernel functional form and its parameters do not vary in set $\mathcal{G}_c$, but only the mean parameter $Mh \in \,]-\infty, \infty[$. For each prior GP, a posterior GP can be inferred. This results in a set of posteriors and a corresponding set of mean estimates $\hat{\mu}(x)$, of which the upper and lower mean estimates $\underline{\hat{\mu}}(x)$, $\overline{\hat{\mu}}(x)$ can be derived analytically. To this very end, let $k_\theta(x,x')$

be a kernel function as defined in [18]. The finitely positive semi-definite matrix $\boldsymbol{K}_n$ is then formed by applying $k_\theta(x, x')$ on the training data vector $x \in \mathcal{X}$:

$$\boldsymbol{K}_n = [k_\theta(x_i, x'_j)]_{ij}. \tag{2}$$

Following [11], we call $\boldsymbol{K}_n$ base kernel matrix. Note that $\boldsymbol{K}_n$ is restricted only to be finitely positive semi-definite and not to have diagonal elements of 1. In statistical terms, $\boldsymbol{K}_n$ is a covariance matrix and not necessarily a correlation matrix. Hence, the variance $I\sigma^2$ is included. Diverging from [11], we only consider target functions without explicit noise, thus no "nugget term" $I\sigma^2_{nugget}$ needs to be included in $\boldsymbol{K}_n$.

Now let x be a scalar input of test data, whose $f(x)$ is to be predicted. Then $\boldsymbol{k}_x = [k_\theta(x, x_1), ..., k_\theta(x, x_n)]^T$ is the vector of covariances between x and the training data. Furthermore, define $\boldsymbol{s}_k = \boldsymbol{K}_n^{-1}\mathbb{1}_n$ and $\boldsymbol{S}_k = \mathbb{1}_n^T \boldsymbol{K}_n^{-1}\mathbb{1}_n$. Then [11] shows that upper and lower bounds of the posterior predictive mean function $\hat{\mu}(x)$ for $f(x)$ can be derived. If $\left|\frac{s_k y}{S_k}\right| \leq 1 + \frac{c}{S_k}$, they are:

$$\overline{\hat{\mu}}(x) = \boldsymbol{k}_x^T \boldsymbol{K}_n^{-1}\boldsymbol{y} + (1 - \boldsymbol{k}_x^T \boldsymbol{s}_k)\frac{\boldsymbol{s}_k^T}{\boldsymbol{S}_k}\boldsymbol{y} + c\frac{|1 - \boldsymbol{k}_x^T \boldsymbol{s}_k|}{\boldsymbol{S}_k} \tag{3}$$

$$\underline{\hat{\mu}}(x) = \boldsymbol{k}_x^T \boldsymbol{K}_n^{-1}\boldsymbol{y} + (1 - \boldsymbol{k}_x^T \boldsymbol{s}_k)\frac{\boldsymbol{s}_k^T}{\boldsymbol{S}_k}\boldsymbol{y} - c\frac{|1 - \boldsymbol{k}_x^T \boldsymbol{s}_k|}{\boldsymbol{S}_k} \tag{4}$$

If $\left|\frac{s_k y}{S_k}\right| > 1 + \frac{c}{S_k}$:

$$\overline{\hat{\mu}}(x) = \boldsymbol{k}_x^T \boldsymbol{K}_n^{-1}\boldsymbol{y} + (1 - \boldsymbol{k}_x^T \boldsymbol{s}_k)\frac{\boldsymbol{s}_k^T}{\boldsymbol{S}_k}\boldsymbol{y} + c\frac{1 - \boldsymbol{k}_x^T \boldsymbol{s}_k}{\boldsymbol{S}_k} \tag{5}$$

$$\underline{\hat{\mu}}(x) = \boldsymbol{k}_x^T \boldsymbol{K}_n^{-1}\boldsymbol{y} + (1 - \boldsymbol{k}_x^T \boldsymbol{s}_k)\frac{\boldsymbol{s}_k^T \boldsymbol{y}}{c + \boldsymbol{S}_k} \tag{6}$$

Inspired by multi-objective BO [9], one might think (despite knowing better) of an IGP and a GP as surrogate models for different target functions. A popular approach in multi-objective BO to proposing points based on various surrogate models is to scalarize their predictions by an acquisition function defined *a priori*. The herein proposed generalized lower confidence bound (GLCB) is such an acquisition function, since it combines mean and variance predictions of a precise GP with upper and lower mean estimates of an IGP. In this way, it generalizes the popular lower confidence bound $LCB(\boldsymbol{x}) = \hat{\mu}(\boldsymbol{x}) - \tau \cdot \sqrt{var(\hat{\mu}(\boldsymbol{x}))}$, initially proposed by [6].[7]

Definition 4 (Generalized Lower Confidence Bound (GLCB)). *Let $\boldsymbol{x} \in \mathcal{X}$. As above, let $\overline{\hat{\mu}}(\boldsymbol{x}), \underline{\hat{\mu}}(\boldsymbol{x})$ be the upper/lower mean estimates of an IGP with imprecision c. Let $\hat{\mu}(\boldsymbol{x})$ and $var(\hat{\mu}(\boldsymbol{x}))$ be the mean and variance predictions*

---

[7] Note that from a decision-theoretic point of view, LCB violates the dominance principle. GLCB inherits this property.

of a precise GP. The prior-mean-robust acquisition function generalized lower confidence bound (GLCB) *shall then be*

$$GLCB(\dot{x}) = \hat{\mu}(x) - \tau \cdot \sqrt{var(\hat{\mu}(x))} - \rho \cdot (\overline{\hat{\mu}}(x) - \underline{\hat{\mu}}(x)).$$

By explicitly accounting for the prior-induced imprecision, GLCB generalizes the trade-off between exploration and exploitation: $\tau > 0$ controls the classical "mean vs. data uncertainty" trade-off (degree of risk aversion) and $\rho > 0$ controls the "mean vs. model imprecision" trade-off (degree of ambiguity aversion). Notably, $\overline{\hat{\mu}}(x) - \underline{\hat{\mu}}(x)$ simplifies to an expression only dependent on the kernel vector between x and the training data $k_x = [k_\theta(x, x_1), ..., k_\theta(x, x_n)]^T$, the base kernel matrix K_n (Eq. 2) and the degree of imprecision c, which follows from Eqs. 5 and 6 in case $\left|\frac{s_k y}{S_k}\right| > 1 + \frac{c}{S_k}$:

$$\overline{\hat{\mu}}(x) - \underline{\hat{\mu}}(x) = (1 - k_x^T s_k)\left(\frac{s_k^T}{S_k}y + \frac{c}{S_k} - \frac{s_k^T y}{c + S_k}\right) \tag{7}$$

As can be seen by comparing Eqs. 3 and 4, in case of $\left|\frac{s_k y}{S_k}\right| \leq 1 + \frac{c}{S_k}$, the model imprecision $\overline{\hat{\mu}}(x) - \underline{\hat{\mu}}(x)$ even simplifies further: $\overline{\hat{\mu}}(x) - \underline{\hat{\mu}}(x) = 2c\frac{|1 - k_x^T s_k|}{S_k}$. In this case, the GLCB comes down to $GLCB(x) = \hat{\mu}(x) - \tau \cdot \sqrt{var(\hat{\mu}(x))} - 2 \cdot \rho c\frac{|1 - k_x^T s_k|}{S_k}$ and the two hyperparameters ρ and c collapse to one. In both cases, the surrogate models $\underline{\hat{\mu}}(x)$ and $\overline{\hat{\mu}}(x)$ do not have to be fully implemented. Only K_n and $k_x = [k_\theta(x, x_1), ..., k_\theta(x, x_n)]^T$ need to be computed. GLCB can thus be plugged into standard BO without much additional computational cost.[8] Algorithm 2 describes the procedure.

Algorithm 2. Prior-mean-RObust Bayesian Optimization (PROBO)

1: create an initial design $D = \{(x^{(i)}, \Psi^{(i)})\}_{i=1,...,n_{init}}$ of size n_{init}
2: specify c and ρ
3: **while** termination criterion is not fulfilled **do**
4:     **train** a precise GP on data D and obtain $\hat{\mu}(x)$, $var(\hat{\mu}(x))$
5:     **compute** k_x, s_k and S_k
6:     **if** $\left|\frac{s_k y}{S_k}\right| > 1 + \frac{c}{S_k}$ **then**
7:       $\overline{\hat{\mu}}(x) - \underline{\hat{\mu}}(x) = (1 - k_x^T s_k)\left(\frac{s_k^T}{S_k}y + \frac{c}{S_k} - \frac{s_k^T y}{c + S_k}\right)$
8:     **else** $\overline{\hat{\mu}}(x) - \underline{\hat{\mu}}(x) = 2c\frac{|1 - k_x^T s_k|}{S_k}$
9:     **compute** $GLCB(x) = -\hat{\mu}(x) + \tau \cdot \sqrt{var(\hat{\mu}(x))} + \rho \cdot (\overline{\hat{\mu}}(x) - \underline{\hat{\mu}}(x))$
10:    **propose** x^{new} that optimizes $GLCB(x)$
11:    **evaluate** Ψ on x^{new}
12:    **update** $D \leftarrow D \cup (x^{new}, \Psi(x^{new}))$
13: **end while**
14: **return** $\arg\min_{x \in D} \Psi(x)$ and respective $\Psi(\arg\min_{x \in D} \Psi(x))$

---

[8] Further note that with expensive target functions to optimize, the computational costs of surrogate models and acquisition functions in BO can be regarded as negligible. The computational complexity of PROBO is the same as for BO with GP.

Just like LCB, the generalized LCB balances optimization of $\hat{\mu}(x)$ and reduction of uncertainty with regard to the model's prediction variation $\sqrt{var(\hat{\mu}(\boldsymbol{x}))}$ through τ. What is more, GLCB aims at reducing model imprecision caused by the prior specification, controllable by ρ. Ideally, this would allow returning optima that are robust not only towards classical prediction uncertainty but also towards imprecision of the specified model.

4   Results

We test our method on a univariate target function generated from a data set that describes the quality of experimentally produced graphene, an allotrope of carbon with potential use in semiconductors, smartphones and electric batteries [24]. The data set comprises $n = 210$ observations of an experimental manufacturing process of graphene. A polyimide film, typically Kapton, is irradiated with laser in a reaction chamber in order to trigger a chemical reaction that results in graphene. Four covariates influence the manufacturing process, namely power and time of the laser irradiation as well as gas in and pressure of the reaction chamber [24]. The target variable (to be maximized) is a measure for the quality of the induced graphene, ranging from 0.1 to 5.5. In order to construct a univariate target function from the data set, a random forest was trained on a subset of it (target quality and time, see Fig. 1). The predictions of this random forests were then used as target function to be optimized.

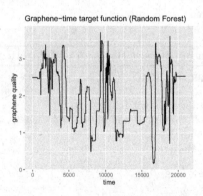

Fig. 1. Univariate target function estimated from graphene data.

We compare GLCB to its classical counterpart LCB as well as to the expected improvement (EI), which is usually considered the most popular acquisition function. It was initially proposed by [14, p. 1–2], disguised as a utility function in a decision problem that captures the expected deviation from the extremum. Let $\psi(\boldsymbol{x})$ be the surrogate model, in our case the posterior predictive GP, and Ψ_{min} the incumbent minimal function value. The expected improvement at point $\boldsymbol{x}$ then is $EI(\boldsymbol{x}) = \mathbb{E}(\max\{\Psi_{min} - \psi(\boldsymbol{x}), 0\})$. For pairwise comparisons of GLCB to LCB and EI, we observe $n = 60$ BO runs with a budget of 90 evaluations and an initial design of 10 data points generated by latin hypercube sampling [13] each. Focus search [4, p. 7] was used as infill optimizer with 1000 evaluations per round and 5 maximal restarts. All experiments were conducted in R version 4.0.3 [17] on a high performance computing cluster using 20 cores (linux gnu). Figure 2 depicts mean optimization paths of BO with GLCB compared to LCB and EI on the graphene-time target function. The paths are shown for three different GLCB settings: $\rho = 1, c = 50$ and $\rho = 1, c = 100$ as well as $\rho = 10, c = 100$. Figure 2 shows that GLCB surpasses LCB (all settings) and EI ($\rho = 10, c = 100$) in late iterations. We also compare GLCB to other acquisition functions and retrieve similar results, except for one purely exploratory and thus degenerated acquisition function, see chapter 6.7.2 in [20].

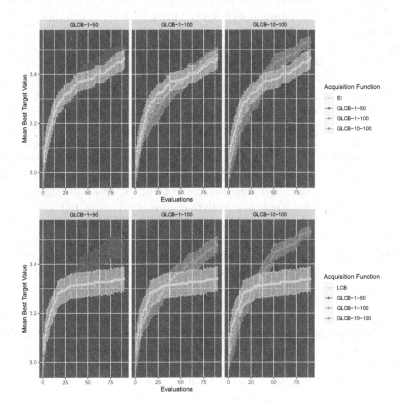

Fig. 2. Benchmarking results from graphene data: Generalized lower confidence bound (GLCB) vs. expected improvement (EI) and lower confidence bound (LCB). Shown are 60 runs per Acquisition Function with 90 evaluations and initial sample size 10 each. Error bars represent 0.95 confidence intervals. GLCB-1-100 means $\rho = 1$ and c = 100; $\tau = 1$ for all GLCBs and LCB.

Further benchmark experiments are conducted on meteorological data, heartbeat time series as well as synthetic functions from [5]. In case of multimodal and wiggly target functions, the results resemble Fig. 2. When optimizing smooth and unimodal functions, however, classical acquisition functions like EI and LCB are superior to GLCB. We suppose that when faced with such simpler problems, the model imprecision is not severe enough to justify additional explorations. For a detailed documentation of these further experiments, we refer the interested reader to chapters 6.6.2, 6.6.3 and 6.6.4 in [20].

5   Discussion

The promising results should not hide the fact that the proposed modification makes the optimizer robust only with regard to possible misspecification of the mean function parameter given a constant trend. Albeit the sensitivity analysis conducted in Sect. 2 demonstrated its importance, the mean parameter is clearly not the only influential component of the GP prior in BO. For instance, the functional form of the kernel also plays a major role, see Table 1. The question of how to specify this prior component is discussed in [7,10]. Apart from this, it is important to note that PROBO depends on a subjectively specified degree of imprecision c. It does not account for any imaginable prior mean (the model would become vacuous, see Sect. 3). What is more, it may be difficult to interpret c and thus specify it in practical applications. However, our method still offers more generality than a precise choice of the mean parameter. Specifying c corresponds to a weaker assumption than setting precise mean parameters.

Notwithstanding such deliberations concerning PROBO's robustness and generality, the method simply converges faster than BO when faced with multimodal and non-smooth target functions. The latter make up an arguably considerable part of problems not only in hyperparameter-tuning, but also in direct applications of BO such as in engineering [8] or drug discovery [16].

The herein proposed method opens several venues for future work. An extension to other Bayesian surrogate models seems feasible, since there is a variety of prior near-ignorance models. What is more, also non-Bayesian surrogate models like random forest can be altered such that they account for imprecision in their assumptions, see [23] for instance. Generally speaking, imprecise probability (IP) models appear very fruitful in the context of optimization based on surrogate models. They not only offer a vivid framework to represent prior ignorance, as demonstrated in this very paper, but may also be beneficial in applications where prior knowledge is abundant. In such situations, in the case of data contradicting the prior, precise probabilities often fail to adequately represent uncertainty, whereas IP models can handle these prior-data conflicts, see e.g. [25].

References

1. Augustin, T., Coolen, F.P., de Cooman, G., Troffaes, M.C.M.: Introduction to Imprecise Probabilities. Wiley, Chichester (2014)
2. Awal, M.A., Masud, M., Hossain, M.S., Bulbul, A.A., Mahmud, S.M.H., Bairagi, A.K.: A novel Bayesian optimization-based machine learning framework for COVID-19 detection from inpatient facility data. IEEE Access **9**, 10263–10281 (2021)

3. Benavoli, A., Zaffalon, M.: Prior near ignorance for inferences in the k-parameter exponential family. Statistics **49**(5), 1104–1140 (2015)
4. Bischl, B., Richter, J., Bossek, J., Horn, D., Thomas, J., Lang, M.: mlrMBO: a modular framework for model-based optimization of expensive black-box functions. arXiv preprint arXiv:1703.03373 (2017)
5. Bossek, J.: smoof: Single- and multi-objective optimization test functions. R J. **9**(1), 103–113 (2017)
6. Cox, D.D., John, S.: A statistical method for global optimization. In: Proceedings of 1992 IEEE International Conference on Systems, Man, and Cybernetics, pp. 1241–1246. IEEE (1992)
7. Duvenaud, D.: Automatic model construction with Gaussian processes. Ph.D. thesis. University of Cambridge (2014)
8. Frazier, P.I., Wang, J.: Bayesian optimization for materials design. In: Lookman, T., Alexander, F.J., Rajan, K. (eds.) Information Science for Materials Discovery and Design. SSMS, vol. 225, pp. 45–75. Springer, Cham (2016). https://doi.org/10.1007/978-3-319-23871-5_3
9. Horn, D., Wagner, T., Biermann, D., Weihs, C., Bischl, B.: Model-based multi-objective optimization: taxonomy, multi-point proposal, toolbox and benchmark. In: Gaspar-Cunha, A., Henggeler Antunes, C., Coello, C.C. (eds.) EMO 2015. LNCS, vol. 9018, pp. 64–78. Springer, Cham (2015). https://doi.org/10.1007/978-3-319-15934-8_5
10. Malkomes, G., Garnett, R.: Automating Bayesian optimization with Bayesian optimization. Adv. Neural. Inf. Process. Syst. **31**, 5984–5994 (2018)
11. Mangili, F.: A prior near-ignorance Gaussian process model for nonparametric regression. In: ISIPTA 2015: Proceedings of the 9th International Symposium on Imprecise Probability: Theories and Applications, pp. 187–196 (2015)
12. Mangili, F.: A prior near-ignorance Gaussian process model for nonparametric regression. Int. J. Approx. Reason. **78**, 153–171 (2016)
13. McKay, M.D., Beckman, R.J., Conover, W.J.: A comparison of three methods for selecting values of input variables in the analysis of output from a computer code. Technometrics **42**(1), 55–61 (2000)
14. Močkus, J.: On Bayesian methods for seeking the extremum. In: Marchuk, G.I. (ed.) Optimization Techniques 1974. LNCS, vol. 27, pp. 400–404. Springer, Heidelberg (1975). https://doi.org/10.1007/3-540-07165-2_55
15. Nguyen, V.: Bayesian optimization for accelerating hyper-parameter tuning. In: 2019 IEEE Second International Conference on Artificial Intelligence and Knowledge Engineering (AIKE), pp. 302–305. IEEE (2019)
16. Pyzer-Knapp, E.O.: Bayesian optimization for accelerated drug discovery. IBM J. Res. Dev. **62**(6), 2:1-2:7 (2018)
17. R Core Team: R: a language and environment for statistical computing. R Foundation for Statistical Computing, Vienna, Austria (2020)
18. Rasmussen, C.E.: Gaussian processes in machine learning. In: Bousquet, O., von Luxburg, U., Rätsch, G. (eds.) ML -2003. LNCS (LNAI), vol. 3176, pp. 63–71. Springer, Heidelberg (2004). https://doi.org/10.1007/978-3-540-28650-9_4
19. Insua, D.R., Ruggeri, F.: Robust Bayesian Analysis. Springer, New York (2000)
20. Rodemann, J.: Robust generalizations of stochastic derivative-free optimization. Master's thesis. LMU Munich (2021)
21. Schmidt, A.M., Conceição, M.F.G., Moreira, G.A.: Investigating the sensitivity of Gaussian processes to the choice of their correlation function and prior specifications. J. Stat. Comput. Simul. **78**(8), 681–699 (2008)

22. Snoek, J., Larochelle, H., Adams, R.P.: Practical Bayesian optimization of machine learning algorithms. Adv. Neural. Inf. Process. Syst. **25**, 2951–2959 (2012)
23. Utkin, L., Kovalev, M., Meldo, A., Coolen, F.: Imprecise extensions of random forests and random survival forests. In: International Symposium on Imprecise Probabilities: Theories and Applications, pp. 404–413. PMLR (2019)
24. Wahab, H., Jain, V., Tyrrell, A.S., Seas, M.A., Kotthoff, L., Johnson, P.A.: Machine-learning-assisted fabrication: Bayesian optimization of laser-induced graphene patterning using in-situ Raman analysis. Carbon **167**, 609–619 (2020)
25. Walter, G., Augustin, T.: Imprecision and prior-data conflict in generalized Bayesian inference. J. Stat. Theory Pract. **3**(1), 255–271 (2009)

Additional Out-Group Search for JADE

Yuichi Miyahira$^{(\boxtimes)}$ and Akira Notsu

Osaka Prefecture University, 1-1 Gakuen-cho, Naka-ku, Sakai, Osaka 599-8531, Japan
miyahira@hi.cs.osakafu-u.ac.jp

Abstract. JADE is a method to adaptively select parameters using probability distribution, and shows good searching accuracy and speed. However, the search outside the solution group is not considered as well as other methods. Though it is possible to forcibly increase the search outside the solution group by dividing the solution group or adding random search, the search speed lowers. In this study, referring to the Nelder-Mead method, the algorithm for improving the accuracy without reducing the search speed as much as possible by adding the solution group outside search to JADE is proposed. Concretely, when a fixed condition is satisfied, one point of the search point carries out the group outside search. In order to prevent the search speed from decreasing even in the case of a high dimension, the number of search points for out-of-group search is set not to increase even when the dimension increases. The effectiveness of the proposed method is confirmed by numerical experiments.

Keywords: Differential evolution · JADE · Nelder-Mead

1 Introduction

Difference Evolution [1] is an optimizing algorithm without gradients that has good search performance and is applicable to a variety of applications. DE is a relatively simple algorithm with the advantage that it has only three control parameters: mutation coefficient F, crossover rate CR, and population size N. However, due to the small number of parameters, the accuracy of the search is affected by the parameters. Therefore, it is necessary to set the appropriate parameter value according to the problem to be handled and the search situation. To solve this problem, SaDE [2], jDE [3], SHADE [4], JADE [5,6] have been proposed.

JADE is a method for efficiently adapting the parameters of DE to the environment by sampling them from a Cauchy or Gaussian distribution adapted to the environment. Specifically, the mean value in the probability distribution used to generate the mutation coefficient F and the crossover rate CR is adjusted according to the success value of each parameter, allowing each parameter to

Supported by JSPS KAKENHI Grant Number JP18K11473.

K. Honda et al. (Eds.): IUKM 2022, LNAI 13199, pp. 105–116, 2022.
https://doi.org/10.1007/978-3-030-98018-4_9

be adjusted automatically. This allows the appropriate parameters to be used, increasing the accuracy of the search.

As with other methods, out-group search is not considered much in JADE, and performance degrades if a solution is missed outside the group after the search has progressed. In our basic experiment [7], we confirmed that searching can be improved by adding the probability of selecting $F = 1.5$, which is an enhanced parameter of out-group search that is not normally set in DE. Therefore, we believe that the performance can be further improved by adding out-group search to JADE.

In this study, referring to Nelder-Mead method [8], we propose an algorithm to add out-group search to JADE and to improve the accuracy without decreasing the search speed as much as possible. If the distance of the solution with the best and worst solutions is more than half of the distance from the end of the search point group, it is judged that there is a large bias in the direction of the update of the search point group, and the outside of the solution group is searched. The composition of this paper is as follows. In Sect. 2, JADE and Nelder-Mead method which are handled in this study are explained. In Sect. 3, we propose and explain JADE with the addition of out-group search. In Sect. 4, the results of numerical experiments using the proposed method are discussed, and in Sect. 5, future problems are summarized.

2   Base Algorithm

2.1   JADE

JADE is one of the improved methods of DE, and the mutation coefficient F and the crossover rate CR are sampled from the probability distribution and used. The Cauchy distribution is used for the mutation coefficient, and the Gaussian distribution is used for the crossover rate, and the adjustment of each distribution is carried out based on the proportion in which the solution was renewed in the past. In JADE, parameters are generated for each individual. As a mutation strategy, "DE/current-to-$p$best" is used. The generation method of each parameter is shown below.

For each generation, the mutation factor F_i of each individual x_i is generated according to the Cauchy distribution of the positional parameter μ_F and the scaling parameter $\sigma_F = 0.1$ as follows:

$$F_i \sim C(\mu_F, \sigma_F)$$

F_i is regenerated if $F_i \leq 0$, and truncated to 1 if $F_i \geq 1$. The positional parameter μ_F is initialized at 0.5 and updated for each generation as follows:

$$\mu_F = (1 - c) \cdot \mu_F + c \cdot \frac{\sum_{F \in S_F} F^2}{\sum_{F \in S_F} F}$$

c is a constant for $(0, 1]$ and the recommended value is 0.1. S_F is the set of mutation coefficients that have successfully updated the solution in that generation.

Similarly, for each generation, the crossover rate CR_i of each individual x_i is generated according to a Gaussian distribution with a mean of μ_{CR} and a standard deviation of $\sigma_{CR} = 0.1$ as follows:

$$CR_i \sim N(\mu_{CR}, \sigma_{CR}^2)$$

CR_i is truncated to the interval $[0, 1]$. The average μ_{CR} is initialized at 0.5 and updated for each generation as follows:

$$CR_i \sim N(\mu_{CR}, \sigma_{CR}^2)$$

S_N is the number of times the solution is updated successfully in each generation, and S_{CR} is the set of crossover rates CR when the solution is updated successfully.

The following describes the mutation strategy "DE/current-to-pbest" used in JADE. In this strategy, the mutation vector $v_{i,g}$ for each individual $x_{i,g}$ of each generation g is generated by:

$$v_{i,g} = x_{i,g} + F_i \cdot (x_{best,g}^p - x_{i,g}) + F_i \cdot (x_{r1,g} - x_{r2,g})$$

F_i is the mutation coefficient of each individual x_i, and $x_{best,g}^p$ is an individual selected from the top $100p\%$ individuals. Also, $x_{r1,g}$ and $x_{r2,g}$ are two points randomly selected from the search points other than x_i so that they do not overlap. Here p is a constant in $(0, 1)$, and the recommended value is 0.1. In addition, there is a way to use an archive in this strategy. In this case, the mutation vector $v_{i,g}$ is generated by the following equation:

$$v_{i,g} = x_{i,g} + F_i \cdot (x_{best,g}^p - x_{i,g}) + F_i \cdot (x_{r1,g} - \tilde{x}_{r2,g})$$

$\tilde{x}_{r2,g}$ is an individual selected at random from the aggregate of past failed individuals stored in the archive and the aggregate of current solution populations. The archive is initially empty and adds failed individuals at the end of each generation update.

The next search point is generated by crossing over this mutation vector with the individual as follows based on the crossover rate:

$$x_{i,g+1,j} = \begin{cases} v_{i,g,j} \, (w \leq CR) \\ x_{i,g,j} \, (otherwise) \end{cases}$$

w is uniform random number between 0 and 1, and $x_{i,g,j}$ means the j-th element of the i-th individual in generation g.

2.2 Nelder-Mead Method

The Nelder-Mead method is a kind of optimizing algorithm without using gradient information. By giving $D + 1$ search points in D dimension space and repeating reflection, expansion and contraction for them, the optimum solution can be searched. The algorithm of function minimization in the Nelder-Mead method is shown below.

Step 0. For $D + 1$ search points, $f(\boldsymbol{x}_1) \leq f(\boldsymbol{x}_2) \leq \cdots \leq f(\boldsymbol{x}_{D+1})$ is set in the order of the values of the objective function f, and if the end condition is not satisfied, **Step1** is assumed, and if the end condition is satisfied, $\boldsymbol{x}_1$ is assumed as the solution.

Step 1. Using the centroids $\boldsymbol{x}_c$ of $\boldsymbol{x}_1 \ldots \boldsymbol{x}_n$, the reflection point $\boldsymbol{x}_{ref}$ of $\boldsymbol{x}_{D+1}$ is determined by the following equation:

$$\boldsymbol{x}_{ref} = \boldsymbol{x}_c + \alpha(\boldsymbol{x}_c - \boldsymbol{x}_{D+1})$$

Step 2.

> **case 1** If $f(\boldsymbol{x}_1) \leq f(\boldsymbol{x}_{ref}) < f(\boldsymbol{x}_D)$, replace $\boldsymbol{x}_{D+1}$ with $\boldsymbol{x}_{ref}$ and go to **Step0**.
>
> **case 2** If $f(\boldsymbol{x}_{ref}) < f(\boldsymbol{x}_1)$, the expansion point, which is the point where the reflection point is further extended, is obtained as follows:

$$\boldsymbol{x}_{exp} = \boldsymbol{x}_c + \gamma(\boldsymbol{x}_{ref} - \boldsymbol{x}_c)$$

> If $f(\boldsymbol{x}_{exp}) \leq f(\boldsymbol{x}_{ref})$, replace $\boldsymbol{x}_{D+1}$ with $\boldsymbol{x}_{exp}$ and go to **Step0**, otherwise replace $\boldsymbol{x}_{D+1}$ with $\boldsymbol{x}_{ref}$ and go to **Step0**.
>
> **case 3** If $f(\boldsymbol{x}_D) \leq f(\boldsymbol{x}_{ref})$
>
>> **case 3–1** If $f(\boldsymbol{x}_{ref}) < f(\boldsymbol{x}_{D+1})$, the contraction point is obtained as follows:

$$\boldsymbol{x}_{con} = \boldsymbol{x}_c + \beta(\boldsymbol{x}_{ref} - \boldsymbol{x}_c)$$

>> Then go to **Step3**.
>>
>> **case 3–2** Otherwise, the contraction point is obtained as follows:

$$\boldsymbol{x}_{con} = \boldsymbol{x}_c + \beta(\boldsymbol{x}_{D+1} - \boldsymbol{x}_c)$$

>> Then go to **Step3**.

Step 3. If $f(\boldsymbol{x}_{con}) < \min\{f(\boldsymbol{x}_{ref}), f(\boldsymbol{x}_{D+1})\}$, replace $\boldsymbol{x}_{D+1}$ with $\boldsymbol{x}_{con}$ and go to **Step0**, otherwise go to **Step4**.

Step 4. Shrink all individual i to point $\boldsymbol{x}_1$ as follows:

$$\boldsymbol{x}_i = \boldsymbol{x}_1 + \delta(\boldsymbol{x}_i - \boldsymbol{x}_1)$$

Then go to **Step0**.

3   Proposed Method

In the proposed method, one of the search points of JADE is made to search outside the group according to the situation. By this, the search is made so that the solution is not missed, when the function in which there are multiple local solutions is searched. In addition, escape from the local solution by the search outside the group is expected, when it falls into the local solution. And, the lowering of the search speed by the increase of the outside group search is prevented by limiting the point of the outside group search to one point.

Concretely, if the distance between the search point with the best solution and the search point with the worst solution is more than half of the distance from the end to the end of the search point group, one search point searches the point by extending a 2^n vector from the search point with the worst solution to the search point with the best solution. The value of n is determined by sampling from a geometric distribution. The success probability of the geometric distribution p is defined as $1/G^{p_{update}}$ by the update rate of the solution p_{update} and the generation G. This p_{update} is updated in the same way as JADE with an initial value of 0 (Fig. 1).

$$\text{if } ||\boldsymbol{x}_{\text{best}} - \boldsymbol{x}_{\text{worst}}|| > ||\boldsymbol{x}_{\text{max}} - \boldsymbol{x}_{\text{min}}||/2$$
$$n = \text{GeometricDistribution}(1/G^{p_{update}})$$
$$\boldsymbol{x}_{\text{last}} = \boldsymbol{x}_{\text{worst}} + (\boldsymbol{x}_{\text{best}} - \boldsymbol{x}_{\text{worst}}) * 2^n$$

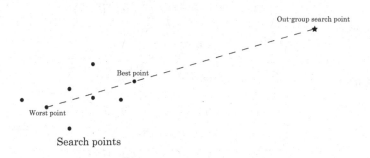

Fig. 1. Out-group search

4   Numerical Experiments and Results

JADE and the proposed method (JADE+) are compared for 16 benchmark functions shown in Table 1. In the comparison experiment, calculation up to 1000 generations was tried 1000 times for each function at a search point of 10. Figure 2 through 24 show the average of 1000 times of the best solution for each generation, with the evaluation value on the vertical axis and the generation number on the horizontal axis. Figures 2, 3, 4, 5, 6, 7, 8, 9, 10, 11, 12, 13, 14, 15, 16 and 17 show the results of the experiment with the dimension of 2, and Figs. 18, 19, 20, 21, 22, 23 and 24 show the results of the experiment with the dimension of 10. In order to deal with the minimization problem, we show that the lower the graph is, the better the solution is. Tables 2 and 3 summarize the mean ± standard deviation of the final solutions for each function. Table 2 shows the case where the dimension is 2, and Table 3 shows the case where the dimension is 10. In Tables 2 and 3, the methods which obtained good solutions and obtained solutions are shown in bold type.

Table 1. Test functions used in the experiment

Function	Equation · Domain · Optimum solution	Shape (D=2)
F1: Sphere	$f(x_1,\cdots,x_D) = \sum_{j=1}^{D} x_j^2$ $-1 \le x_j \le 1$ $f_{min}(0,\cdots,0) = 0$	
F2: Rosenbrock	$f(x_1,\cdots,x_D) = \sum_{j=1}^{D-1}(100(x_{j+1}-x_j^2)^2 + (x_j-1)^2)$ $-5 \le x_j \le 5$ $f_{min}(1,\cdots,1) = 0$	
F3: Booth	$f(x_1,x_2) = (x_1 + 2x_2 - 7)^2 + (2x_1 + x_2 - 5)^2$ $-10 \le x_j \le 10$ $f_{min}(1,-3) = 0$	
F4: Matyas	$f(x_1,x_2) = 0.26(x_1^2 + x_2^2) - 0.48x_1x_2$ $-10 \le x_j \le 10$ $f_{min}(0,0) = 0$	
F5: Easom	$f(x_1,x_2) = -\cos(x_1)\cos(x_2)\exp(-((x_1-\pi)^2 + (x_2-\pi)^2))$ $-100 \le x_j \le 100$ $f_{min}(\pi,\pi) = -1$	
F6: Rastrigin	$f(x_1,\cdots,x_D) = 10D + \sum_{j=1}^{D}(x_j^2 - 10\cos(2\pi x_j))$ $-5 \le x_j \le 5$ $f_{min}(0,\cdots,0) = 0$	
F7: Ackley	$f(x_1,\cdots,x_D) = 20 - 20\exp\left(-0.2\sqrt{\frac{1}{D}\sum_{j=1}^{D} x_j^2}\right)$ $+e - \exp\left(\frac{1}{D}\sum_{j=1}^{D}\cos(2\pi x_j)\right)$ $-32.768 \le x_j \le 32.768$ $f_{min}(0,\cdots,0) = 0$	
F8: Levi N.13	$f(x_1,x_2) = \sin^2(3\pi x_1) + (x_1-1)^2(1+\sin^2(3\pi x_2))$ $+(x_2-1)^2(1+\sin^2(2\pi x_2))$ $-10 \le x_j \le 10$ $f_{min}(1,1) = 0$	
F9: Bukin N.6	$f(x_1,x_2) = 100\sqrt{|x_2 - 0.01x_1^2|} + 0.01|x_1 + 10|$ $-15 \le x_1 \le 5$ $-3 \le x_2 \le 3$ $f_{min}(-10,1) = 0$	
F10: Beale	$f(x_1,x_2) = (1.5 - x_1 + x_1x_2)^2 + (2.25 - x_1 + x_1x_2^2)^2$ $+(2.625 - x_1 + x_1x_2^3)^2$ $-4.5 \le x_j \le 4.5$ $f_{min}(3,0.5) = 0$	
F11: Goldstein-Price	$f(x_1,x_2) = (1 + (x_1 + x_2 + 1)^2(19 - 14x_1$ $+3x_1^2 - 14x_2 + 6x_1x_2 + 3x_2^2))(30 + (2x_1 - 3x_2)^2$ $(18 - 32x_1 + 12x_1^2 + 48x_2 - 36x_1x_2 + 27x_2^2))$ $-2 \le x_1,x_2 \le 2$ $f_{min}(0,-1) = 3$	
F12: Schaffer N.2	$f(x_1,x_2) = 0.5 + \frac{\sin^2(x_1^2 - x_2^2) - 0.5}{(1+0.001(x_1^2+x_2^2))^2}$ $-100 \le x_1,x_2 \le 100$ $f_{min}(0,0) = 0$	
F13: Five-well potential	$f(x_1,x_2) = (1 - \frac{1}{1+0.05(x_1^2+(x_2-10)^2)} - \frac{1}{1+0.05((x_1-10)^2+x_2^2)}$ $-\frac{1.5}{1+0.03((x_1+10)^2+x_2^2)} - \frac{2}{1+0.05((x_1-5)^2+(x_2+10)^2)}$ $-\frac{1}{1+0.1((x_1+5)^2+(x_2+10)^2)})(1 + 0.0001(x_1^2 + x_2^2)^{1.2})$ $-20 \le x_1 \le 20$ $-20 \le x_2 \le 20$ $f_{min}(4.92,-9:89) = -1.4616$	
F14: Griewank	$f(x_1,\ldots,x_D) = 1 + \frac{1}{4000}\sum_{j=1}^{D} x_j^2 - \prod_{j=1}^{D}\cos(\frac{x_j}{\sqrt{j}})$ $-600 \le x_j \le 600$ $f_{min}(0,\ldots,0) = 0$	
F15: Xin-She Yang	$f(x_1,\ldots,x_D) = (\sum_{j=1}^{D}|x_j|)\exp(-\sum_{j=1}^{D}\sin(x_j^2))$ $-2\pi \le x_j \le 2\pi$ $f_{min}(0,\ldots,0) = 0$	
F16: Styblinski-Tang	$f(x_1,\ldots,x_D) = \frac{\sum_{j=1}^{D} x_j^4 - 16x_j^2 + 5x_j}{2}$ $-5 \le x_1 \le 4$ $-3 \le x_2 \le 4$ $-39.16617D \le f(-2.903534,\ldots,-2.903534) \le -39.16616D$ $f_{min}(-2.903534,\ldots,-2.903534) \approx -39.166165D$	

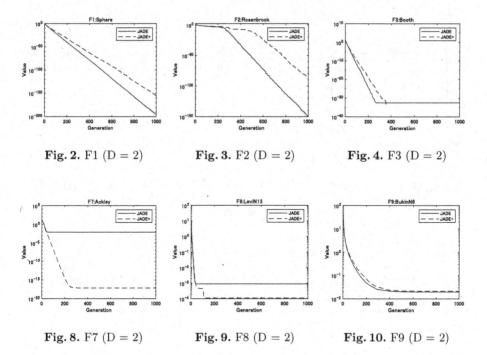

Fig. 2. F1 (D = 2)   **Fig. 3.** F2 (D = 2)   **Fig. 4.** F3 (D = 2)

Fig. 8. F7 (D = 2)   **Fig. 9.** F8 (D = 2)   **Fig. 10.** F9 (D = 2)

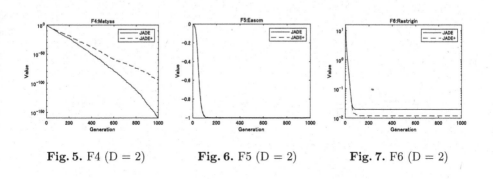

Fig. 5. F4 (D = 2)   **Fig. 6.** F5 (D = 2)   **Fig. 7.** F6 (D = 2)

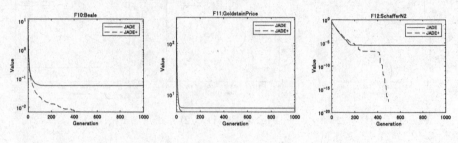

Fig. 11. F10 (D = 2)   Fig. 12. F11 (D = 2)   Fig. 13. F12 (D = 2)

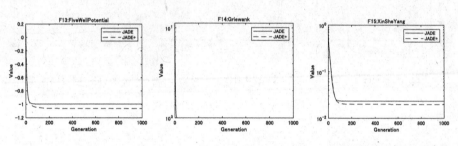

Fig. 14. F13 (D = 2)   Fig. 15. F14 (D = 2)   Fig. 16. F15 (D = 2)

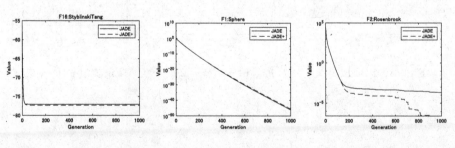

Fig. 17. F16 (D = 2)   Fig. 18. F1 (D = 10)   Fig. 19. F2 (D = 10)

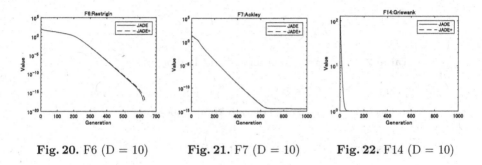

Fig. 20. F6 (D = 10)   **Fig. 21.** F7 (D = 10)   **Fig. 22.** F14 (D = 10)

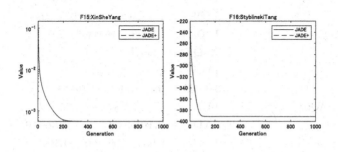

Fig. 23. F15 (D = 10)   **Fig. 24.** F16 (D = 10)

We begin with a discussion of the two-dimensional case, Figs. 2 through 17. The graphs show that the proposed method has a better solution than JADE in Fig. 7 (Rastrigin function), Fig. 8 (Ackley function), Fig. 9 (Levi N. 13 function), Fig. 11 (Beale function), Fig. 12 (Goldstein-Price function), Fig. 13 (SchafferN2 function), Fig. 14 (Five-well potential function), Fig. 16 (Xin-She Yang function), and Fig. 17 (Styblinski-Tang function), and the convergence speed is almost the same. From this fact, it is proven that the accuracy heightens without lowering the search speed by carrying out the group outside search in one search point. In Fig. 9 (Levi N. 13 function) and Fig. 13 (SchafferN2 function), it can be seen that the proposed method escaped from the point where it almost converged once, and reached a better solution. This suggests that the search outside the group works well to escape from the local solution. However, in Fig. 2 (Sphere function), Fig. 3 (Rosenbrock function) and Fig. 5 (Matyas function), the search speed of the solution is inferior to JADE, and it can be read that the useless group outside search leads to the lowering of the search speed in the simple unimodal function.

We then discuss the 10 dimensional case, Figs. 18 through 24. First, as shown in Fig. 19, the result of F2 was improved. The results of other functions were not

Table 2. Comparison of mean and standard deviation (D = 2)

Function	Method	Mean ± Standard deviation
F1:Sphere	**JADE**	**5.7656e−196 ± 0**
	JADE+	8.7776e−156 ± 1.9808e−154
F2:Rosenbrock	**JADE**	**1.3679e−150 ± 4.3256e−149**
	JADE+	2.9785e−87 ± 9.4188e−86
F3:Booth	JADE	3.1554e−33 ± 9.9784e−32
	JADE+	**0 ± 0**
F4:Matyas	**JADE**	**1.3784e−160 ± 4.0084e−159**
	JADE+	7.5967e−96 ± 2.4023e−94
F5:Easom	JADE	−0.998 ± 0.044695
	JADE+	**−1 ± 0**
F6:Rastrigin	JADE	0.019899 ± 0.1463
	JADE+	**0.01194 ± 0.10839**
F7:Ackley	JADE	0.0077398 ± 0.14117
	JADE+	**7.1054e−18 ± 1.588e−16**
F8:LeviN13	JADE	0.00087899 ± 0.0097929
	JADE+	**0.00010987 ± 0.0034745**
F9:BukinN6	**JADE**	**0.019676 ± 0.014354**
	JADE+	0.021124 ± 0.026538
F10:Beale	JADE	0.057155 ± 0.20082
	JADE+	**0.0068586 ± 0.072006**
F11:GoldsteinPrice	JADE	5.43 ± 12.4936
	JADE+	**4.566 ± 10.2735**
F12:SchafferN2	JADE	3.1266e−06 ± 9.8872e−05
	JADE+	**0 ± 0**
F13:FiveWellPotential	JADE	−0.99589 ± 0.32489
	JADE+	**−1.0621 ± 0.29633**
F14:Griewank	**JADE**	**0.99975 ± 2.2327e−14**
	JADE+	**0.99975 ± 2.2327e−14**
F15:XinSheYang	JADE	0.023408 ± 0.084715
	JADE+	**0.019781 ± 0.078336**
F16:StyblinskiTang	JADE	−77.0742 ± 4.1255
	JADE+	**−77.3569 ± 3.6402**

Table 3. Comparison of mean and standard deviation (D = 10)

Function	Method	Mean ± Standard deviation
F1:Sphere	**JADE**	**7.3367e–47 ± 1.3492e–45**
	JADE+	1.9139e–46 ± 3.7947e–45
F2:Rosenbrock	JADE	0.00022373 ± 0.0097558
	JADE+	**2.5623e–07 ± 1.1459e–05**
F6:Rastrigin	**JADE**	**0 ± 0**
	JADE+	**0 ± 0**
F7:Ackley	**JADE**	**3.503e–15 ± 4.1751e–16**
	JADE+	3.5136e–15 ± 3.7065e–16
F14:Griewank	**JADE**	**0.99975 ± 3.6647e–14**
	JADE+	**0.99975 ± 3.6647e–14**
F15:XinSheYang	**JADE**	**0.00056607 ± 1.4686e–14**
	JADE+	**0.00056607 ± 2.2173e–14**
F16:StyblinskiTang	**JADE**	**−391.5415 ± 1.2981**
	JADE+	−391.5274 ± 1.3717

so different from those of the conventional method. In this experiment, it can be said that the result as intended was obtained, because the search frequency of the outside region did not increase, even if the dimension increased, so that the performance in the high dimension would not be lowered. And, it was proven that it was good to add outer region search like the proposed method, because the region near the optimum solution was lined in F2.

Table 2 shows that the proposed method improved the accuracy of the solution to 11 functions out of 16. For these 11 functions, the proposed method gives better values for both mean and standard deviation. Since all the functions whose accuracy is improved are multimodal functions, the proposed method is more stable and gives better solution in multimodal functions, and it is proven that the search outside the group is effective for the multimodal function. In the sphere function, Rosenbrock function and Matyas function which are the unimodal function, the reason why the accuracy is inferior to JADE is the lowering of the search speed by the out-of-group search, and there seems to be a large room of the improvement on the judging method of whether to carry out the out-group search or not.

Next, we discuss Table 3. In the proposed method, one of the search points searches outside the population. Therefore, as the dimension increases, the effect of out-of-group search decreases and the difference between JADE and JADE+ disappears. In Table 3, we can see that the results are similar between JADE and JADE+.

5   Conclusion

In this study, we propose a method to add search outside the group referring to the Nelder-Mead method to JADE which is adaptive differential evolution. A comparison experiment between the proposed method and the conventional method was carried out using 16 benchmark functions, and it succeeded in improving the accuracy of the solution in many functions. Especially, search outside the group works effectively in the multimodal function, and escape from the local solution is also observed. However, since there were some cases in which the search speed was inferior to JADE in the unimodal function, it is also necessary to examine how to decide whether to search outside the group.

Acknowledgment. This work was supported in part by JSPS KAKENHI Grant Number JP18K11473.

References

1. Storn, R., Price, K.: Differential evolution - a simple and efficient heuristic for global optimization over continuous spaces. J. Glob.Optim. **11**, 341–359 (1997)
2. Qin, A.K., Suganthan, P.N.: Self-adaptive differential evolution algorithm for numerical optimization. IEEE Congr. Evol. Comput. **2**, 1785–1791 (2005)
3. Brest, J., Greiner, S., Boskovic, B., Mernik, M., Zumer, V.: Self-adapting control parameters in differential evolution: a comparative study on numerical benchmark problems. IEEE Trans. Evol. Comput. **10**(6), 646–657 (2006)
4. Tanabe, R., Fukunaga, A.: Success-history based parameter adaptation for differential evolution. In: Proceedings of the 2013 IEEE Congress on Evolutionary Computation, pp. 71–78 (2013)
5. Zhang, J., Sanderson, A.C.: JADE: adaptive differential evolution with optional external archive. IEEE Trans. Evol. Comput. **13**(5), 945–958 (2009)
6. Zhang, J., Sanderson, A.C.: JADE: self-adaptive differential evolution with fast and reliable convergence performance. In: Proceedings IEEE Congress on Evolutionary Computation, Singapore, pp. 2251–2258 (2007)
7. Miyahira, Y., Notsu, A., Honda, K., Ubukata, S.: Adaptive selection of parameters in differential evolution by bandit algorithm. In: Proceeding of the 65th Annual Conference of the Institute of Systems, Control and Information Engineers, pp. 1020–1026 (2021)
8. Nelder, J.A., Mead, R.: A simplex method for function minimization. Comput. J. **7**, 308–313 (1965)
9. Singer, S., Singer, S.: Efficient implementation of the Nelder-Mead search algorithm. Appl. Numer. Anal. Comput. Math. **1**(2), 524–534 (2004)
10. Hansen, N., Ostermeier, A.: Completely derandomized self-adaptation in evolution strategies. Evol. Comput. **9**(2), 159–195 (2001)
11. Storn, R., Price, K.: Differential evolution - a simple and efficient adaptive scheme for global optimization over continuous spaces. J. Glob. Optim. **23**(1) (1995)

Confidence Intervals for Mean of Delta Two-Parameter Exponential Distribution

Wansiri Khooriphan, Sa-Aat Niwitpong[(✉)], and Suparat Niwitpong

Faculty of Applied Science, Department of Applied Statistics,
King Mongkut's University of Technology North Bangkok, Bangkok 10800, Thailand
Sa-aat.n@sci.kmutnb.ac.th

Abstract. The two-parameter exponential distribution is widely used for many applications in real life, and the data can include zero observations. The mean, which represents the center of a population, is one of the parameters of interest. Herein, we propose confidence intervals for the mean of a delta two-parameter exponential distribution based on parametric bootstrapping (PB), standard bootstrapping (SB), the generalized confidence interval (GCI), and the method of variance estimates recovery (MOVER). The performances of the proposed confidence intervals were evaluated by using coverage probabilities and average lengths via Monte Carlo simulations. The results indicate that GCI can be recommended for small-to-moderate sample sizes whereas PB is appropriate for large sample sizes.

Keywords: Parametric bootstrap · Standard bootstrap · Generalized confidence interval · Method of variance estimates recovery · Delta two parameter exponential distribution

1   Introduction

In applied statistics, the two-parameter exponential distribution has been widely used for many applications in real life, such as lifetime, survival, and reliability analyses [1]. Thus, defining confidence intervals for estimating its parameters is important for statistical inference on this distribution [2]. Confidence intervals provide better information with respect to the population than point estimation [3]. Thus, many researchers have constructed confidence intervals for the parameters of a two-parameter exponential distribution. For example, Sangnawakij and Niwitpong [4] proposed confidence intervals for the coefficient of variation of a two-parameter exponential distribution by using the method of variance of estimates recovery (MOVER), the generalized confidence interval (GCI), and the asymptotic confidence interval. Of these, MOVER performs well in terms of coverage probability when data only consist of positive values. Sangnawakij and Niwitpong [5] constructed confidence intervals for the ratio of the coefficients of variation of two-parameter exponential distributions by using MOVER and GCI;

© Springer Nature Switzerland AG 2022
K. Honda et al. (Eds.): IUKM 2022, LNAI 13199, pp. 117–129, 2022.
https://doi.org/10.1007/978-3-030-98018-4_10

the latter provided the best performance in terms of acceptable coverage probability and the shortest length. Thangjai and Niwitpong [2] provided confidence intervals for the weighted coefficients of variation of two-parameter exponential distributions by using adjusted MOVER, GCI, and the large sample method and found that GCI can be recommended for all situations studied. Thangjai and Niwitpong [6] created confidence intervals for the difference between the signal-to-noise ratios of two-parameter exponential distributions by using GCI, MOVER, LS, and parametric bootstrapping (PB). Here, the PB approach provided better coverage probabilities than the others for all of the scenarios studied.

A data series can include zero observations in various situations. Aitchison [7] focused on the characteristics of a data series including zero observations and determined that the probability of having zero observation is $0 < \delta < 1$ whereas the remaining probability $1 - \delta$ is used to describe the positive observations. Later, Aitchison and Brown [8] introduced the delta-lognormal distribution where the number of zero observations can be viewed as a random variable with a binomial distribution and the positive observations comprise a random variable from a lognormal distribution.

The mean of a random variable is an average value that is weighted according to the probability distribution. It is a useful parameter employed as a measure of the center of a population [3]. The aim of the current study is to propose confidence intervals for the mean of a delta two-parameter exponential distribution based on PB, SB, GCI, and MOVER.

2   Methods

Let X_i be a random variable following two parameters exponential distribution (λ, β) distribution with scale parameter λ and location parameter β. The probability density function can be derived as follows:

$$f(x; \lambda, \beta) = \begin{cases} \frac{1}{\lambda} exp(-\frac{(x-\beta)}{\lambda}); & \lambda > 0, \beta \in \mathbb{R} \\ 0; & \text{otherwise.} \end{cases} \tag{1}$$

The maximum likelihood estimators of β and λ are respectively given by

$$\hat{\beta} = X_{(1)} \tag{2}$$

$$\hat{\lambda} = \bar{X} - X_{(1)} \tag{3}$$

Suppose that the population of interest contains both zero and non-zero observations denoted by $n_{(0)}$ and $n_{(1)}$, respectively, where $n = n_{(0)} + n_{(1)}$. The zero observations follow a binomial distribution $n_{(0)} \sim Bin(n, \delta)$ whereas the non-zero observations follow a two-parameter exponential distribution. Let $X = (X_1, X_2, ..., X_n)$ be a random sample from a delta two-parameter exponential distribution denoted by $\Delta(\delta, \lambda, \beta)$. Its distribution function can be derived as

$$G(x_i; \delta, \lambda, \beta) = \begin{cases} \delta; & x = 0, \\ \delta + (1 - \delta)F(x; \lambda, \beta); & x > 0 \end{cases} \tag{4}$$

$F(x; \lambda, \beta)$ is a two-parameter exponential distribution with cumulative distribution function $\widehat{\delta} = n_{(0)}/n$. The mean of X is

$$E(X) = \theta = (1 - \delta) \cdot (\lambda + \beta) \qquad (5)$$

The methods for constructing the confidence interval for θ are proposed in the following section.

2.1 Bootstrap Confidence Intervals

A bootstrap sample denoted by $x_1^*, x_2^*, ..., x_n^*$ is of size n and drawn with replacement from the original sample. The corresponding bootstrap for $\widehat{\theta}$ is denoted as $\widehat{\theta}^*$. We have assumed that 5,000 bootstrap samples are taken, and thus 5,000 bootstrap $\widehat{\theta}$'s will be obtained and can be ordered from the smallest to the largest, denoted by $\widehat{\theta}_{(1)}^*, \widehat{\theta}_{(2)}^*, ..., \widehat{\theta}_{(5,000)}^*$.

Parametric Bootstrap (PB). In this study, θ is the parameter of interest. Subsequently, Eq. (5) is transformed by using log function

$$\vartheta = \ln(1 - \delta) + \ln(\lambda + \beta) \qquad (6)$$

The bootstrap for the mean of a delta two-parameter exponential distribution can be written as

$$\widehat{\theta}^* = \ln(1 - \widehat{\delta}^*) + \ln(\widehat{\lambda}^* + \widehat{\beta}^*) \qquad (7)$$

The $100(1 - \alpha)\%$ PB interval is given by

$$CI_{PB} = [\widehat{\theta}^*(\alpha/2), \widehat{\theta}^*(1 - \alpha/2)] \qquad (8)$$

Standard Bootstrap (SB). From the 5,000 bootstrap estimates $\widehat{\theta}_i^*$, for $i = 1, 2, ..., 5,000$ calculate the sample average as

$$\bar{\theta}^* = \frac{1}{5,000} \sum_{i=1}^{5,000} \widehat{\theta}_i^* \qquad (9)$$

and the sample standard deviation as

$$S_\theta^* = \left[\frac{1}{4,999} \sum_{i=1}^{5,000} (\widehat{\theta}_i^* - \bar{\theta}^*)^2 \right]^{\frac{1}{2}} \qquad (10)$$

The $100(1 - \alpha)\%$ SB interval is given by

$$CI_{SB} = [\bar{\theta}^* + S_\theta^* Z_{\frac{\alpha}{2}}, \bar{\theta}^* + S_\theta^* Z_{1-\frac{\alpha}{2}}] \qquad (11)$$

The confidence intervals for the mean (θ) can be obtained by using Algorithm 1.

Algorithm 1. Bootstrapping

1: For a given sample from a delta two-parameters exponential distribution (x), compute $\bar{x}$, $\hat{\delta}$, $\hat{\lambda}$ and $\hat{\beta}$.
2: Generate x^* from x.
3: Compute $\bar{x}^*$, $\hat{\delta}^*$, $\hat{\lambda}^*$ and $\hat{\beta}^*$.
4: Compute $\hat{\theta}^*$ from Eq. (7).
5: Repeat Steps 2–4 5,000 times and obtain an array of $\hat{\theta}^*$.
6: Compute the 95% confidence intervals for $\hat{\theta}^*$ from Eqs. (8) and (11).
7: Repeat Steps 1–6 15,000 times to compute the coverage probabilities and average lengths.

2.2 Generalized Confidence Interval (GCI)

The GCI for constructing confidence intervals was first presented by Weerahandi [9]. This method is based on the generalized pivotal quantity (GPQ) concept. Recall that the MLEs of the parameters are $\hat{\beta} = X_{(1)}$, $\hat{\lambda} = \bar{X} - X_{(1)}$ and $\hat{\delta} = n_{(0)}/n$. The GPQ for δ proposed by Anirban [10] and Wu and Hsieh [11] is given by

$$R_\delta = \sin^2 \left[\arcsin \sqrt{\hat{\delta}} - \frac{Z}{2\sqrt{n}} \right] \tag{12}$$

where $Z = 2\sqrt{n} \left(\arcsin \sqrt{\hat{\delta}} - \arcsin \sqrt{\delta} \right) \xrightarrow{D} N(0,1)$ as $n \to \infty$. We known that $\hat{\beta}$ and $\hat{\lambda}$ are independent [12] so the respective pivots of $\hat{\beta}$ and $\hat{\lambda}$ [13] can be derived as

$$W_1 = \frac{2n(\hat{\beta} - \beta)}{\lambda} \sim \chi_2^2 \tag{13}$$

$$W_2 = \frac{2n\hat{\lambda}}{\lambda} \sim \chi_{2n-2}^2 \tag{14}$$

where χ_2^2 and χ_{2n-2}^2 denote Chi-squared distributions with 2 and $2n-2$ degrees of freedom, respectively.

The GPQ of β is given by:

$$R_\beta = x_{(1)} - \frac{W_1 \hat{\lambda}}{W_2} \tag{15}$$

The GPQ of λ is given by:

$$R_\lambda = \frac{2n\hat{\lambda}}{W_2} \tag{16}$$

Thus, the GPQ of θ is given by:

$$R_\theta = \ln(1 - R_\delta) + \ln(R_\lambda + R_\beta) \tag{17}$$

The $100(1 - \alpha)\%$ GCI interval is given by

$$CI_{GCI} = [R_\theta(\alpha/2), R_\theta(1 - \alpha/2)] \tag{18}$$

Algorithm 2. GCI

1: For a given sample from a delta two-parameters exponential distribution (x), compute $\bar{x}$, $\hat{\delta}$, $\hat{\lambda}$ and $\hat{\beta}$.
2: Generate W_1 from Chi-square distribution with degrees of freedom 2.
3: Generate W_2 from Chi-square distribution with degrees of freedom $2n - 2$.
4: Compute R_β from Eq. (15).
5: Compute R_λ from Eq. (16).
6: Compute R_θ from Eq. (17).
7: Repeat Steps 2–6 5,000 times and obtain an array of R_θ.
8: Compute the 95% confidence intervals for θ from Eq. (18).
9: Repeat Steps 1–8 15,000 times to compute the coverage probabilities and average lengths.

2.3 Method of Variance Estimates Recovery (MOVER)

The idea behind MOVER based on the central limit theorem was first proposed by Donner and Zou [14]. In this study, we focus only on the confidence interval for $\hat{\theta}_1 + \hat{\theta}_2$. The MOVER to construct the confidence interval for $\hat{\theta}_1 + \hat{\theta}_2$ defined as

$$CI_M = [L_M, U_M] \tag{19}$$

where

$$L_M = (\hat{\theta}_1 + \hat{\theta}_2) - \sqrt{(\hat{\theta}_1 - l_1)^2 + (\hat{\theta}_2 - l_2)^2}$$
$$U_M = (\hat{\theta}_1 + \hat{\theta}_2) + \sqrt{(u_1 - \hat{\theta}_1)^2 + (u_2 - \hat{\theta}_2)^2}$$

Recall that the parameter of interest is $\vartheta = \ln(1 - \delta) + \ln(\lambda + \beta)$. We set $\theta_1 = \ln(1 - \delta) = \delta'$ and $\theta_2 = \ln(\lambda + \beta)$. The CI for δ was examined by Zou et al. [15]. The $100(1 - \alpha)\%$ CI for $\ln(1 - \delta)$ as

$$CI_{\ln(1-\delta)} = [l_1, u_1] \tag{20}$$

where

$$l_1 = \ln\left[\left(\hat{\delta}' + \frac{T_{1-\frac{\alpha}{2}}^2}{2n} - \sqrt{\frac{\hat{\delta}'(1-\hat{\delta}')}{n} + \frac{T_{1-\frac{\alpha}{2}}^2}{4n^2}}\right) \Big/ \left(1 + T_{1-\frac{\alpha}{2}}^2/n\right)\right]$$

$$u_1 = \ln\left[\left(\hat{\delta}' + \frac{T_{1-\frac{\alpha}{2}}^2}{2n} + \sqrt{\frac{\hat{\delta}'(1-\hat{\delta}')}{n} + \frac{T_{1-\frac{\alpha}{2}}^2}{4n^2}}\right) \Big/ \left(1 + T_{1-\frac{\alpha}{2}}^2/n\right)\right]$$

Note that $T = \frac{n_{(1)} - n\delta'}{\sqrt{n\delta'(1-\delta')}} \overset{d}{\sim} N(0,1)$

Sangnawakij and Niwitpong [4] proposed the $100(1 - \alpha)\%$ CI for $\lambda + \beta$ as

$$CI_{\lambda+\beta} = [l_{\lambda+\beta}, u_{\lambda+\beta}] \tag{21}$$

where

$$l_{\lambda+\beta} = \bar{x} - \sqrt{\left[\hat{\lambda} - \frac{n\hat{\lambda}}{Z_{\frac{\alpha}{2}}\sqrt{n-1}+(n-1)}\right]^2 + \left[\frac{\hat{\lambda}}{n}\ln(\frac{\alpha}{2})\right]^2}$$

$$u_{\lambda+\beta} = \bar{x} + \sqrt{\left[\frac{n\hat{\lambda}}{-Z_{\frac{\alpha}{2}}\sqrt{n-1}+(n-1)} - \hat{\lambda}\right]^2 + \left[\frac{\hat{\lambda}}{n}\ln(1-\frac{\alpha}{2})\right]^2}$$

Then the $100(1-\alpha)\%$ CI for $\ln(\lambda+\beta)$ is given by

$$CI_{\ln(\lambda+\beta)} = [l_2, u_2] \tag{22}$$

where

$$l_2 = \ln\left(\bar{x} - \sqrt{\left[\hat{\lambda} - \frac{n\hat{\lambda}}{Z_{\frac{\alpha}{2}}\sqrt{n-1}+(n-1)}\right]^2 + \left[\frac{\hat{\lambda}}{n}\ln(\frac{\alpha}{2})\right]^2}\right)$$

$$u_2 = \ln\left(\bar{x} + \sqrt{\left[\frac{n\hat{\lambda}}{-Z_{\frac{\alpha}{2}}\sqrt{n-1}+(n-1)} - \hat{\lambda}\right]^2 + \left[\frac{\hat{\lambda}}{n}\ln(1-\frac{\alpha}{2})\right]^2}\right)$$

Thus the $100(1-\alpha)\%$ MOVER interval for ϑ is given by

$$CI_{MOVER} = [l_{MOVER}, u_{MOVER}] \tag{23}$$

where

$$L_{MOVER} = (\hat{\theta}_1 + \hat{\theta}_2) - \sqrt{(\hat{\theta}_1 - l_1)^2 + (\hat{\theta}_2 - l_2)^2}$$

$$U_{MOVER} = (\hat{\theta}_1 + \hat{\theta}_2) + \sqrt{(\hat{\theta}_1 - u_1)^2 + (\hat{\theta}_2 - u_2)^2}$$

Note that $Z = \frac{\hat{\mu}-\mu}{\sqrt{\sigma^2/n}} \overset{d}{\sim} N(0,1)$

T and Z are independent random variables.

3   Simulation Studies and Results

A simulation study with 15,000 replications (M) and 5,000 repetitions (m) for PB, SB and GCI with a nominal confidence level of 0.95 was conducted. Sample size n was set as 10, 20, 30, 50, 100 or 200; δ as 0.1, 0.2, 0.5, 0.8 or 0.9; scale parameter λ as 1 or 2 and location parameter β as 0. The performances of the confidence intervals were assessed by comparing their coverage probabilities and average lengths using Monte Carlo simulation. In each scenario, the best-performing confidence interval had a coverage probability close to or greater than 0.95 and the shortest average length. The coverage probability and average length results for the nominal 95% two-sided confidence intervals for the mean of a delta two-parameter exponential distribution are reported in Table 1.

It can be seen that GCI performed well for small-to-moderate sample sizes and small δ whereas the PB and SB confidence intervals performed well for large sample sizes and large δ. However, PB obtained narrower average lengths

Table 1. The coverage probabilities and (Average lengths) of nominal 95% two-sided confidence intervals for mean of delta two parameters exponential distribution

n	λ	δ	Coverage probability (Average length)			
			PB	SB	GCI	MOVER
10	1	0.1	0.8840	0.8740	**0.9572**	0.6010
			(1.0672)	(1.0847)	(1.4059)	(0.5233)
		0.2	0.8841	0.8762	**0.9487**	0.5905
			(1.0549)	(1.0738)	(1.3020)	(0.4823)
		0.5	0.8926	0.8781	0.9298	0.5977
			(0.9099)	(0.9350)	(0.9706)	(0.4026)
		0.8	0.8114	0.8088	0.8429	0.6018
			(0.4915)	(0.5536)	(0.5516)	(0.3085)
		0.9	0.3429	0.7141	0.7487	0.3919
			(0.2013)	(0.2930)	(0.3471)	(0.2205)
	2	0.1	0.8813	0.8722	**0.9506**	0.6017
			(2.1444)	(2.1799)	(2.8160)	(1.0488)
		0.2	0.8867	0.8756	**0.9498**	0.5853
			(2.1076)	(2.1444)	(2.6079)	(0.9687)
		0.5	0.8865	0.8712	0.9200	0.5897
			(1.8195)	(1.8681)	(1.9393)	(0.8071)
		0.8	0.8136	0.8129	0.8461	0.6129
			(0.9819)	(1.1055)	(1.1023)	(0.6184)
		0.9	0.3559	0.7187	0.7518	0.4035
			(0.4044)	(0.5890)	(0.6975)	(0.4430)
20	1	0.1	0.9137	0.9091	**0.9507**	0.8032
			(0.8104)	(0.8170)	(0.9058)	(1.0279)
		0.2	0.9253	0.9168	**0.9462**	0.7859
			(0.7976)	(0.8044)	(0.8460)	(0.9102)
		0.5	0.9332	0.9211	0.9410	0.7725
			(0.6918)	(0.7009)	(0.6478)	(0.5858)
		0.8	0.8951	0.8747	0.8919	0.7811
			(0.4385)	(0.4551)	(0.3970)	(0.2881)
		0.9	0.8173	0.8136	0.8425	0.6482
			(0.2497)	(0.2847)	(0.2709)	(0.1832)
	2	0.1	0.9158	0.9094	**0.9481**	0.8031
			(1.6216)	(1.6348)	(1.8106)	(2.0545)
		0.2	0.9271	0.9217	**0.9492**	0.7961
			(1.6023)	(1.6165)	(1.6963)	(1.8260)
		0.5	0.9323	0.9199	0.9364	0.7814
			(1.3904)	(1.4087)	(1.3015)	(1.1771)

(*continued*)

Table 1. (*continued*)

n	λ	δ	Coverage probability (Average length)			
			PB	SB	GCI	MOVER
		0.8	0.8972	0.8804	0.8941	0.7833
			(0.8731)	(0.9075)	(0.7892)	(0.5737)
		0.9	0.8129	0.8119	0.8372	0.6483
			(0.5061)	(0.5770)	(0.5485)	(0.3715)
30	1	0.1	0.9302	0.9250	**0.9520**	0.8363
			(0.6730)	(0.6767)	(0.7143)	(0.7633)
		0.2	0.9370	0.9305	**0.9495**	0.8276
			(0.6665)	(0.6705)	(0.6726)	(0.6830)
		0.5	**0.9494**	0.9426	**0.9446**	0.8107
			(0.5838)	(0.5890)	(0.5221)	(0.4489)
		0.8	0.9229	0.9055	0.9103	0.7916
			(0.3803)	(0.3894)	(0.3225)	(0.2259)
		0.9	0.8784	0.8544	0.8749	0.7063
			(0.2407)	(0.2570)	(0.2247)	(0.1487)
	2	0.1	0.9334	0.9274	**0.9504**	0.8426
			(1.3522)	(1.3595)	(1.4346)	(1.5330)
		0.2	0.9358	0.9296	**0.9474**	0.8210
			(1.3336)	(1.3417)	(1.3423)	(1.3631)
		0.5	**0.9497**	0.9401	0.9442	0.8023
			(1.1652)	(1.1757)	(1.0442)	(0.8978)
		0.8	0.9303	0.9129	0.9187	0.8093
			(0.7602)	(0.7781)	(0.6448)	(0.4524)
		0.9	0.8868	0.8644	0.8809	0.7124
			(0.4825)	(0.5149)	(0.4508)	(0.2983)
50	1	0.1	0.9402	0.9371	**0.9492**	0.8729
			(0.5345)	(0.5364)	(0.5433)	(0.5559)
		0.2	**0.9501**	**0.9454**	**0.9512**	0.8583
			(0.5267)	(0.5286)	*(0.5100)*	(0.4976)
		0.5	**0.9631**	**0.9573**	**0.9469**	0.8449
			(0.4632)	(0.4658)	*(0.4006)*	(0.3327)
		0.8	**0.9508**	0.9373	0.9287	0.8012
			(0.3084)	(0.3125)	(0.2498)	(0.1694)
		0.9	0.9234	0.9050	0.9045	0.7311
			(0.2110)	(0.2177)	(0.1761)	(0.1123)
	2	0.1	0.9412	0.9372	**0.9503**	0.8704
			(1.0687)	(1.0727)	(1.0843)	(1.1097)

(*continued*)

Table 1. (*continued*)

n	λ	δ	Coverage probability (Average length)			
			PB	SB	GCI	MOVER
		0.2	**0.9472**	0.9437	**0.9491**	0.8571
			(1.0515)	(1.0555)	*(1.0199)*	(0.9948)
		0.5	**0.9575**	**0.9525**	0.9426	0.8364
			(0.9224)	(0.9274)	(0.7991)	(0.6636)
		0.8	**0.9506**	0.9395	0.9297	0.8071
			(0.6174)	(0.6257)	(0.4994)	(0.3386)
		0.9	0.9187	0.8991	0.8984	0.7241
			(0.4173)	(0.4301)	(0.3488)	(0.2221)
100	1	0.1	**0.9517**	**0.9496**	**0.9503**	0.9067
			(0.3837)	(0.3845)	*(0.3781)*	(0.3751)
		0.2	**0.9542**	**0.9526**	**0.9486**	0.8896
			(0.3774)	(0.3782)	*(0.3552)*	(0.3368)
		0.5	**0.9703**	**0.9662**	0.9432	0.8555
			(0.3332)	(0.3342)	(0.2803)	(0.2273)
		0.8	**0.9719**	**0.9629**	0.9381	0.7981
			(0.2267)	(0.2283)	(0.1762)	(0.1162)
		0.9	**0.9512**	0.9368	0.9199	0.7323
			(0.1586)	(0.1607)	(0.1239)	(0.0756)
	2	0.1	**0.9484**	0.9455	**0.9482**	0.9003
			(0.7673)	(0.7690)	*(0.7559)*	(0.7498)
		0.2	**0.9602**	**0.9586**	**0.9526**	0.8921
			(0.7567)	(0.7582)	*(0.7122)*	(0.6754)
		0.5	**0.9716**	**0.9665**	**0.9466**	0.8596
			(0.6656)	(0.6677)	*(0.5603)*	(0.4547)
		0.8	**0.9704**	**0.9621**	0.9387	0.7980
			(0.4517)	(0.4548)	(0.3513)	(0.2315)
		0.9	**0.9538**	**0.9397**	0.9221	0.7326
			(0.3173)	(0.3217)	(0.2476)	(0.1510)
200	1	0.1	**0.9541**	**0.9547**	**0.9500**	0.9199
			(0.2733)	(0.2737)	*(0.2650)*	(0.2590)
		0.2	**0.9598**	**0.9593**	**0.9462**	0.9107
			(0.2697)	(0.2701)	*(0.2500)*	(0.2339)
		0.5	**0.9769**	**0.9754**	**0.9499**	0.8746
			(0.2373)	(0.2378)	*(0.1968)*	(0.1579)
		0.8	**0.9803**	**0.9761**	**0.9450**	0.7890
			(0.1628)	(0.1634)	*(0.1243)*	(0.0807)

(*continued*)

Table 1. (*continued*)

n	λ	δ	Coverage probability (Average length)			
			PB	SB	GCI	MOVER
		0.9	**0.9727**	**0.9647**	0.9358	0.7318
			(0.1163)	(0.1171)	(0.0877)	(0.0521)
	2	0.1	**0.9585**	**0.9577**	0.9532	0.9235
			(0.5464)	(0.5471)	*(0.5302)*	(0.5181)
		0.2	**0.9626**	**0.9616**	**0.9480**	0.9074
			(0.5381)	(0.5389)	*(0.4991)*	(0.4671)
		0.5	**0.9766**	**0.9743**	**0.9471**	0.8675
			(0.4750)	(0.4759)	*(0.3939)*	(0.3161)
		0.8	**0.9793**	**0.9758**	**0.9479**	0.7996
			(0.3252)	(0.3264)	*(0.2479)*	(0.1609)
		0.9	**0.9737**	**0.9651**	0.9365	0.7392
			(0.2322)	(0.2339)	(0.1751)	(0.1041)

than SB. The coverage probabilities obtained by MOVER were lower than the nominal confidence level in all cases, which is probably because the coverage probability does not depend on the scale parameter (λ) and so is unaffected by its value. Many researchers have investigated and used GCI and PB methods for construct confidence intervals for the parameters of various distributions [2, 4–6]. Indeed, GCI provides satisfactory and more accurate confidence intervals for the weighted coefficients of variation and the ratio of coefficients of variation of two-parameter exponential distributions than other methods. Moreover, confidence intervals constructed by using the PB approach are recommended for the difference between the signal-to-noise ratios of two-parameter exponential distributions.

4 An Empirical Application

In this section, the performances of the confidence intervals were compared by using real datasets comprising sulfur dioxide emissions reported by the Division of Air Quality and Noise Management Bureau, Pollution Control Department, Thailand. The sulfur dioxide data were obtained from 11 August to 11 September 2021 in Phuket province, Thailand, and from 21 August to 21 September 2021 in Songkhla province, Thailand. First, fitting of the data to four distributions (normal, Cauchy, gamma, and exponential) was compared by using the minimum Akaike information criterion (AIC) and Bayesian information criterion (BIC). AIC and BIC are defined as $AIC = -2\ln L + 2k$ and $BIC = -2\ln L + 2k\ln(n)$. The results in Tables 2 and 3 show that the lowest AIC and BIC values for Phuket province (146.7457 and 148.2115, respectively) and Songkhla province

(96.0579 and 97.4252 respectively) were for the exponential model, which is thus the most suitable distribution.

Table 2. AIC and BIC results of SO_2 data from Phuket province

Models	Normal	Cauchy	Gamma	Exponential
AIC	197.0046	179.8331	148.7457	146.7457
BIC	199.9361	182.7646	151.6772	148.2115

Table 3. AIC and BIC results of SO_2 data from Songkhla province

Models	Normal	Cauchy	Gamma	Exponential
AIC	139.1536	134.1444	98.0579	96.0579
BIC	141.8882	136.8790	100.7925	97.4252

The sulfur dioxide data from Phuket province are $\bar{x} = 4.91, x_{(1)} = 1, n = 32, n_{(1)} = 23, n_{(0)} = 9$ with the MLEs for δ, β, λ and θ being $\hat{\delta} = 0.28, \hat{\beta} = 1, \hat{\lambda} = 3.91$ and $\hat{\theta} = 3.53$, respectively. The 95% two-sided confidence intervals for θ were calculated, as reported in Table 4. The summary statistics for the sulfur dioxide data from Songkhla province are $\bar{x} = 3.37, x_{(1)} = 1, n = 29, n_{(1)} = 16, n_{(0)} = 13$ with the MLEs for δ, β, λ and θ being $\hat{\delta} = 0.45, \hat{\beta} = 1, \hat{\lambda} = 2.38$ and $\hat{\theta} = 1.86$, respectively. The 95% two-sided confidence intervals for θ were calculated, as reported in Table 5.

Table 4. The 95% two-sided confidence intervals for mean of SO_2 data from Phuket province

Methods	Confidence intervals for θ		Length of intervals
	Lower	Upper	
PB	2.0313	5.4063	3.3750
SB	1.8178	5.2476	3.4298
GCI	2.4565	5.1151	2.6586
MOVER	1.8018	4.2551	2.4533

According to the simulation in the previous section, in case of $n = 30, \beta = 1, \delta = 0.2$, the GCI method obtained a coverage probability close to the nominal confidence level of 0.95. For $n = 30, \beta = 1, \delta = 0.5$, PB, SB, and GCI provided

Table 5. The 95% two-sided confidence intervals for mean of SO_2 data from Songkhla province

Methods	Confidence intervals for θ		Length of intervals
	Lower	Upper	
PB	1.0000	2.7931	1.7931
SB	0.9539	2.7648	1.8109
GCI	1.1893	2.8145	1.6252
MOVER	0.9741	2.2957	1.3216

appropriate confidence intervals for the mean because their coverage probabilities were close to the nominal confidence level of 0.95. Meanwhile, the average lengths using MOVER were the shortest but its coverage probabilities were less than the nominal level. Thus, GCI is the best method for constructing confidence intervals for the mean of the sulfur dioxide data because it provided coverage probabilities close to 0.95 and shorter average lengths than PB and SB.

5   Conclusions

The objective of this study was to construct confidence intervals for the mean of a delta two-parameter exponential distribution by using PB, SB, GCI, and MOVER. From the coverage probability and average length results obtained via Monte Carlo simulation and by using real data following an exponential distribution, GCI can be recommended for small-to-moderate sample size cases whereas PB is appropriate for large sample size cases. Future researchers may also be extended to the case of difference between means of delta two-parameter exponential distributions.

References

1. Elgmati, E.A., Gregni, N.B.: Quartile method estimation of two-parameter exponential distribution data with outliers. Int. J. Stat. Probab. **5**(5), 12–15 (2016)
2. Thangjai, W., Niwitpong, S.A.: Confidence intervals for the weighted coefficients of variation of two-parameter exponential distributions. Cogent Math. **4**(1), 1315880 (2017)
3. Casella, G., Berger, R.L.: Statistical inference, 2nd edn. Pacific Grove, Pacific Grove, CA (2002)
4. Sangnawakij, P., Niwitpong, S.A.: Confidence intervals for coefficients of variation in two-parameter exponential distributions. Commun. Stati. - Simul. Comput. **46**(8), 6618–6630 (2017)
5. Sangnawakij, P., Niwitpong, S.A., Niwitpong, S.: Confidence intervals for the ratio of coefficients of variation in the two-parameter exponential distributions. In: IUKM (2016)

6. Thangjai, W., Niwitpong, S.A., Niwitpong, S.: Adjusted generalized confidence intervals for the common coefficient of variation of several normal populations. Commun. Stat.-Simul. Comput. **49**(1), 194–206 (2020)

7. Aitchison, J.: On the distribution of a positive random variable having a discrete probability mass at the origin. J. Am. Stat. Assoc. **50**(271), 901–908 (1955)

8. Aitchison, J., Brown, J.A.C.: The lognormal distribution: with special reference to its uses in economics. J. Am. Stat. Assoc. (1963)

9. Weerahandi, S.: Generalized confidence intervals. J. Am. Stat. Assoc. **88**(423), 899–905 (1993)

10. Anirban, D.: Asymptotic theory of statistics and probability theory of statistics and probability. J. Am. Stat. Assoc.(2008)

11. Wu, W.H., Hsieh, H.N.: Generalized confidence interval estimation for the mean of delta-lognormal distribution: an application to New Zealand trawl survey data. J. Appl. Stat. **41**(7), 1471–1485 (2014)

12. Li, J., Song, W., Shi, J.: Parametric bootstrap simultaneous confidence intervals for differences of means from several two-parameter exponential distributions. Stat. Probab. Lett. **106**, 39–45 (2015)

13. Roy, A., Mathew, T.: A generalized confidence limit for the reliability function of a two-parameter exponential distribution. J. Stat. Plan. Infer. **128**(2), 509–517 (2005)

14. Donner, A., Zou, G.: Closed-form confidence intervals for functions of the normal mean and standard deviation. Stat. Methods Med. Res. **21**(4), 347–359 (2010)

15. Zou, G.Y., Taleban, J., Huo, C.Y.: Confidence interval estimation for lognormal data with application to health economics. Comput. Stat. Data Anal. **53**(11), 3755–3764 (2009)

An Analysis to Treat the Degeneracy of a Basic Feasible Solution in Interval Linear Programming

Zhenzhong Gao[✉] and Masahiro Inuiguchi[✉]

Osaka University, Toyonaka, Osaka 560-8531, Japan
zhenzhong@inulab.sys.es.osaka-u.ac.jp, inuiguti@sys.es.osaka-u.ac.jp

Abstract. When coefficients in the objective function cannot be precisely determined, the optimal solution is fluctuated by the realisation of coefficients. Therefore, analysing the stability of an optimal solution becomes essential. Although the robustness analysis of an optimal basic solution has been developed successfully so far, it becomes complex when the solution contains degeneracy. This study is devoted to overcoming the difficulty caused by the degeneracy in a linear programming problem with interval objective coefficients. We focus on the tangent cone of a degenerate basic feasible solution since the belongingness of the objective coefficient vector to its associated normal cone assures the solution's optimality. We decompose the normal cone by its associated tangent cone to a direct union of subspaces. Several propositions related to the proposed approach are given. To demonstrate the significance of the decomposition, we consider the case where the dimension of the subspace is one. We examine the obtained propositions by numerical examples with comparisons to the conventional techniques.

Keywords: Interval linear programming · Degeneracy · Polyhedral convex cone · Tangent cone · Basic space

1 Introduction

Linear programming addresses enormous real-world problems. The conventional LP techniques assume that all coefficients are precisely determined. However, this assumption cannot always be guaranteed. Sometimes the coefficients can only be imprecisely known with ranges or distributions due to measurement limitation, noise and insufficient knowledge. Since the imprecise coefficients may fluctuate the solution's optimality, a decision-maker is usually interest in analysing its stability.

Researchers have studied the problem for decades. An approach called *sensitivity analysis* [1] that utilises *shadow price* can analyse the maximum variation on a single coefficient. To treat the case of multiple coefficients, Bradley, Hax, and

This work was supported by JSPS KAKENHI Grant Number JP18H01658.

K. Honda et al. (Eds.): IUKM 2022, LNAI 13199, pp. 130–142, 2022.
https://doi.org/10.1007/978-3-030-98018-4_11

Magnanti [1] solved a convex cone by the *100 Percent Rule*, which is also called the *optimality assurance cone* in this paper. Then one only needs to check the belongingness of the imprecise coefficients to this convex cone. To represent the imprecise coefficients, researchers utilise several methods such as interval [5,16], fuzzy [7,9], and probability distribution [10,11,14,15]. For example, the *necessary optimality* [9] is widely utilised in the interval case if a feasible solution is optimal for all realisations derived by the interval coefficients. The *tolerance approach* [2,16,17] can address it straightforwardly if the feasible set is constant.

Despite the usefulness of the optimality assurance cone, we cannot always solve it directly by the simplex method. When the feasible solution is non-basic or degenerate, it becomes problematic [6]. To handle it, researchers aim to separate the non-zero part of the solution instead of focusing on its basis. Some remarkable techniques have emerged, such as *support set invariancy* and *optimal partition invariancy* [6]. However, they only concentrate on the non-basic situation, which is called *dual degeneracy* in this paper. On the other hand, *primal degeneracy* is merely considered a particular case and treated by *variational analysis* [12] and *convex analysis* [13] theoretically. The reason is that the optimal solution and optimal value would not change even the basic index set varies. However, when solving the optimal assurance cone of a basic feasible solution, the variance of the basic index set causes troubles. If we list all combinations violently, the computational burden will become tremendous for a large-scale problem [6]. Hence, the study of the *primal degeneracy* is vital.

In this paper, we study the optimality assurance cone by its counterpart *tangent cone* [12,13] in the view of linear algebra. We start by reviewing the interval linear programming and introduce the necessary optimality of a feasible solution in the next section. After identifying the difference between the primal and dual degeneracy, we focus on the primal one. We consider solving the tangent cone of a feasible basic solution and decompose the derived optimality assurance cone into a union of subspaces with equivalence. To simplify our analysis, we assume that the dimension of the subspace is only 1, i.e. the cardinality of the non-zero variable set is strictly 1 less than the basic index set. We finally give numerical examples to show that our approach can treat the problem with no loop or iteration.

2   Preliminaries

2.1   The Linear Programming

The linear programming (LP) problem in this paper follows the standard form as

$$\text{minimize } \boldsymbol{c}^{\mathrm{T}}\boldsymbol{x}, \text{ subject to } A\boldsymbol{x} = \boldsymbol{b}, \ \boldsymbol{x} \geq \boldsymbol{0}, \tag{1}$$

where $\boldsymbol{x} \in \mathbb{R}^n$ is the decision variable vector, while $A \in \mathbb{R}^{m \times n}$, $\boldsymbol{b} \in \mathbb{R}^m$ and $\boldsymbol{c} \in \mathbb{R}^n$ are the coefficient matrix, right-hand-side coefficient vector and objective coefficient vector, respectively.

Since the simplex method needs a basic index set $\mathbb{I}_B$ with $\text{Card}(\mathbb{I}_B) = m$, we have to consider *basic* feasible solutions. Therefore, let $x_B^* \in \mathbb{R}^m$ and $x_N^* \in \mathbb{R}^{n-m}$ denote the basic and non-basic sub-vectors of x^* separated by $\mathbb{I}_B$, respectively. Then we can also separate A with $A_B \in \mathbb{R}^{m \times m}$ and $A_N \in \mathbb{R}^{m \times (n-m)}$, and c with $c_B \in \mathbb{R}^m$ and $b_N \in \mathbb{R}^{n-m}$ accordingly.

Since $\mathbb{I}_B$ is solved by the simplex method, A_B should be non-singular. Therefore, we have the proposition for the optimality of a basic feasible solution:

Proposition 1. *A basic feasible solution x^* is optimal if and only if the following conditions are valid:*

$$c_N - A_N^{\mathrm{T}} A_B^{-\mathrm{T}} c_B \geq 0, \tag{2}$$
$$A_B^{-1} b \geq 0, \tag{3}$$

where the optimal solution is $x_B^* = A_B^{-1} b$, $x_N^* = 0$ with the optimal value being $c_B^{\mathrm{T}} A_B^{-1} b$.

2.2 The Interval Linear Programming

Since the coefficients in an LP problem cannot always be guaranteed to be precise in reality, *interval linear programming* (ILP) considers utilising intervals to represent the imprecise coefficients. A typical ILP problem is written as

$$\text{minimize } \gamma^{\mathrm{T}} x, \text{ subject to } \varLambda x = \varphi, \ x \geq 0, \tag{4}$$

where x represents the decision variable vector, but $\varLambda \subseteq \mathbb{R}^{m \times n}$, $\varphi \subseteq \mathbb{R}^m$ and $\gamma \subseteq \mathbb{R}^n$ are the interval subsets composed of the imprecise A, b and c, respectively. Therefore, an ILP problem can be regarded as a combination of multiple conventional LP problems, called *scenarios* [4]. Hence, the robustness analysis of a solution equals to analysing its all scenarios.

However, Proposition 1 only guarantees the invariance of $\mathbb{I}_B$ instead of the optimal solution x^* due to $x_B^* = A_B^{-1} b$. Since the imprecision in constraints is difficult (see [3,4]) to study, we assume the ILP problem always has a constant feasible set, i.e. $\varLambda$ and φ are singletons containing A and b, respectively. Hence, the ILP problem becomes

$$\text{minimize } \gamma^{\mathrm{T}} x, \text{ subject to } Ax = b, \ x \geq 0, \tag{5}$$

where $\gamma \in \varPhi := \{(c_1, \ldots, c_n)^{\mathrm{T}} : c_i^{\mathrm{L}} \leq c_i \leq c_i^{\mathrm{U}}, \ i = 1, \ldots, n\} \subseteq \mathbb{R}^n$. c_i^{L} and c_i^{U} are the lower and upper bounds of the interval $\varPhi_i$, $i = 1, \ldots, n$, respectively.

To analyse the optimality of a feasible solution in an ILP problem (5), we utilise *possible and necessary optimality* [9]:

Definition 1 (possible and necessary optimality). *Let $\varPhi$ defined in Problem (5) denote an interval hyper-box composed of γ and let x^* be a feasible solution, then x^* is possibly optimal for $\varPhi$ if $\exists \gamma \in \varPhi$ that x^* is optimal, and x^* is necessarily optimal for $\varPhi$ if $\forall \gamma \in \varPhi$ that x^* is optimal.*

To check the necessary optimality of $\boldsymbol{x}^*$, we use the *optimality assurance cone* [8] defined as

Definition 2 (optimality assurance cone). *Let $\boldsymbol{x}^*$ be a feasible solution. Then the* optimality assurance cone, *denoted as $\mathscr{S}^O(\boldsymbol{x}^*)$, is defined by*

$$\mathscr{S}^O(\boldsymbol{x}^*) := \left\{ \boldsymbol{c} \in \mathbb{R}^n : \boldsymbol{c}^T \boldsymbol{x}^* = \min\{\boldsymbol{c}^T \boldsymbol{x} : A\boldsymbol{x} = \boldsymbol{b}, \ \boldsymbol{x} \geq \boldsymbol{0}\} \right\}. \tag{6}$$

Since a decision-maker usually does not prefer a possibly optimal solution, we focus on the necessary optimality, which can be checked by the lemma below:

Lemma 1. *A feasible solution $\boldsymbol{x}^*$ is* necessarily optimal *if and only if $\Phi \subseteq \mathscr{S}^O(\boldsymbol{x}^*)$.*

However, Eq. (6) is not applicable for solving the *optimality assurance cone*. Fortunately, if $\boldsymbol{x}^*$ is a *non-degenerate basic feasible* solution, we can utilise Proposition 1 to get an equivalent result as the *100 Percent Rule* [1] did:

Proposition 2. *Let $\boldsymbol{x}^*$ be a non-degenerate basic feasible solution to the ILP Problem (5). Then a convex cone defined by $\boldsymbol{x}^*$, denoted as $\mathscr{M}^O(\boldsymbol{x}^*)$, is equivalent to $\mathscr{S}^O(\boldsymbol{x}^*)$. Namely,*

$$\mathscr{M}^O(\boldsymbol{x}^*) := \left\{ \boldsymbol{c} \in \mathbb{R}^n : \boldsymbol{c}_N - A_N^T A_B^{-T} \boldsymbol{c}_B \geq \boldsymbol{0} \right\} = \mathscr{S}^O(\boldsymbol{x}^*). \tag{7}$$

With the condition in Proposition 2, Lemma 1 equals to the following one:

Lemma 2. *A non-degenerate basic feasible solution $\boldsymbol{x}^*$ is* necessarily optimal *if and only if $\Phi \subseteq \mathscr{M}^O(\boldsymbol{x}^*)$.*

By Lemma 2, the necessary optimality can be checked straightforwardly by *tolerance approach* [2,16,17]. Since the only difference between Lemma 1 and 2 is whether $\boldsymbol{x}^*$ is *non-degenerate* and *basic*, the problem becomes difficult when there exists degeneracy. Hence, the key is how to correctly solve $\mathscr{S}^O(\boldsymbol{x}^*)$ in an efficient way, which becomes the main topic in the following content.

3 Degeneracy and Optimality Assurance Cone

3.1 Difference Between Dual and Primal Degeneracy

Before proposing our approach, we need to illustrate what is the *degeneracy* that has been mentioned in previous sections by the following examples.

Example 1. Let us consider the following LP problem:

$$\begin{aligned}
\text{minimize} \quad & c_1 x_1 + c_2 x_2, \\
\text{subject to} \quad & 3x_1 + 4x_2 + x_3 = 42, \\
& 3x_1 + x_2 + x_4 = 24, \\
& x_2 + x_5 = 9, \\
& x_i \geq 0, \ i = 1, 2, ..., 5,
\end{aligned}$$

where $c_1 = -3$ and $c_2 = -4$. Solve this problem.

By the simplex method, we obtain the tabular as

Basis	x_1	x_2	x_3	x_4	x_5	RHS
x_2	0	1	1/3	−1/3	0	6
x_1	1	0	−1/9	4/9	0	6
x_5	0	0	−1/3	1/3	1	3
−z	0	0	1	$\boxed{0}$	0	42

where the optimal solution is $x^* = (6, 6, 0, 0, 3)^T$. However, since the last row of x_4 position being 0, we can re-pivot the tabular as:

Basis	x_1	x_2	x_3	x_4	x_5	RHS
x_2	0	1	0	0	1	9
x_1	1	0	1/3	0	−4/3	2
x_4	0	0	−1	1	3	9
−z	0	0	1	0	$\boxed{0}$	42

This time $x^* = (2, 9, 0, 9, 0)^T$, where the optimal value maintains to be −42. However, if we modify Example 1 as the following one:

Example 2. Reconsider Example 1, if $c_1 = -3$ and $c_2 = -2$ and there exists an extra constraint $x_1 + x_2 + x_6 = 12$, solve this problem.

By the simplex method, we obtain the tabular as

Basis	x_1	x_2	x_3	x_4	x_5	x_6	RHS
x_2	0	1	1/3	−1/3	0	0	6
x_1	1	0	−1/9	4/9	0	0	6
x_5	0	0	−1/3	1/3	1	0	3
x_6	0	0	−2/9	−1/9	0	1	$\boxed{0}$
−z	0	0	1/3	2/3	0	0	30

where the optimal solution is $x^* = (6, 6, 0, 0, 3, 0)^T$. However, since $x_6 = 0$, we can also pivot the tabular as:

Basis	x_1	x_2	x_3	x_4	x_5	x_6	RHS
x_2	0	1	0	−1/2	0	3/2	6
x_1	1	0	0	1/2	0	−1/2	6
x_5	0	0	0	2/3	1	−3/2	3
x_3	0	0	1	1/2	0	−9/2	$\boxed{0}$
−z	0	0	0	1/2	0	3/2	30

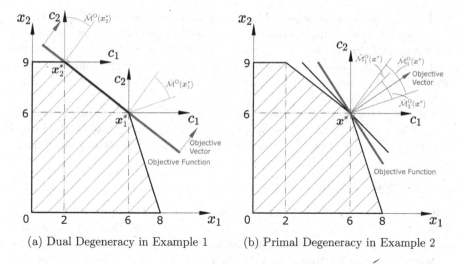

(a) Dual Degeneracy in Example 1   (b) Primal Degeneracy in Example 2

Fig. 1. Primal and dual degeneracy

Unlike Example 1, both optimal solution and optimal value maintain to be the same even the basis changes, which means the situation of Example 2 is different even the basis in both examples change.

To explain both examples illustratively, we project them into x_1-x_2 coordinate in Fig. 1 since $c_i = 0$, $i \neq 1, 2$. $\mathscr{M}_i^O(x_j^*)$ denotes the ith projection of the optimality assurance cone of x_j^*.

Subfig. 1a shows the result of Example 1, where we find two optimal basic solutions. We show that the optimality assurance cones of x_1^* and x_2^* are independent, and can be solved once the optimal solution is determined. Moreover, it shows that any vertex on the line of x_1^* and x_2^* can be the optimal solution.

However, for the result of Example 2 shown in Subfig. 1b, the situation becomes different. At first, there exists only one optimal solution x^* with 3 active constraints on it, where only 2 of them are needed. Consequently, we have 3 potential optimality assurance cones as $\mathscr{M}_1^O(x^*)$, $\mathscr{M}_2^O(x^*)$ and $\mathscr{M}_3^O(x^*)$, where the union of them is what we want.

Example 1 and 2 show two different degeneracies. When considering them by the simplex method, we find that the degeneracy in Example 1 is connected with the objective coefficients, while in Example 2 is the right-hand-side coefficients. Hence, we identify them as *dual degeneracy* (Example 1) and *primal degeneracy* (Example 2) by Proposition 1, and state that for a basic feasible solution x^*,

- there exists no *dual degeneracy* if $c_N - A_N^T A_B^{-T} c_B > 0$, and
- there exists no *primal degeneracy* if $A_B^{-1} b > 0$,

and if x^* satisfies both, we define it as *non-degenerate basic feasible*.

Conventionally, *dual* and *primal degeneracy* are usually treated as the same question for both enabling the basic index set to change. Moreover, most of studies only focused on the *dual* one when a feasible solution is non-basic.

However, the rationales of *dual* and *primal degeneracy* are completely different. In the view of linear algebra, when the objective coefficient vector c is not independent from the rows of matrix A, *dual degeneracy* happens. On the other hand, *primal degeneracy* has no relation with the objective function, and is usually caused by the over-constraints on the solution. Therefore, it is necessary to do respective discussion.

Since the *dual degeneracy* only enables us to choose a non-basic solution, we can handle such trouble by simply choosing a basic one. Instead, the *primal degeneracy* does not diminish. Hence, we concentrate on the *primal degeneracy*.

3.2 Analysis of Primal Degeneracy

In Example 2, we show what is *primal degeneracy* by the simplex tabular and figure. We find that, despite with computational burden, the simplex method can always find the correct optimality assurance cone. Therefore, by utilising the *support set* [6] of a basic feasible solution x^*, which is the index set $\mathbb{I}_P(x^*) := \{i : x_i^* > 0\}$, we can treat the problem by the following theorem:

Theorem 1. *Let $\mathbb{I}_P(x^*) \subseteq \{1, 2, \ldots, n\}$ denote the support set of a basic feasible solution x^* in Problem (5) with $\mathrm{Card}(\mathbb{I}_P(x^*)) \leq m$, where m is the number of constraints. Then the optimality assurance cone of x^* is*

$$\mathcal{M}^O(x^*) = \bigcup_{\substack{\mathbb{I}_{B^i} \supseteq \mathbb{I}_P(x^*), \\ A_{B^i}^{-1} b \geq 0}} \left\{ c \in \mathbb{R}^n : c_{N^i} - A_{N^i}^T A_{B^i}^{-T} c_{B^i} \geq 0 \right\} = \mathscr{S}^O(x^*), \quad (8)$$

where $\mathbb{I}_{B^i}$ with $\mathrm{Card}(\mathbb{I}_{B^i}) = m$ is the index set that determines A_{B^i} and c_{B^i}.

Proof. By the definition of support set, we have the result as

$$\{x^*\} = \bigcap_{\mathbb{I}_{B^i} \supseteq \mathbb{I}_P(x^*)} \left\{ x \in \mathbb{R}^n : A_{B^i} x_{B^i} + A_{N^i} x_{N^i} = b, \ x_{B^i} \geq 0, \ x_{N^i} = 0 \right\},$$

$$= \bigcap_{\substack{\mathbb{I}_{B^i} \supseteq \mathbb{I}_P(x^*), \\ A_{B^i}^{-1} b \geq 0}} \left\{ x \in \mathbb{R}^n : x_{B^i} = A_{B^i}^{-1} b, \ x_{N^i} = 0 \right\}.$$

For any $\mathbb{I}_{B^i}$ satisfying $\mathbb{I}_{B^i} \supseteq \mathbb{I}_P(x^*)$ and $A_{B^i}^{-1} b \geq 0$, its counterpart normal cone calculated as $\{c \in \mathbb{R}^n : c_{N^i} - A_{N^i}^T A_{B^i}^{-T} c_{B^i} \geq 0\}$ makes x^* optimal for any c belonging to it. Hence, we have the result by uniting all of them. $\square$

The key of Theorem 1 is to list all situations where a feasible basic solution maintains to be optimal and unite them. However, since the calculation is linearly related to the combination of the basis, it becomes enormous when the system is in large-scale.

To treat the difficulty in Theorem 1, the following lemma is necessary for our approach:

Lemma 3. *Let x^* be a basic feasible solution in the ILP Problem (5). Then the optimality assurance cone is normal to the tangent cone of the feasible set on x^*.*

Lemma 3 states the relation between the optimality assurance cone with its counterpart, i.e. tangent cone, which is significant since we can simply solve the tangent cone of a convex set on a point. By utilising the support set, we have the following proposition:

Proposition 3. *The tangent cone of the feasible set on a basic feasible solution x^* in the ILP Problem (5) is*

$$\mathscr{T}(x^*) = \left\{ x \in \mathbb{R}^n : A_P x_P + A_Z x_Z = 0, \ x_Z \geq 0 \right\}, \tag{9}$$

where x_P and x_Z are separated by the support set $\mathbb{I}_P$ and its counterpart $\mathbb{I}_Z :=$ $\{1, 2, \ldots, n\} \backslash \mathbb{I}_P$.

Since $\mathscr{T}(x^*)$ in Eq. (9) forms a convex cone, let us review the definition and some useful properties of it.

Definition 3 (Convex Cone). *A subset $\mathscr{C}$ of a vector space $\mathscr{V}$ over an ordered field $\mathscr{F}$ is a cone if for every vector $v \in \mathscr{C}$ and any positive scalar $\alpha \in \mathscr{F}$, $\alpha v \in \mathscr{C}$. Moreover, if for every $v, w \in \mathscr{C}$ and for any positive scalar $\alpha, \beta \in \mathscr{F}$ such that $\alpha v + \beta w \in \mathscr{C}$, then $\mathscr{C}$ is a convex cone.*

Furthermore, the following lemma indicates that we can utilise convex techniques to analyse the tangent cone.

Lemma 4. *The tangent cone of a convex set is convex.*

Since it is known that a convex cone is not a vector space due to the non-negative scalar, we cannot utilise the *basic space* directly. However, we can still use the concept, where the convex cone is *spanned* by a series of vectors. We call these vectors as the *basic vectors* of the convex cone and note that, if the convex cone is polyhedral, e.g. the tangent cone, the number of the basic vectors is finite. Similar to the linear space, we call the left part as the *null space*.

The following theorem illustrates the relation between the basic vectors with the basis of the ILP problem (5).

Theorem 2. *Given an ILP problem (5) with a non-degenerate basic feasible solution x^*, then the null space of the tangent cone $\mathscr{T}(x^*)$ corresponds to the basic index set $\mathbb{I}_B$ of x^*.*

Proof. Since x^* is non-degenerate and basic, $x_B^* = A_B^{-1} b > 0$ is always valid. Hence the tangent cone becomes

$$\mathscr{T}(x^*) = \left\{ x \in \mathbb{R}^n : A_B x_B + A_N x_N = 0, \ x_N \geq 0 \right\}.$$

If we ignore $x_N \geq 0$ and only consider the linear space $\{ x \in \mathbb{R}^n : Ax = 0 \}$, it can be spanned by $(n - m)$ independent vectors due to $A \in \mathbb{R}^{m \times n}$. Since x^* is non-degenerate and basic, $\mathrm{Card}(\boldsymbol{x}_N) = n - m$. Hence, to cover the condition of $\boldsymbol{x}_N \geq \boldsymbol{0}$, we can utilise it directly as the basic vectors. Hence, the basic vectors of $\mathscr{T}(\boldsymbol{x}^*)$ corresponds to the non-basic part of $\boldsymbol{x}^*$, which is equivalent to the condition that the null space of $\mathscr{T}(\boldsymbol{x}^*)$ corresponds to $\mathbb{I}_B$ of $\boldsymbol{x}^*$ □

Theorem 2 indicates the fact that, once we determine the basic vectors of the tangent cone of a basic feasible solution, the corresponding basis is known. Hence, we can extend this property to the following proposition:

Proposition 4. *For the tangent cone $\mathscr{T}(\boldsymbol{x}^*)$ defined in Proposition 3, all combinations of choosing $(n - m)$ entries from $\mathbb{I}_Z$ can span the tangent cone.*

By Theorem 1 and 2, Proposition 4 is obvious. However, the method in Proposition 4 has the same computational complexity as Theorem 1, indicating that our calculation speed would not improve with purely changing to the realm of convex cone.

To simplify the procedure, we only consider the dimension of the subspace to be 1, i.e. 1-dimensional degeneracy that $\mathrm{Card}(\mathbb{I}_P) = m - 1$. Therefore, there exist an extra entry in $\mathbb{I}_Z$, giving an extra constraint in spanning the tangent cone. It is because we need to choose $(n - m)$ basic vectors from $\mathbb{I}_Z$, but there exist $(n - m + 1)$ entries should be non-negative. Hence, when $(n - m)$ basic vectors are chosen, there always leaves an entry in $\mathbb{I}_Z$ that should be non-negatively spanned by the chosen basic vectors.

However, to treat such problem, we can firstly use the following modification to make all coefficients in the extra constraint be non-negative:

Proposition 5. *Let $\boldsymbol{x}^*$ be a basic feasible solution with 1-dimensional degeneracy for the ILP Problem (5) and let $\mathbb{I}_P$ denote its support set. Then there always exists an extra constraint that can be written with all coefficients non-negative:*

$$\sum_{i \in \mathbb{I}_Z^k} k_i (\boldsymbol{x}_Z)_i = \sum_{j \in \mathbb{I}_Z^l} l_j (\boldsymbol{x}_Z)_j, \tag{10}$$

where $\mathbb{I}_Z^k \cup \mathbb{I}_Z^l = \mathbb{I}_Z$ and $\mathrm{Card}(\mathbb{I}_Z^k) \leq \mathrm{Card}(\mathbb{I}_Z^l)$. k_i and l_j are non-negative scalars.

It is easy to understand that when there exists no primal degeneracy, $\mathbb{I}_Z^k$ is empty by Theorem 2. Moreover, it also hints the following lemma:

Lemma 5. *If $\mathrm{Card}(\mathbb{I}_Z^k) \leq 1$ in Proposition 5, then $\mathbb{I}_Z^l$ is the index set that corresponds to the basic vectors of the tangent cone.*

The rationality of Lemma 5 is that, once a variable can be expressed by other non-negative variables multiplied with non-negative scalars, it becomes non-negative. So it is no longer necessary to consider the non-negative constraint any more and hence can be abandoned.

To explain Lemma 5 more illustratively, let us use Example 2 again. In Example 2, we solve the optimal solution as $\boldsymbol{x}^* = (6, 6, 0, 0, 3, 0)^{\mathrm{T}}$, which gives $\mathbb{I}_P = \{1, 2, 5\}$ and $\mathbb{I}_Z = \{3, 4, 6\}$. Hence we can write its tangent cone as

$$\mathscr{T}(\boldsymbol{x}^*) = \Big\{ A\boldsymbol{x} = 0 \text{ with } x_3, x_4, x_6 \geq 0 \Big\}.$$

Since $A \in \mathbb{R}^{4\times6}$, there should only exist 2 entries in the basic space. Therefore, we need to pick out v_1 and v_2 from x_3, x_4 and x_6. As Proposition 5 indicates, we remove x_1, x_2 and x_5 from $Ax = 0$, which give the following extra constraint:

$$x_4 = 9x_6 - 2x_3 \Rightarrow 9x_6 = x_4 + 2x_3$$

After modification to make all scalars to be non-negative, it shows that x_6 should be removed. Hence, the correct basic space should be formed by x_3 and x_4, which indicates the correct basic index set $\mathbb{I}_B = \{1, 2, 5, 6\}$. The result corresponds to the conventional analysis in Example 2.

However, if $\mathrm{Card}(\mathbb{I}_Z^k) \geq 2$, we cannot treat the problem simply by Lemma 5. Instead, we can pick every entry in $\mathbb{I}_Z^k$ as the one that should be removed, which results a series of basic vectors. Then we can form the tangent cone by their union. Mathematically, we have the following proposition:

Proposition 6. *Let the extra constraint of x_Z be written in the form of Eq. (10) with $\mathrm{Card}(\mathbb{I}_Z^k) \leq \mathrm{Card}(\mathbb{I}_Z^l)$ and $\mathbb{I}_Z^k \cup \mathbb{I}_Z^l = \mathbb{I}_Z$, and k_i and l_j are all non-negative scalars. Then the index of the basic vectors of the tangent cone is the union of the following sets:*

$$\mathbb{I}_Z \backslash \{i\}, \ i \in \mathbb{I}_Z^k. \tag{11}$$

It is easy to see that if $\mathrm{Card}(\mathbb{I}_Z^k) \leq 1$, then Proposition 6 is equivalent to Lemma 5. To illustrate Proposition 6, let us consider a brief example.

Example 3. Let us consider the following LP problem:

$$\text{minimize } c_1 x_1 + c_2 x_2 + c_3 x_3$$
$$\begin{aligned}
\text{subject to } -2x_1 + x_3 + x_4 &= -4 &\text{(i)}\\
2x_2 + x_3 + x_5 &= 8 &\text{(ii)}\\
2x_1 + x_3 + x_6 &= 8 &\text{(iii)}\\
-2x_2 + x_3 + x_7 &= -4 &\text{(iv)}\\
x_i \geq 0, \ i = 1, 2, \ldots, 7,
\end{aligned}$$

where we assume $c_1 = 2$, $c_2 = 1$, $c_3 = -10$. Solve this problem.

By the simplex method with the following tabular, the degenerate optimal solutions is $x^* = (3, 3, 2, 0, 0, 0, 0)^{\mathrm{T}}$, indicating that $\mathbb{I}_P = \{1, 2, 3\}$.

Basis	x_1	x_2	x_3	x_4	x_5	x_6	x_7	RHS
x_1	1	0	0	$-1/2$	$1/4$	0	$1/4$	3
x_3	0	0	1	0	$1/2$	0	$1/2$	2
x_6	0	0	0	1	-1	1	-1	0
x_2	0	1	0	0	$1/4$	0	$-1/4$	3
$-z$	0	0	0	1	$17/4$	0	$19/4$	11

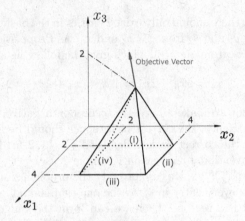

Fig. 2. Degeneracy in Example 3

Hence, we can list 4 situations and take the union of them by Theorem 1. Since $c_i = 0$, $i = 4, 5, 6, 7$, we can simply project the system to $\mathbb{R}^3$ and write the optimality assurance cone as

$$\mathcal{M}^O(\check{x}^*) = \bigcup_{i=4,5,6,7} \{\check{c} \in \mathbb{R}^3 : G^i \check{c} \geq 0\},$$

where $\check{x}^* = (x_1^*, x_2^*, x_3^*)^T$ and $\check{c} = (c_1, c_2, c_3)^T$, and

$$G^4 = \begin{bmatrix} 1/4 & -1/4 & -1/2 \\ -1/2 & 0 & 0 \\ 1/4 & 1/4 & -1/2 \end{bmatrix}, \qquad G^5 = \begin{bmatrix} 1/4 & -1/4 & -1/2 \\ -1/4 & -1/4 & -1/2 \\ 0 & 1/2 & 0 \end{bmatrix},$$

$$G^6 = \begin{bmatrix} 1/2 & 0 & 0 \\ -1/4 & -1/4 & -1/2 \\ -1/4 & 1/4 & -1/2 \end{bmatrix}, \qquad G^7 = \begin{bmatrix} 1/4 & 1/4 & -1/2 \\ 0 & -1/2 & 0 \\ -1/4 & 1/4 & -1/2 \end{bmatrix},$$

where G^{i+3} denote the situation that ignore (i_{th}) constraint. However, in our approach by Proposition 5, we have the extra constraint as $x_4 + x_6 = x_5 + x_7$, which indicates that we only need to take the union of G^4 and G^6, or the union of G^5 and G^7, i.e.

$$\mathcal{M}^O(\check{x}^*) = \bigcup_{i=4,6} \{\check{c} \in \mathbb{R}^3 : G^i \check{c} \geq 0\} = \bigcup_{i=5,7} \{\check{c} \in \mathbb{R}^3 : G^i \check{c} \geq 0\} \qquad (12)$$

Moreover, if we draw the projection $\check{x}$ in x_1-x_2-x_3 coordinate, we have Fig. 2, where the feasible set is the space in the tetrahedron. Then result shown by Eq. (12) is obvious.

4 Conclusion and Future Work

In this paper, we proposed a linear algebraic approach to treating the primal degeneracy in ILP problem with imprecise objective coefficients, since we always need to solve the optimality assurance cone explicitly when analysing the robustness of a basic feasible solution in the ILP problem.

In accomplishing our goal, we first consider the tangent cone of the basic feasible solution instead of listing all bases by the simplex method. Since the tangent cone is normal to the optimality assurance cone and is always polyhedral and convex, we modify the concept of basic space in linear subspace and apply it to the tangent cone. We show that once we can span the tangent cone by its basic space, we find the correct basis of the corresponding problem, which would lead to the correct optimality assurance cone. To illustrate and validate our technique, we give numerical examples.

However, since we assume that there exists only one degeneracy, the analysis is not complete. Moreover, when degeneracy becomes multiple, it is necessary to have some algorithms for forming the correct basic space. Another incomplete section is that we assume the feasible solution is basic even there exists dual degeneracy. Therefore, we would take dual degeneracy into consideration in our next step.

References

1. Bradley, S.P., Hax, A.C., Magnanti, T.L.: Applied Mathematical Programming. Addison-Wesley Publishing Company (1977)
2. Filippi, C.: A fresh view on the tolerance approach to sensitivity analysis in linear programming. Eur. J. Oper. Res. **167**(1), 1–19 (2005)
3. Garajová, E., Hladík, M.: On the optimal solution set in interval linear programming. Comput. Optim. Appl. **72**(1), 269–292 (2018). https://doi.org/10.1007/s10589-018-0029-8
4. Garajová, E., Hladík, M., Rada, M.: Interval linear programming under transformations: optimal solutions and optimal value range. CEJOR **27**(3), 601–614 (2018). https://doi.org/10.1007/s10100-018-0580-5
5. Hladík, M.: Computing the tolerances in multiobjective linear programming. Optim. Methods Softw. **23**(5), 731–739 (2008)
6. Hladík, M.: Multiparametric linear programming: support set and optimal partition invariancy. Eur. J. Oper. Res. **202**(1), 25–31 (2010)
7. Inuiguchi, M.: Necessity measure optimization in linear programming problems with fuzzy polytopes. Fuzzy Sets Syst. **158**(17), 1882–1891 (2007)
8. Inuiguchi, M., Gao, Z., Carla, O.H.: Robust optimality analysis of non-degenerate basic feasible solutions in linear programming problems with fuzzy objective coefficients, submitted
9. Inuiguchi, M., Sakawa, M.: Possible and necessary optimality tests in possibilistic linear programming problems. Fuzzy Sets Syst. **67**(1), 29–46 (1994)
10. Kall, P., Mayer, J., et al.: Stochastic Linear Programming, vol. 7. Springer, Heidelberg (1976)
11. Kondor, I., Pafka, S., Nagy, G.: Noise sensitivity of portfolio selection under various risk measures. J. Banking Finan. **31**(5), 1545–1573 (2007)

12. Rockafellar, R.T., Wets, R.J.B.: Variational Analysis, vol. 317. Springer, Heidelberg (2009)
13. Rockafellar, R.T.: Convex Analysis. Princeton University Press (2015)
14. Shamir, R.: Probabilistic analysis in linear programming. Stat. Sci. **8**(1), 57–64 (1993). http://www.jstor.org/stable/2246041. ISSN 08834237
15. Todd, M.J.: Probabilistic models for linear programming. Math. Oper. Res. **16**(4), 671–693 (1991)
16. Wendell, R.E.: The tolerance approach to sensitivity analysis in linear programming. Manag. Sci. **31**(5), 564–578 (1985)
17. Wondolowski, F.R., Jr.: A generalization of Wendell's tolerance approach to sensitivity analysis in linear programming. Decis. Sci. **22**(4), 792–811 (1991)

Job-Satisfaction Enhancement in Nurse Scheduling: A Case of Hospital Emergency Department in Thailand

Pavinee Rerkjirattikal[1,2] (ID), Raveekiat Singhaphandu[1,2] (ID), Van-Nam Huynh[1] (ID), and Sun Olapiriyakul[2(✉)] (ID)

[1] School of Knowledge Science, Japan Advanced Institute of Science and Technology, Ishikawa 923-1211, Japan
{p.rerkjirattikal,r.singhaphandu,huynh}@jaist.ac.jp
[2] School of Manufacturing Systems and Mechanical Engineering, Sirindhorn International Institute of Technology, Thammasat University, Pathum Thani 12120, Thailand
suno@siit.tu.ac.th

Abstract. A well-designed nurse schedule can improve nurses' job satisfaction, organizational commitment, and intention to stay. To generate effective scheduling outcomes, the simultaneous consideration of staffing costs, workload, individual preferences, and fairness must be made. However, the integration of these aspects into the scheduling model is still lacking in practice. This study develops a bi-objective mixed-integer linear programming approach for nurse scheduling that minimizes the total staffing cost while maximizing nurses' preference-based satisfaction. The proposed model allows the nurses' shift and day-off preferences to be fulfilled while ensuring equitable workloads and cost-effectiveness. The model is validated using actual data collected from a public hospital emergency department in Thailand with approximately 800 beds capacity. Our results highlight the performance of the proposed model in terms of cost, job satisfaction, and fairness compared to the manually-made schedule.

Keywords: Nurse Scheduling Problem · Workload · Individual preferences · Job satisfaction · Fairness

1 Introduction

Hospitals operate around the clock in rendering medical care to patients. There is a need for medical personnel, especially nurses, to work under prolonged and strenuous shifts. Such shift work results in increased risk of excessive fatigue [1], sleep deprivation [2], and work-life imbalance [3]. These factors are known to induce job dissatisfaction, and turnover intention among nurses [4], the common causes of nurse shortage. The positive correlation between job satisfaction on nurse retention has been addressed by many studies [5–8]. The scheduling of

© Springer Nature Switzerland AG 2022
K. Honda et al. (Eds.): IUKM 2022, LNAI 13199, pp. 143–154, 2022.
https://doi.org/10.1007/978-3-030-98018-4_12

nurses without a careful workload consideration can lead to excessive work hours that may not only have negative effects on the health and well-being of nurses but also the service quality and patients' safety [9]. To address all these concerns, the management has to implement proper scheduling measures that improve nurses' working conditions and job satisfaction. To achieve a high job satisfaction level among nurses, hospital management must thoroughly comprehend and consider the influential attributes contributing to job satisfaction. Many factors have been investigated by the previous studies, such as work conditions [10], work schedule [11], job autonomy [12], and fairness [13]. The consideration of these factors during the scheduling process can result in positive scheduling outcomes, including improved workload distribution, and preferred and equitable work schedules.

Significant research efforts have been made to solve Nurse Scheduling Problem (NSP). Mathematical techniques enable the determination of optimal nurse-shift assignments that fulfill operational objectives while complying with hospital regulations and staffing policies. To date, the research that combines cost, workload balance, shift-preference fulfillment, and scheduling fairness has not been well-addressed. To bridge this gap and to promote fairness and job satisfaction in nurse scheduling, we propose a bi-objective NSP model that minimizes the staffing cost and maximizes the fulfillment of nurses' shift and day-off preferences. The fairness of workload and preferred assignment allocation is also considered. The proposed model is validated using a case study of a public hospital emergency room in Thailand.

The rest of this paper is organized as follows. Section 2 provides a review of related literature. Then, the mathematical model formulation is presented in Sect. 3. Section 4 describes the details of the illustrative case study used for model validation. Section 5 presents the experimental results and analysis. Finally, concluding remarks and future works are summarized in Sect. 6.

2   Literature Review

The research on NSP has been well-documented in the literature for its complexity and practical challenges. Many nurse scheduling approaches with multiple scheduling features are designed to improve the work conditions in actual hospital cases [14,15]. The consideration of job satisfaction in an NSP has received more attention in recent years. The integration of individual preferences to improve the job satisfaction of nurses has been extensively addressed according to the NSP literature. In general, the scheduling objectives and constraints are formulated, and the individual preferences of nurses can be accounted for in terms of preferred shifts and days off [16–18], weekend day-off [19,20], and co-workers [21]. Aside from individual preferences, fairness is another desirable scheduling outcome commonly considered by the NSP literature. Michael et al. [22] developed an NSP that balances the number of preferred days off assigned to nurses. Youssef and Senbel [23] formulated an NSP that includes nurses' shift and day-off preferences as the soft constraints that can be relaxed while minimizing the soft constraint violation. Lin et al. [24] developed an NSP algorithm to balance nurses' preferred shifts and days off assigned to nurses. Rerkjirattikal et al. [25] proposed an NSP model that balances the workload, and preferable shift and day-off assignments among the nurses in a hospital's intensive care unit. More examples of scheduling approaches that consider schedule preferences and fairness for job satisfaction improvement can also be found in the workforce scheduling literature [26,27].

The previous studies provide fundamental guidelines of how job satisfaction and fairness can be integrated into an NSP. To further improve their application value, two significant research gaps are addressed here in this study. Firstly, there is a need to consider scheduling fairness based on a more comprehensive accounting of nurses' workloads and individual preferences. The fairness consideration in the current NSP literature is usually built based on a single aspect of fairness that either offers workload or satisfaction balance [28–31]. The outcomes based on single-aspect fairness may not be a good representative of schedules that promote job satisfaction, especially in the long run. Secondly, the perspective of cost consideration is still lacking in the existing literature. The viewpoint of cost for practical application of job satisfaction- and fairness-enhanced NSP needs to be further examined. An example can be found in Hamid et al. [32]. They developed an NSP that optimizes both staffing cost and nurses' job satisfaction under workload balancing constraints.

While job satisfaction is essential, the practical use of NSP relies heavily on economic aspects. The inclusion of wages and other costs to investigate the cost tradeoff for improving nurses' work conditions is still an open research area. This research aims to fill these gaps by proposing a bi-objective NSP model that minimizes the total staffing cost and maximizes the minimum total preference score among all nurses. The preference score is based on individual shift and day-off preferences. This study expects to produce scheduling outcomes that satisfy nurses in terms of both preferable assignment and equitable workload. Thus far, only a few studies consider fairness while satisfying individual preferences. The practicality of the proposed model is also examined using an actual case study, bridging the theoretical and practical aspects of NSP research as indicated by Petrovic [33]. Our model is solved using the ε-constraint method to determine the optimal cost solution with a maximum preference score. Based on the NSP literature, the use of the ε-constraint method, so-called pre-emptive optimization is suitable when the priority of each objective is given as demonstrated in Di Martinelly and Meskens [34] and Hamid et al. [32].

3 Mathematical Model Formulation

In this study, a bi-objective nurse scheduling model is proposed using the pre-emptive mixed-integer linear programming (MILP) approach. The two scheduling goals are to minimize the staffing cost and maximize the fulfillment of nurses' shift and day-off preferences. Without loss of generality, the assumptions and notations used in the model formulation are summarized below.

Assumptions

- The planning horizon is 28 days (1 month). Each workday contains multiple shifts of the same length.
- Nurses are classified into levels based on their experience. In each shift, the total number of nurses and nurses with a particular skill level must meet the requirements.
- The number of regular shifts for nurses in a month is known. All shifts worked in excess of the employee's regular shifts of work per month will be regarded as overtime shifts.
- The total amount of shifts assigned to each nurse must not exceed the limit.
- Each nurse must receive at least the minimum allowable day-offs per week.
- Any night shift cannot be followed by a morning shift of the next day.
- There can be no more than three consecutive night shifts.
- In case a double-shift workday is allowed, there can be no more than two consecutive double-shift workdays.

Indices

$\mathcal{N}$ Set of nurses; $\mathcal{N} = \{1, 2, \ldots, N\}$
$\mathcal{S}$ Set of shifts in a workday; $\mathcal{S} = \{1, 2, \ldots, S\}$
$\mathcal{K}$ Set of nurse skill levels; $\mathcal{K} = \{1, 2, \ldots, K\}$
$\mathcal{D}$ Set of days in planning horizon; $\mathcal{D} = \{1, 2, \ldots, D\}$

Input Parameters

R_{sd} The total number of nurses required in shift s on the day d
RL_{sk} The minimum number of nurse with skill level k required in shift s
N_k A set of nurses that belong to skill level k; $\mathcal{N} = N_1 \cup N_2 \cup \cdots \cup N_K$
SK_{nk} A binary parameter equals 1 if nurse n belongs to skill level k; 0 otherwise.
SP_{ns} The preference score of nurse n towards working in shift s; $SP_{ns} \in \{1, \ldots, Q\}$
DP_{nd} The preference score of nurse n towards taking a day-off on day d; $DP_{nd} \in \{1, \ldots, Q\}$
Q_{nd} A binary parameter equals 1 if nurse n requests to take a day-off on day d; 0 otherwise.
$CReg_s$ Cost of assigning a regular shift during shift s to a nurse
CO_s Cost of assigning an overtime shift during shift s to a nurse
DS The maximum number of shifts can be assigned to a nurse per day.
REG The regulated number of regular shifts per month equals the total days in a month subtracted by number of weekends and holidays
TS The maximum total shifts can be assigned to a nurse per month
DO The minimum number of days off a nurse must receive per week
Gap_{TPC} The allowable gap between the total preference score (TPC) of all nurses
Gap_{WL} The allowable gap between the total shifts assigned (WL) to all nurses
$BigM$ A large positive number used for formulating conditional linear equations

Decision Variables

$X_{nsd} = 1$ if nurse n is assigned to shift s on day d; 0 otherwise.
$Y_{nd} = 1$ if nurse n is assigned to take a day-off on day d; 0 otherwise.

Auxiliary Variables

For ease of understanding, auxiliary variables derived from the value of decision variables used in objective functions and constraints are listed below.

TPC_n The total preference score of nurse n calculated by the summation of the total shift and day off preference scores

$$TPC_n = \sum_{s=1}^{S}\sum_{d=1}^{D}(X_{nsd} \cdot SP_{ns}) + \sum_{d=1}^{D}(Y_{nd} \cdot DP_{nd}) \qquad \forall n \in \mathcal{N} \quad (1)$$

TPC_{min} The minimum total preference score among all nurses

$$TPC_{min} = \min_{n \in \mathcal{N}}\{TPC_n\} \qquad (2)$$

WL_n The total shifts assigned to nurse n.

$$WL_n = \sum_{s=1}^{S}\sum_{d=1}^{D}X_{nsd} \qquad \forall n \in \mathcal{N} \qquad (3)$$

OT_n The total overtime shifts assigned to nurse n.

$$OT_n = WL_n - REG \qquad \forall n \in \mathcal{N} \qquad (4)$$

The job satisfaction-enhanced NSP model consists of two objectives as follows.

1) Minimize the total staffing cost equals a summation of total regular shift cost and total overtime shift cost.

$$\min \sum_{n=1}^{N}(\sum_{s=1}^{S}((WL_n - OT_n) \cdot CReg_s)) + \sum_{n=1}^{N}(\sum_{s=1}^{S}(OT_n \cdot CO_s)) \qquad (5)$$

2) Maximize the minimum total preference score among all nurses.

$$\max\ TPC_{min} \qquad (6)$$

   subject to

$$\sum_{n=1}^{N}X_{nsd} = R_{sd} \qquad \forall s \in \mathcal{S}; d \in \mathcal{D} \qquad (7)$$

$$\sum_{n=1}^{N}(X_{nsd} \cdot SK_{nk}) \geq RL_{sk} \qquad \forall s \in \mathcal{S}; d \in \mathcal{D}; k \in \mathcal{K} \qquad (8)$$

$$\sum_{s=1}^{S} X_{nsd} \leq DS \qquad \forall n \in \mathcal{N}; d \in \mathcal{D} \qquad (9)$$

$$\sum_{d=d}^{d+6} Y_{nd} \geq DO \qquad \forall n \in \mathcal{N}; d \in \{1, 8, 15, 22\} \qquad (10)$$

$$WL_n \leq TS \qquad \forall n \in \mathcal{N} \qquad (11)$$

$$\sum_{s=1}^{S} X_{nsd} \leq BigM \cdot (1 - Y_{nd}) \qquad \forall n \in \mathcal{N}; d \in \mathcal{D} \qquad (12)$$

$$\sum_{s=1}^{S} X_{nsd} + Y_{nd} \geq 1 \qquad \forall n \in \mathcal{N}; d \in \mathcal{D} \qquad (13)$$

$$Q_{nd} \leq Y_{nd} \qquad \forall n \in \mathcal{N}; d \in \mathcal{D} \qquad (14)$$

$$|TPC_n - TPC_{n'}| \leq Gap_{TPC} \qquad \forall n \in \mathcal{N}; n \neq n' \qquad (15)$$

$$|WL_n - WL_{n'}| \leq Gap_{WL} \qquad \forall n \in \mathcal{N}; n \neq n' \qquad (16)$$

$$X_{n,s=S,d} + X_{n,s=1,d+1} \leq 1 \qquad \forall n \in \mathcal{N}; d \in \{\mathcal{D} - 1\} \qquad (17)$$

$$\sum_{j=Night}^{} \sum_{d=d}^{d+3} X_{nsd} \leq 3 \qquad \forall n \in \mathcal{N}; d \in \{\mathcal{D} - 3\} \qquad (18)$$

$$\sum_{j=1}^{J} \sum_{d=d}^{d+2} X_{nsd} \leq 5 \qquad \forall n \in \mathcal{N}; d \in \{\mathcal{D} - 2\} \qquad (19)$$

Constraint (7) regulates the total number of nurses assigned to any shift must equal the requirements. Constraint (8) ensures the number of nurses in each skill level assigned to any shift meets minimum requirements. Constraint (9) restricts the number of daily shifts assigned to nurses. Constraint (10) ensures that nurses receive at least a certain amount of days off per week. Constraint (11) regulates that nurses' total number of shifts must not exceed the limit. Constraints (12) and (13) enforce that no shift assignment is made on any day-off. Constraint (14) ensures that the requested day off of nurses is fulfilled. Constraints (15) and (16) limit the differences between total preference score (TPC_n) and total shift assignments (WL_n) between nurse n and any other nurses to maintain scheduling fairness. Constraint (17) restricts that no morning shift of the following day can be assigned after any night shift. Constraints (18) and (19) limit the number of consecutive night shifts and double-shift workdays, respectively. Note that Constraint (19) can be excluded if double-shift workdays are not allowed.

An ε-constraint or so-called pre-emptive optimization technique is used to solve the proposed NSP model. The technique suits for solving multi-objective models with objectives ranked in the order of importance. In pre-emptive optimization, the model is solved iteratively under each objective function. The objective value obtained from each iteration becomes the bound in the subsequent iterations. In this study, the primary objective is the total staffing cost, and the nurses' job satisfaction is secondary. The model is firstly solved under cost minimization objective (5) subject to Constraints (7)–(19) to obtain the optimal cost ($Cost^*$). Then, the model is solved under the maximization of the minimum total preference score (6) with respect to Constraints (7)–(20), where (20) is an additional constraint imposed to bound the total staffing cost under the total preference score objective as follows,

$$\sum_{n=1}^{N}(\sum_{s=1}^{S}((WL_n - OT_n) \cdot CReg_s)) + \sum_{n=1}^{N}(\sum_{s=1}^{S}(OT_n \cdot CO_s)) \leq Cost^* \tag{20}$$

4   Illustrative Case Study

The case study used for model validation is the Emergency Department (ED) at the Thammasat University Hospital (TUH), a large-scale public hospital located in Pathum Thani, Thailand. The data collection processes, including field questionnaire survey and interview with the head nurse, were conducted during March–June 2021. There are 40 nurses, working under a 3-shift rotation plan: Morning (8 AM–4 PM), Afternoon (4 PM–12 AM), and Night (12 AM–8 AM) shifts. Nurses are divided into five skill levels, with level 5 being the most experienced. There are 10, 11, 7, 9, and 3 nurses with skill levels 1–5, respectively.

For the current scheduling practices, the head nurse manually creates a shift assignment schedule subject to the hospital regulations and the requested day-offs without considering any individual preferences. The scheduling task requires about 3–5 days to complete depending on the degree of request conflicts. In terms of fairness, the head nurse tried to distribute the workload as evenly as possible. The use of a fairness indicator is still lacking. A questionnaire survey was conducted to obtain nurses' preferences data regarding the preferred shifts and days off. The nurses were asked to rank the most to least preferred working shifts and three most to least days of the week to take days off. The collected data and parameters derived from hospital regulations summarized in Table 1 are used to validate the proposed model. In this hospital case study, the costs of assigning nurses to each shift are different. The night shift has the highest pay rate. The cost of overtime shifts is also the same as that of regular shifts. The cost of assigning nurses with different experience levels is the same. The total number of nurses required across the three shifts is different but the same across days. The right-most column of Table 1 is the minimum requirement of the number of nurses with skill levels 1–5, respectively. In this study, the nurses' shifts (SP_{ns}) and day-off preferences (DP_{nd}) scores are given for first, second, third-most preferred slots as 3, 2, and 1, respectively. In DP_{nd}, 0 point is given to the day-of-weeks that are not preferred.

Table 1. The regulation-related parameters

Parameters	Value	Parameters	Value	Parameters	Value
$CReg_s = CO_s(\$)$		R_{nd}		RL_{sk}	
Morning	24.76	Morning	13	Morning	3, 3, 2, 2, 1
Afternoon	33.31	Afternoon	12	Afternoon	3, 3, 2, 2, 0
Night	33.92	Night	9	Night	2, 2, 1, 1, 0
REG	16	DS	2	Gap_{TPC}	5
TS	26	DO	1	Gap_{WL}	5

5   Result and Discussion

The proposed satisfaction-enhanced NSP is solved using GUROBI optimizer version 9.1.2 coded in Python and a 2.3 GHz Dual-Core Intel Core i5-8300H operating system. The model can be solved to optimality of both objectives in less than a minute. The example of nurse schedule outputs is shown in Table 2.

Table 2. An example of nurse schedule output

Nurses	Day 1	Day 2	Day 3	Day 4	Day 5	...	Day 27	Day 28
1	O	M	M	M	M	...	M	O
2	M/A	O	A/N	A	A/N	...	O	O
...	...	...	...	...	...	...		...
40	M/N	O	M	M	M/N	...	O	M

M - Morning shift, A - Afternoon shift, N - Night shift, O - Day-off.

As summarized in Table 3, it can be observed that the obtained results show a good improvement in solutions, compared to the manual scheduling results in terms of 1) total staffing cost, 2) workload assignments, and 3) total preference score. When using the proposed model, the total staffing cost is reduced by 10.6% or about $7,000 for the entire scheduling period of one month. In the current schedule, nurses work on average 27 shifts with the standard deviation of 4.44 and the range between the maximum and minimum shifts assigned to nurses of as high as 19. Considering the workload, nurses' workload decreases to 24 shifts on average in the proposed model. The standard deviation and range decrease to 1.58 and 5 shifts, indicating fewer and more evenly distributed workloads assigned among nurses. Furthermore, the comparison of the use of fairness constraints is made to demonstrate the improved fairness outcomes when the constraints are enforced. It is found that the fairness constraints effectively balance the workload and preferred assignments. However, the result does not seem noticeable since the gap value used in the case study is relatively large. Under tighter gap values, the usefulness of the fairness constraints becomes apparent.

Figure 1 compares the workload distributions between the manual scheduling and the optimal solution. In terms of the total preference score, the proposed model can achieve a good result. The range of preference scores among nurses is from 82 to 83, indicating the fairness in satisfying individual preferred shifts and days off. On this aspect, the performance of manual scheduling cannot be assessed because there was no preference consideration. Based on the result, it can be seen that the proposed model can provide a cost-effective and fair scheduling solution. With the solving time of less than a minute, the scheduling can be much more responsive to any last-minute changes to the input parameters.

Table 3. Comparison of KPIs between the proposed NSP and current schedule

KPIs	Current schedule	The proposed model	
		With fairness constraints	Without fairness constraints
Total staffing cost	$ 65,835.45	$58,830.94	$58,830.94
Total shifts (WL_n)			
Average (SD)	27 (4.44)	24 (1.58)	23.8 (1.61)
Min-Max (Range)	17–36 (19)	21–26 (5)	20–26 (6)
Total preference score (TPC_n)			
Average (SD)	N/A	82.1 (0.3)	82.2 (0.46)
Min-Max (Range)	N/A	82–83 (1)	82–84 (2)

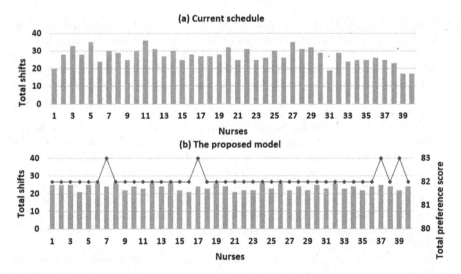

Fig. 1. A comparison of workload distribution between the current schedule and the proposed model

6   Conclusion

In this paper, for the first time, a nurse scheduling approach that considers multiple job satisfaction factors and cost-effectiveness has been developed. The bi-objective nurse scheduling model with cost and job satisfaction objectives is formulated. The proposed model aims to improve nurses' fairness perception about the distribution of workload and preferred shifts and days off. Management can also examine the cost of fairness enhancement and make decisions accordingly. This study also validates the practicality of the proposed model by providing an analysis of an illustrative case study of an emergency department (ED) at a large-scale public hospital in Thailand. Based on the result, it can be seen that the proposed model can provide a more cost-effective and fair scheduling solution compared to manual scheduling. With the solving time of less than a minute, the scheduling can also be much more responsive to any last-minute changes to the input parameters. This research is the preliminary step in formulating a nurse scheduling-and-rescheduling framework. The improving direction is to account for uncertainties related to important scheduling parameters such as demand and absenteeism of nurses.

Acknowledgement. First, the authors would like to express our gratitude to the Thammasat University Hospital for granting permission to collect data. Second, we would like to thank the head nurse and nurses working in the emergency department for facilitating the data collection and participating in the field survey. Last, this study was supported by Thammasat University Research Fund, Contract No. TUFT 052/2563.

References

1. Min, A., Hong, H.C., Son, S., Lee, T.: Sleep, fatigue and alertness during working hours among rotating-shift nurses in Korea: an observational study. J. Nurs. Manag. **29**(8), 2647–2657 (2021)
2. Ferri, P., Guadi, M., Marcheselli, L., Balduzzi, S., Magnani, D., Di Lorenzo, R.: The impact of shift work on the psychological and physical health of nurses in a general hospital: a comparison between rotating night shifts and day shifts. Risk Manag. Healthcare Policy **9**, 203–211 (2016)
3. Navajas-Romero, V., Ariza-Montes, A., Hernández-Perlines, F.: Analyzing the job demands-control-support model in work-life balance: a study among nurses in the European context. Int. J. Environ. Res. Public Health **17**(8), 2847 (2020)
4. Lee, E., Jang, I.: Nurses' fatigue, job stress, organizational culture, and turnover intention: a culture-work-health model. West. J. Nurs. Res. **42**(2), 108–116 (2020)
5. Gebregziabher, D., Berhanie, E., Berihu, H., Belstie, A., Teklay, G.: The relationship between job satisfaction and turnover intention among nurses in Axum comprehensive and specialized hospital Tigray, Ethiopia. BMC Nurs. **19**(1), 79 (2020)
6. Shin, Y., Park, S.H., Kim, J.K.: A study on relationship among organizational fairness, motivation, job satisfaction, intention to stay of nurses. J. Korea Contents Assoc. **14**(10), 596–609 (2014)

7. Dewi, N.M.U.K., Januraga, P.P., Suarjana, K.: The relationship between nurse job satisfaction and turnover intention: a private hospital case study in Bali, Indonesia. In: Proceedings of the 4th International Symposium on Health Research (ISHR 2019). Atlantis Press, Paris, France (2020)
8. Fasbender, U., Van der Heijden, B.I.J.M., Grimshaw, S.: Job satisfaction, job stress and nurses' turnover intentions: the moderating roles of on-the-job and off-the-job embeddedness. J. Adv. Nurs. **75**(2), 327–337 (2019)
9. Iqbal, S., Iram, M.: Determinants of medication errors among nurses in public sector. Pak. J. Nurs. Midwifery **2**(1), 271–276 (2018)
10. Albashayreh, A., Al Sabei, S.D., Al-Rawajfah, O.M., Al-Awaisi, H.: Healthy work environments are critical for nurse job satisfaction: implications for Oman. Int. Nurs. Rev. **66**(3), 389–395 (2019)
11. Rizany, I., Hariyati, R.T.S., Afifah, E., Rusdiyansyah: The impact of nurse scheduling management on nurses' job satisfaction in army hospital: a cross-sectional research. SAGE Open **9**(2), 1–9 (2019). https://doi.org/10.1177/2158244019856189
12. Choi, S., Kim, M.: Effects of structural empowerment and professional governance on autonomy and job satisfaction of the Korean nurses. J. Nurs. Manag. **27**(8), 1664–1672 (2019)
13. Topbaş, E., et al.: The effect of perceived organisational justice on job satisfaction and burnout levels of haemodialysis nurses. J. Ren. Care **45**(2), 120–128 (2019)
14. Zanda, S., Zuddas, P., Seatzu, C.: Long term nurse scheduling via a decision support system based on linear integer programming: a case study at the University Hospital in Cagliari. Comput. Ind. Eng. **126**(September), 337–347 (2018)
15. Svirsko, A.C., Norman, B.A., Rausch, D., Woodring, J.: Using mathematical modeling to improve the emergency department nurse-scheduling process. J. Emerg. Nurs. **45**(4), 425–432 (2019)
16. Burke, E.K., Li, J., Qu, R.: A Pareto-based search methodology for multi-objective nurse scheduling. Ann. Oper. Res. **196**(1), 91–109 (2012)
17. Wright, P.D., Mahar, S.: Centralized nurse scheduling to simultaneously improve schedule cost and nurse satisfaction. Omega (United Kingdom) **41**(6), 1042–1052 (2013)
18. Lin, R.C., Sir, M.Y., Sisikoglu, E., Pasupathy, K., Steege, L.M.: Optimal nurse scheduling based on quantitative models of work-related fatigue. IIE Trans. Healthcare Syst. Eng. **3**(1), 23–38 (2013)
19. Becker, T., Steenweg, P.M., Werners, B.: Cyclic shift scheduling with on-call duties for emergency medical services. Health Care Manag. Sci. **22**(4), 676–690 (2018). https://doi.org/10.1007/s10729-018-9451-9
20. Huang, L., Ye, C., Gao, J., Shih, P.C., Mngumi, F., Mei, X.: Personnel scheduling problem under hierarchical management based on intelligent algorithm. Complexity **2021**, 14 p. (2021). https://doi.org/10.1155/2021/6637207. Article ID 6637207
21. Hamid, M., Tavakkoli-Moghaddam, R., Golpaygani, F., Vahedi-Nouri, B.: A multi-objective model for a nurse scheduling problem by emphasizing human factors. Proc. Inst. Mech. Eng. Part H: J. Eng. Med. **234**, 179–199 (2020)
22. Michael, C., Jeffery, C., David, C.: Nurse preference rostering using agents and iterated local search. Ann. Oper. Res. **226**(1), 443–461 (2014). https://doi.org/10.1007/s10479-014-1701-8
23. Youssef, A., Senbel, S.: A bi-level heuristic solution for the nurse scheduling problem based on shift-swapping. In: 2018 IEEE 8th Annual Computing and Communication Workshop and Conference (CCWC), pp. 72–78. IEEE, January 2018

24. Lin, C.C., Kang, J.R., Chiang, D.J., Chen, C.L.: Nurse scheduling with joint normalized shift and day-off preference satisfaction using a genetic algorithm with immigrant scheme. Int. J. Distrib. Sens. Netw. **2015**, 10 p. (2015). https://doi.org/10.1155/2015/595419. Article ID 595419

25. Rerkjirattikal, P., Huynh, V.N., Olapiriyakul, S., Supnithi, T.: A goal programming approach to nurse scheduling with individual preference satisfaction. Math. Prob. Eng. **2020**, 11 p. (2020). https://doi.org/10.1155/2020/2379091. Article ID 2379091

26. Rerkjirattikal, P., Olapiriyakul, S.: Overtime assignment and job satisfaction in noise-safe job rotation scheduling. In: Seki, H., Nguyen, C.H., Huynh, V.-N., Inuiguchi, M. (eds.) IUKM 2019. LNCS (LNAI), vol. 11471, pp. 26–37. Springer, Cham (2019). https://doi.org/10.1007/978-3-030-14815-7_3

27. Soriano, J., Jalao, E.R., Martinez, I.A.: Integrated employee scheduling with known employee demand, including breaks, overtime, and employee preferences. J. Ind. Eng. Manag. **13**(3), 451 (2020)

28. Nasiri, M.M., Rahvar, M.: A two-step multi-objective mathematical model for nurse scheduling problem considering nurse preferences and consecutive shifts. Int. J. Serv. Oper. Manag. **27**(1), 83–101 (2017)

29. Ang, B.Y., Lam, S.W.S., Pasupathy, Y., Ong, M.E.H.: Nurse workforce scheduling in the emergency department: a sequential decision support system considering multiple objectives. J. Nurs. Manag. **26**(4), 432–441 (2018)

30. Chiaramonte, M., Caswell, D.: Rerostering of nurses with intelligent agents and iterated local search. IIE Trans. Healthcare Syst. Eng. **6**(4), 213–222 (2016)

31. Fügener, A., Pahr, A., Brunner, J.O.: Mid-term nurse rostering considering cross-training effects. Int. J. Prod. Econ. **196**, 176–187 (2018)

32. Hamid, M., Barzinpour, F., Hamid, M., Mirzamohammadi, S.: A multi-objective mathematical model for nurse scheduling problem with hybrid DEA and a ugmented ϵ-constraint method?: A case study. J. Ind. Syst. Eng. **11**, 98–108 (2018)

33. Petrovic, S.: "You have to get wet to learn how to swim" applied to bridging the gap between research into personnel scheduling and its implementation in practice. Ann. Oper. Res. **275**(1), 161–179 (2019)

34. Di Martinelly, C., Meskens, N.: A bi-objective integrated approach to building surgical teams and nurse schedule rosters to maximise surgical team affinities and minimise nurses'. Int. J. Prod. Econ. **191**, 323–334 (2017)

Analysis of Medical Data Using Interval Estimators for Common Mean of Gaussian Distributions with Unknown Coefficients of Variation

Warisa Thangjai[1] , Sa-Aat Niwitpong[2] , and Suparat Niwitpong[2]([✉])

[1] Department of Statistics, Faculty of Science, Ramkhamhaeng University,
Bangkok 10240, Thailand
[2] Department of Applied Statistics, Faculty of Applied Science, King Mongkut's
University of Technology North Bangkok, Bangkok 10800, Thailand
{sa-aat.n,suparat.n}@sci.kmutnb.ac.th

Abstract. The common mean of Gaussian distributions is a parameter of interest when analyzing medical data. In practice, the population coefficient of variation (CV) is unknown because the population mean and variance are unknown. In this study, the common mean of Gaussian distributions with unknown CVs is considered and four new interval estimators for it using generalized confidence interval (GCI), large sample (LS), adjusted method of variance estimates recovery (adjusted MOVER), and standard bootstrap (SB) approaches are proposed. Furthermore, the proposed interval estimators are compared with a previously reported one based on the GCI approach. Monte Carlo simulation was used to evaluate the performances of the interval estimators based on their coverage probabilities and average lengths, while, medical datasets were used to illustrate the efficacy of these approaches. Our findings show that the interval estimator based on the GCI approach for the common mean of Gaussian distributions with unknown CVs provided the best performance in terms of coverage probability for all sample sizes. However, the adjusted MOVER and SB approaches can be considered as an alternative when the sample size is large ($n_i \geq 100$).

Keywords: Adjusted MOVER approach · CV · GCI approach · Mean · SB approach

1 Introduction

The population coefficient of variation (CV), which is free from a unit of measurement, is defined as the ratio of the population standard deviation to the population mean, $\tau = \sigma/\mu$, and has been widely applied in many fields, e.g., agriculture, biology, and environmental and physical sciences. Estimating a known CV has been suggested by many scholars. For example, Gerig and Sen [1] used Canadian migratory bird survey data from 1969 and 1970 while assuming that the CV for

© Springer Nature Switzerland AG 2022
K. Honda et al. (Eds.): IUKM 2022, LNAI 13199, pp. 155–166, 2022.
https://doi.org/10.1007/978-3-030-98018-4_13

each province was known. Meanwhile, estimating the mean of Gaussian distributions with a known CV has also been studied extensively (e.g., Searls [2] and Niwitpong [3]). However, the CV is unknown when the population mean and variance have been estimated and thus, needs to be estimated too. Srivastava [4] and Srivastava [5] proposed an estimator for the normal population mean with an unknown CV and indicated that it is more efficient than a previously reported sample mean estimator. He later presented a uniformly minimum variance unbiased estimator of the efficiency ratio and compared its usefulness to estimate an unknown CV with an existing estimator (Srivastava and Singh [6]). Sahai [7] provided an estimator for the normal mean with unknown CV and studied it along the same lines as Srivastava [4] and Srivastava [5] estimators. Meanwhile, Thangjai et al. [8] presented confidence intervals for the normal mean and the difference between two normal means with unknown CVs. In addition, Thangjai et al. [9] proposed the Bayesian confidence intervals for means of normal distributions with unknown CVs.

In practice, samples are collected at different time points, and the problem of estimating common parameters under these circumstances has been widely studied by several researchers. Krishnamoorthy and Lu [10] proposed the generalized variable approach for inference on the common mean of normal distributions. Lin and Lee [11] developed a new generalized pivotal quantity based on the best linear unbiased estimator for constructing confidence intervals for the common mean of normal distributions. Tian [12] presented procedures for inference on the common CV of normal distributions. Tian and Wu [13] provided the generalized variable approach for inference on the common mean of log-normal distributions. Thangjai et al. [14] investigated a new confidence interval for the common mean of normal distributions using the adjusted method of variance estimates recovery (adjusted MOVER) approach. Finally, the estimator of Srivastava [4] is well established for constructing confidence intervals for the common mean of Gaussian distributions with unknown CVs.

Interval estimators for the common mean of Gaussian distributions with unknown CVs have been proposed in several medical science studies, such as the common percentage of albumin in human plasma proteins from four sources (Jordan and Krishnamoorthy [15]) and quality assurance in medical laboratories for the diagnostic determinations of hemoglobin, red blood cells, the mean corpuscular volume, hematocrit, white blood cells, and platelets in normal and abnormal blood samples (Tian [12] and Fung and Tsang [16]).

Herein, the concepts in Thangjai et al. [8] and Thangjai and Niwitpong [17] are extended to k populations to construct new interval estimators for the common mean of Gaussian distributions with unknown CVs. The approaches to construct these interval estimators: the generalized confidence interval (GCI), large sample (LS), adjusted MOVER, and standard bootstrap (SB) are compared with the GCI approach of Lin and Lee [11]. The GCI approach first introduced by Weerahandi [18] has been used successfully to construct interval estimators (e.g., Krishnamoorthy and Lu [10], Lin and Lee [11], Tian [12], Tian and Wu [13], Ye et al. [19]). The LS approach using the central limit theorem (along with a GCI approach) was first proposed by Tian and Wu [13] to construct confidence intervals for the common mean of log-normal distributions. The adjusted MOVER approach motivated by Zou and Donner [20] and Zou et al. [21] was extended by Thangjai et al. [14] and Thangjai and Niwitpong [17] to construct an interval estimator for a common parameter.

2   Preliminaries

In this section, the lemma and theorem are explained to estimate the interval estimators for common Gaussian mean with unknown CVs.

Let $X = (X_1, X_2, ..., X_n)$ be a random variable from the Gaussian distribution with mean μ and variance σ^2. The population CV is $\tau = \sigma/\mu$. Let $\bar{X}$ and S^2 be sample mean and sample variance for X, respectively. The CV estimator is $\hat{\tau} = S/\bar{X}$.

Following Srivastava [4] and Thangjai et al. [8], the Gaussian mean estimator when the CV is unknown, $\hat{\theta}$, is

$$\hat{\theta} = \frac{n\bar{X}}{n + \frac{S^2}{\bar{X}^2}}. \tag{1}$$

According to Thangjai et al. [8], the mean and variance of the mean estimator with unknown CV are

$$E\left(\hat{\theta}\right) = \left(\frac{\mu}{1 + \left(\frac{\sigma^2}{n\mu^2 + \sigma^2}\right)\left(1 + \frac{2\sigma^4 + 4n\mu^2\sigma^2}{(n\mu^2 + \sigma^2)^2}\right)}\right)$$
$$* \left(1 + \frac{\left(\frac{n\sigma^2}{n\mu^2 + \sigma^2}\right)^2 \left(\frac{2}{n} + \frac{2\sigma^4 + 4n\mu^2\sigma^2}{(n\mu^2 + \sigma^2)^2}\right)}{\left(n + \left(\frac{n\sigma^2}{n\mu^2 + \sigma^2}\right)\left(1 + \frac{2\sigma^4 + 4n\mu^2\sigma^2}{(n\mu^2 + \sigma^2)^2}\right)\right)^2}\right) \tag{2}$$

and

$$Var\left(\hat{\theta}\right) = \left(\frac{\mu}{1 + \left(\frac{\sigma^2}{n\mu^2 + \sigma^2}\right)\left(1 + \frac{2\sigma^4 + 4n\mu^2\sigma^2}{(n\mu^2 + \sigma^2)^2}\right)}\right)^2$$
$$* \left(\frac{\sigma^2}{n\mu^2} + \frac{\left(\frac{n\sigma^2}{n\mu^2 + \sigma^2}\right)^2 \left(\frac{2}{n} + \frac{2\sigma^4 + 4n\mu^2\sigma^2}{(n\mu^2 + \sigma^2)^2}\right)}{\left(n + \left(\frac{n\sigma^2}{n\mu^2 + \sigma^2}\right)\left(1 + \frac{2\sigma^4 + 4n\mu^2\sigma^2}{(n\mu^2 + \sigma^2)^2}\right)\right)^2}\right). \tag{3}$$

Consider k independent Gaussian distributions with a common mean with unknown CVs. Let $X_i = (X_{i1}, X_{i2}, ..., X_{in_i})$ be a random variable from the i-th Gaussian distribution with the common mean μ and possibly unequal variances σ_i^2 as follows: $X_{ij} \sim N(\mu, \sigma_i^2)$; $i = 1, 2, ..., k$, $j = 1, 2, ..., n_i$.

For the i-th sample, let $\bar{X}_i$ and $\bar{x}_i$ be sample mean and observed sample mean of X_i, respectively. And let S_i^2 and s_i^2 be sample variance and observed sample variance of X_i, respectively. According to Thangjai et al. [8], the estimator of Srivastava [4] is well established. The estimator is given by

$$\hat{\theta}_i = \frac{n_i \bar{X}_i}{n_i + \frac{S_i^2}{(\bar{X}_i)^2}}; i = 1, 2, ..., k. \tag{4}$$

This paper is interested in constructing confidence intervals for the common Gaussian mean with unknown CVs, based on Graybill and Deal [22], defined as follows:

$$\hat{\theta} = \sum_{i=1}^{k} \frac{\hat{\theta}_i}{\widetilde{Var}\left(\hat{\theta}_i\right)} \Bigg/ \sum_{i=1}^{k} \frac{1}{\widetilde{Var}\left(\hat{\theta}_i\right)}, \tag{5}$$

where $\widetilde{Var}\left(\hat{\theta}_i\right)$ denotes the estimator of $Var\left(\hat{\theta}_i\right)$ which is defined in Eq. (3) with μ_i and σ_i^2 replaced by $\bar{x}_i$ and s_i^2, respectively.

2.1   GCI

Definition 1. Let $X = (X_1, X_2, ..., X_n)$ be a random variable from a distribution $F(x|\delta)$, where $x = (x_1, x_2, ..., x_n)$ be an observed sample, $\delta = (\theta, \nu)$ is a unknown parameter vector, θ is a parameter of interest, and ν is a nuisance parameters. Let $R = R(X; x, \delta)$ be a function of X, x and δ. The random quantity R is called a generalized pivotal quantity if it satisfies the following two properties; see Weerahandi [18]:

(i) The probability distribution of R is free of unknown parameters.
(ii) The observed value of R does not depend on the vector of nuisance parameters.

The $100(\alpha/2)$-th and $100(1-\alpha/2)$-th percentiles of R are the lower and upper limits of $100(1-\alpha)\%$ two-sided GCI.

Following Thangjai et al. [8], the generalized pivotal quantities of σ_i^2, μ_i, and θ_i based on the i-th sample are defined as follows:

$$R_{\sigma_i^2} = \frac{(n_i - 1) s_i^2}{V_i}. \tag{6}$$

$$R_{\mu_i} = \bar{x}_i - \frac{Z_i}{\sqrt{U_i}} \sqrt{\frac{(n_i - 1) s_i^2}{n_i}} \tag{7}$$

and

$$R_{\theta_i} = \frac{n_i R_{\mu_i}}{n_i + \frac{R_{\sigma_i^2}}{(R_{\mu_i})^2}}, \tag{8}$$

where V_i denotes a chi-squared distribution with $n_i - 1$ degrees of freedom, Z_i denotes a standard normal distribution, and U_i denotes a chi-squared distribution with $n_i - 1$ degrees of freedom.

According to Tian and Wu [13], the generalized pivotal quantity for the common Gaussian mean with unknown CVs is a weighted average of the generalized pivotal quantity. That is given by

$$R_\theta = \sum_{i=1}^{k} \frac{R_{\theta_i}}{R_{Var(\hat{\theta}_i)}} \Big/ \sum_{i=1}^{k} \frac{1}{R_{Var(\hat{\theta}_i)}}, \tag{9}$$

where $R_{Var(\hat{\theta}_i)}$ is defined in Eq. (3) with μ_i and σ_i^2 replaced by R_{μ_i} and $R_{\sigma_i^2}$, respectively.

Hence, the R_θ is the generalized pivotal quantity for θ and is satisfied the conditions (i) and (ii) in Definition 1. Then the common Gaussian mean with unknown CVs can be constructed from R_θ.

Therefore, the $100(1 - \alpha)\%$ two-sided confidence interval for the common Gaussian mean with unknown CVs based on the GCI approach is

$$CI_{GCI} = [L_{GCI}, U_{GCI}] = [R_\theta(\alpha/2), R_\theta(1 - \alpha/2)], \tag{10}$$

where $R_\theta(\alpha/2)$ and $R_\theta(1 - \alpha/2)$ denote the $100(\alpha/2)$-th and $100(1 - \alpha/2)$-th percentiles of R_θ, respectively.

2.2 LS Confidence Interval

According to Graybill and Deal [22] and Tian and Wu [13], the LS estimate of the Gaussian mean with unknown CV is a pooled estimated estimator of the Gaussian mean with unknown CV defined as in Eq. (5), where $\hat{\theta}_i$ is defined in Eq. (4) and $\widetilde{Var}\left(\hat{\theta}_i\right)$ denotes the estimator of $Var\left(\hat{\theta}_i\right)$ which is defined in Eq. (3) with μ_i and σ_i^2 replaced by $\bar{x}_i$ and s_i^2, respectively.

The distribution of $\hat{\theta}$ is approximately Gaussian distribution when the sample size is large. Then the quantile of the Gaussian distribution is used to construct confidence interval for θ. Therefore, the $100(1 - \alpha)\%$ two-sided confidence interval for the common Gaussian mean with unknown CVs based on the LS approach is

$$CI_{LS} = [L_{LS}, U_{LS}]$$
$$= [\hat{\theta} - z_{1-\alpha/2}\sqrt{1 \Big/ \sum_{i=1}^{k} \frac{1}{\widetilde{Var}\left(\hat{\theta}_i\right)}}, \hat{\theta} + z_{1-\alpha/2}\sqrt{1 \Big/ \sum_{i=1}^{k} \frac{1}{\widetilde{Var}\left(\hat{\theta}_i\right)}}], \tag{11}$$

where $z_{1-\alpha/2}$ denotes the $(1 - \alpha/2)$-th quantile of the standard normal distribution.

2.3   Adjusted MOVER Confidence Interval

Now recall that Z is a standard normal distribution with the mean 0 and variance 1, defined as follows:

$$Z = \frac{\bar{X} - \mu}{\sqrt{\widehat{Var}\left(\hat{\theta}\right)}} \sim N(0,1). \tag{12}$$

The confidence interval for mean of Gaussian distribution is

$$CI_\mu = [l, u] = [\bar{x} - z_{1-\alpha/2}\sqrt{\widehat{Var}\left(\hat{\theta}\right)}, \bar{x} + z_{1-\alpha/2}\sqrt{\widehat{Var}\left(\hat{\theta}\right)}]. \tag{13}$$

For $i = 1, 2, ..., k$, the lower limit (l_i) and upper limit (u_i) for the normal mean μ_i based on the i-th sample can be defined as

$$l_i = \bar{x}_i - z_{1-\alpha/2}\sqrt{\widehat{Var}\left(\hat{\theta}_i\right)} \tag{14}$$

and

$$u_i = \bar{x}_i + z_{1-\alpha/2}\sqrt{\widehat{Var}\left(\hat{\theta}_i\right)}, \tag{15}$$

where $\widehat{Var}\left(\hat{\theta}_i\right)$ denotes the estimator of $Var\left(\hat{\theta}_i\right)$ which is defined in Eq. (3) and $z_{1-\alpha/2}$ denotes the $(1-\alpha/2)$-th quantile of the standard normal distribution.

According to Thangjai et al. [14] and Thangjai and Niwitpong [17], the common mean with unknown CVs is weighted average of the mean with unknown CV $\hat{\theta}_i$ based on k individual samples. The common mean with unknown CVs has the following form

$$\hat{\theta} = \sum_{i=1}^{k} \frac{\hat{\theta}_i}{\widehat{Var}\left(\hat{\theta}_i\right)} \bigg/ \sum_{i=1}^{k} \frac{1}{\widehat{Var}\left(\hat{\theta}_i\right)}, \tag{16}$$

where $\hat{\theta}_i$ is defined in Eq. (4), $\widehat{Var}\left(\hat{\theta}_i\right) = \frac{1}{2}(\frac{(\hat{\theta}_i - l_i)^2}{z_{\alpha/2}^2} + \frac{(u_i - \hat{\theta}_i)^2}{z_{\alpha/2}^2})$, and l_i and u_i are defined in Eqs. (14) and (15), respectively.

Therefore, the $100(1 - \alpha)\%$ two-sided confidence interval for the common Gaussian mean with unknown CVs based on the adjusted MOVER approach is

$$CI_{AM} = [L_{AM}, U_{AM}]$$

$$= [\hat{\theta} - z_{1-\alpha/2}\sqrt{1 \bigg/ \sum_{i=1}^{k} \frac{z_{\alpha/2}^2}{(\hat{\theta}_i - l_i)^2}}, \hat{\theta} + z_{1-\alpha/2}\sqrt{1 \bigg/ \sum_{i=1}^{k} \frac{z_{\alpha/2}^2}{(u_i - \hat{\theta}_i)^2}}], \tag{17}$$

where $\hat{\theta}$ is defined in Eq. (16), and $z_{\alpha/2}$ and $z_{1-\alpha/2}$ denote the $(\alpha/2)$-th and $(1 - \alpha/2)$-th quantiles of the standard normal distribution, respectively.

2.4 SB Confidence Interval

Let $X_i^* = (X_{i1}^*, X_{i2}^*, ..., X_{in_i}^*)$ be a bootstrap sample with replacement from $X_i = (X_{i1}, X_{i2}, ..., X_{in_i})$ and let $\bar{X}_i^*$ and S_i^{2*} be mean and variance of X_i^*, respectively. Let $x_i^* = (x_{i1}^*, x_{i2}^*, ..., x_{in_i}^*)$ be an observed value of $X_i^* = (X_{i1}^*, X_{i2}^*, ..., X_{in_i}^*)$ and let $\bar{x}_i^*$ and s_i^{2*} be mean and variance of x_i^*, respectively. The estimates of $\hat{\theta}_i^*$ and $Var(\hat{\theta}_i^*)$ are

$$\hat{\theta}_i^* = \frac{n_i \bar{X}_i^*}{n_i + \frac{S_i^{2*}}{(\bar{X}_i^*)^2}} \tag{18}$$

and

$$Var\left(\hat{\theta}_i^*\right) = \left(\frac{\mu_i^*}{1 + \left(\frac{\sigma_i^{2*}}{n_i(\mu_i^*)^2 + \sigma_i^{2*}}\right)\left(1 + \frac{2\sigma_i^{4*} + 4n_i(\mu_i^*)^2\sigma_i^{2*}}{(n_i(\mu_i^*)^2 + \sigma_i^{2*})^2}\right)}\right)^2$$
$$* \left(\frac{\sigma_i^{2*}}{n_i(\mu_i^*)^2} + \frac{\left(\frac{n_i\sigma_i^{2*}}{n_i(\mu_i^*)^2 + \sigma_i^{2*}}\right)^2\left(\frac{2}{n_i} + \frac{2\sigma_i^{4*} + 4n_i(\mu_i^*)^2\sigma_i^{2*}}{(n_i(\mu_i^*)^2 + \sigma_i^{2*})^2}\right)}{\left(n_i + \left(\frac{n_i\sigma_i^{2*}}{n_i(\mu_i^*)^2 + \sigma_i^{2*}}\right)\left(1 + \frac{2\sigma_i^{4*} + 4n_i(\mu_i^*)^2\sigma_i^{2*}}{(n_i(\mu_i^*)^2 + \sigma_i^{2*})^2}\right)\right)^2}\right) .\tag{19}$$

According to Graybill and Deal [22], the common Gaussian mean with unknown CVs is a pooled estimated unbiased estimator of the Gaussian mean with unknown CVs based on k individual samples. The common Gaussian mean with unknown CVs is defined by

$$\hat{\theta}^* = \sum_{i=1}^{k} \frac{\hat{\theta}_i^*}{\widetilde{Var}\left(\hat{\theta}_i^*\right)} \bigg/ \sum_{i=1}^{k} \frac{1}{\widetilde{Var}\left(\hat{\theta}_i^*\right)}, \tag{20}$$

where $\hat{\theta}_i^*$ is defined in Eq. (18) and $\widetilde{Var}\left(\hat{\theta}_i^*\right)$ is the estimator of $Var\left(\hat{\theta}_i^*\right)$ which is defined in Eq. (19) with μ_i^* and σ_i^{2*} replaced by $\bar{x}_i^*$ and s_i^{2*}, respectively.

The B bootstrap statistics are used to construct the sampling distribution for estimating the confidence interval for the common Gaussian mean with unknown CVs. Therefore, the $100(1 - \alpha)\%$ two-sided confidence interval for the common Gaussian mean with unknown CVs based on the SB approach is

$$CI_{SB} = [L_{SB}, U_{SB}] = [\bar{\hat{\theta}}^* - z_{1-\alpha/2}S_{\hat{\theta}*}, \bar{\hat{\theta}}^* + z_{1-\alpha/2}S_{\hat{\theta}*}], \tag{21}$$

where $\bar{\hat{\theta}}^*$ and $S_{\hat{\theta}*}$ are the mean and standard deviation of $\hat{\theta}^*$ defined in Eq. (20) and $z_{1-\alpha/2}$ denotes the $100(1 - \alpha/2)$-th percentile of the standard normal distribution.

Next, we briefly review the GCI of Lin and Lee [11] for the common mean of Gaussian distributions. The generalized pivotal quantity based on the best linear un-biased estimator for the common Gaussian mean μ is

$$R_\mu = \frac{\sum_{i=1}^{k} \frac{n_i \bar{x}_i U_i}{v_i} - Z\sqrt{\sum_{i=1}^{k} \frac{n_i U_i}{v_i}}}{\sum_{i=1}^{k} \frac{n_i U_i}{v_i}}, \tag{22}$$

where Z denotes the standard normal distribution, U_i denotes a chi-squared distribution with $n_i - 1$ degrees of freedom, and $v_i = (n_i - 1)s_i^2$.

Therefore, the $100(1 - \alpha)\%$ two-sided confidence interval for the common Gaussian mean based on the GCI approach of Lin and Lee [20] is

$$C_{LL} = [L_{LL}, U_{LL}] = [R_\mu(\alpha/2), R_\mu(1 - \alpha/2)], \tag{23}$$

where $R_\mu(\alpha/2)$ and $R_\mu(1 - \alpha/2)$ denote the $100(\alpha/2)$-th and $100(1 - \alpha/2)$-th percentiles of R_μ, respectively.

3   Simulation Studies

Monte Carlo simulation was used to estimate the coverage probabilities (CPs) and the average lengths (ALs) of all confidence intervals; those constructed via the GCI, LS, adjusted MOVER, and SB approaches are denoted as CI_{GCI}, CI_{LS}, CI_{AM}, and CI_{SB}, respectively, while the GCI of Lin and Lee [20] is denoted as CI_{LL}. The CP of the $100(1 - \alpha)\%$ confidence level is $c \pm z_{\alpha/2}\sqrt{\frac{c(1-c)}{M}}$, where c is the nominal confidence level and M is the number of simulation runs. At the 95% confidence level, the best performing confidence interval will have a CP in the range $[0.9440, 0.9560]$ with the shortest AL.

Each confidence interval was evaluated at the nominal confidence level of 0.95. The number of populations $k = 2$; and the sample sizes within each population n_1 and n_2 were given in the following table. Without loss of generality (Thangjai et al. [8]), the common mean of Gaussian data within each population was $\mu = 1.0$. The population standard deviations were set at $\sigma_1 = 0.5, 1.0, 1.5, 2.0$ and $\sigma_2 = 1.0$. The CVs were computed by $\tau_i = \sigma_i/\mu$, where $i = 1, 2$. Hence, the ratio of τ_1 to τ_2 was reduced to σ_1/σ_2.

The result of simulations with the number of simulation runs $M = 5,000$ is reported in Table 1. Only CI_{GCI} obtained CPs greater than 0.95 in all cases whereas those of CI_{LS}, CI_{AM}, CI_{SB}, and CI_{LL} were under 0.95. However, the CPs of CI_{LS}, CI_{AM}, and CI_{SB} increased and became close to 0.95 when the sample size was increased. For $n_i \leq 50$, the CPs of CI_{LS}, CI_{AM}, and CI_{SB} tended to decrease when σ_1/σ_2 increased. Moreover, the CPs of CI_{GCI} did not change when σ_1/σ_2 was varied. Hence, CI_{GCI} is preferable for most cases, while CI_{AM} and CI_{SB}, which are easy to use in practice, can be used when the sample size is large ($n_i \geq 100$).

As the sample case (k) increased, CI_{GCI} is preferable when the sample size is small. For a large sample size, CI_{GCI}, CI_{AM}, CI_{SB}, and CI_{LL} performed similarly in terms of CP but the ALs of the CI_{AM}, CI_{SB}, and CI_{LL} were shorter than that of CI_{GCI}.

Table 1. The CPs and ALs of 95% two-sided confidence intervals for the common mean of Gaussian distributions with unknown CVs: 2 sample cases.

n_1	n_2	μ	σ_1/σ_2	CP (AL)				
				CI_{GCI}	CI_{LS}	CI_{AM}	CI_{SB}	CI_{LL}
30	30	1.0	0.5	**0.9532**	0.9338	0.9432	0.9412	0.9450
				(0.3586)	(0.3176)	(0.3313)	(0.3299)	(0.3277)
			1.0	**0.9608**	0.9350	**0.9506**	0.9476	0.9460
				(0.6305)	(0.5297)	(0.5526)	(0.5498)	(0.5136)
			1.5	**0.9580**	0.9232	0.9456	**0.9526**	0.9492
				(0.7689)	(0.6473)	(0.6744)	(0.7130)	(0.6065)
			2.0	**0.9574**	0.8822	0.9140	0.9480	0.9474
				(0.8126)	(0.6707)	(0.6953)	(0.8183)	(0.6563)
50	50	1.0	0.5	0.9496	0.9388	0.9478	0.9418	0.9460
				(0.2706)	(0.2474)	(0.2536)	(0.2501)	(0.2515)
			1.0	**0.9642**	0.9480	**0.9562**	0.9474	0.9480
				(0.4506)	(0.4075)	(0.4177)	(0.4061)	(0.3962)
			1.5	**0.9524**	0.9426	**0.9544**	0.9496	0.9430
				(0.5505)	(0.5020)	(0.5145)	(0.5110)	(0.4658)
			2.0	**0.9508**	0.9116	0.9340	**0.9506**	0.9428
				(0.6042)	(0.5269)	(0.5388)	(0.6109)	(0.5024)
100	100	1.0	0.5	**0.9540**	0.9478	**0.9512**	0.9490	**0.9504**
				(0.1882)	(0.1751)	(0.1772)	(0.1755)	(0.1764)
			1.0	**0.9588**	**0.9520**	**0.9554**	0.9498	0.9492
				(0.2980)	(0.2832)	(0.2867)	(0.2806)	(0.2786)
			1.5	**0.9506**	**0.9570**	**0.9586**	0.9496	0.9494
				(0.3623)	(0.3494)	(0.3536)	(0.3363)	(0.3279)
			2.0	**0.9568**	**0.9550**	**0.9622**	**0.9580**	**0.9506**
				(0.3892)	(0.3748)	(0.3793)	(0.3831)	(0.3536)
200	200	1.0	0.5	**0.9500**	0.9476	**0.9500**	0.9476	0.9482
				(0.1322)	(0.1239)	(0.1247)	(0.1240)	(0.1244)
			1.0	0.9498	0.9478	0.9498	0.9462	0.9474
				(0.2030)	(0.1984)	(0.1997)	(0.1972)	(0.1968)
			1.5	0.9488	0.9490	**0.9514**	0.9418	0.9428
				(0.2543)	(0.2415)	(0.2430)	(0.2334)	(0.2310)
			2.0	**0.9592**	**0.9616**	**0.9638**	**0.9542**	**0.9566**
				(0.2649)	(0.2625)	(0.2641)	(0.2547)	(0.2485)

4   Empirical Application

Empirical application of the proposed confidence intervals to real data were presented and compared with CI_{LL}.

The dataset reported by Fung and Tsang [16] and Tian [12] and used here comprises hemoglobin, red blood cells, the mean corpuscular volume, hematocrit, white blood cells, and platelet values in normal and abnormal blood samples collected by the Hong Kong Medical Technology Association in 1995 and 1996. The summary statistics for 1995 are $\bar{x}_1 = 84.1300$, $s_1^2 = 3.3900$, and $n_1 = 63$, and those for 1996 are $\bar{x}_2 = 85.6800$, $s_2^2 = 2.9460$, and $n_2 = 72$. The means of the Gaussian distributions with unknown CVs are $\hat{\theta}_1 = 84.1294$ and $\hat{\theta}_2 = 85.6795$ for 1995 and 1996, respectively, while the common mean of the Gaussian distributions with unknown CVs is $\hat{\theta} = 85.1962$.

The two datasets fit Gaussian distributions. The 95% two-sided confidence intervals for CI_{GCI}, CI_{LS}, CI_{AM}, and CI_{SB} were [84.1099,85.8884], [61.4262,108.9661], [60.9972,109.3992], and [84.4604,85.6013] with interval lengths of 1.7785, 47.5399, 48.4020, and 1.1409. For comparison, CI_{LL} provided [84.6502,85.3635] with an interval length of 0.7133. Thus, CI_{LL} had the shortest interval length, while CI_{SB} performed the best out of the proposed approaches as its interval length was shorter than those of the other three for $k = 2$.

Therefore, these results confirm our simulation study in the previous section in term of length. In simulation, the GCI of Lin and Lee [20] is the shortest average lengths, but the coverage probabilities are less than the nominal confidence level of 0.95. Furthermore, the coverage probability and length in this example are computed by using only one sample, whereas the coverage probability and average length in the simulation are computed by using 5,000 random samples. Therefore, the GCI of Lin and Lee [20] is not recommended to construct the confidence intervals for common mean of Gaussian distributions with unknown CVs.

5   Discussion and Conclusions

Thangjai et al. [8] proposed confidence intervals for the mean and difference of means of normal distributions with unknown coefficients of variation. In addition, Thangjai et al. [9] presented the Bayesian approach to construct the confidence intervals for means of normal distributions with unknown coefficients of variation. In this paper, we extend the work of Thangjai et al. [8,9] to construct confidence intervals for the common mean of k Gaussian distributions with unknown CVs.

Herein, GCI, LS, adjusted MOVER, and SB approaches to construct interval estimators for the common mean of Gaussian distributions with unknown CVs are presented. Their CPs and ALs were evaluated via a Monte Carlo simulation and compared with the confidence interval based on the GCI approach of Lin and Lee [11]. The results of the simulation studies indicate that the confidence intervals performed similarly based on their CPs for large sample sizes (i.e.,

$n_i \geq 100$). However, the CP of CI_{GCI} was more satisfactory than those of the other confidence intervals. Moreover, the CPs of CI_{AM}, CI_{SB}, and CI_{LL} were close to 0.95 and their ALs were slightly shorter than CI_{GCI} when the sample size was large (i.e., $n_i \geq 100$). Thus, CI_{AM} and CI_{SB} can be considered as an alternative to construct an interval estimator for the common mean of Gaussian distributions with unknown CVs when the sample size is large whereas CI_{LS} is not recommended for small sample sizes (i.e., $n_i < 100$) as its CP is below 0.95. Further research will be conducted to find other approaches for comparison.

Acknowledgments. This research was funded by Faculty of Applied Science, King Mongkut's University of Technology North Bangkok. No. 651118.

References

1. Gerig, T.M., Sen, A.R.: MLE in two normal samples with equal but unknown population coefficients of variation. J. Am. Stat. Assoc. **75**, 704–708 (1980)
2. Searls, D.T.: The utilization of a known coefficient of variation in the estimation procedure. J. Am. Stat. Assoc. **59**, 1225–1226 (1964)
3. Niwitpong, S.: Confidence intervals for the normal mean with a known coefficient of variation. Far East J. Math. Sci. **97**, 711–727 (2015)
4. Srivastava, V.K.: On the use of coefficient of variation in estimating mean. J. Indian Soc. Agric. Stat. **26**, 33–36 (1974)
5. Srivastava, V.K.: A note on the estimation of mean in normal population. Metrika **27**(1), 99–102 (1980). https://doi.org/10.1007/BF01893580
6. Srivastava, V.K., Singh, R.S.: Uniformly minimum variance unbiased estimator of efficiency ratio in estimation of normal population mean. Statist. Probab. Lett. **10**, 241–245 (1990)
7. Sahai, A.: On an estimator of normal population mean and UMVU estimation of its relative efficiency. Appl. Math. Comput. **152**, 701–708 (2004)
8. Thangjai, W., Niwitpong, S., Niwitpong, S.: Confidence intervals for mean and difference of means of normal distributions with unknown coefficients of variation. Mathematics **5**, 1–23 (2017)
9. Thangjai, W., Niwitpong, S.-A., Niwitpong, S.: Bayesian confidence intervals for means of normal distributions with unknown coefficients of variation. In: Huynh, V.-N., Entani, T., Jeenanunta, C., Inuiguchi, M., Yenradee, P. (eds.) IUKM 2020. LNCS (LNAI), vol. 12482, pp. 361–371. Springer, Cham (2020). https://doi.org/10.1007/978-3-030-62509-2_30
10. Krishnamoorthy, K., Lu, Y.: Inference on the common means of several normal populations based on the generalized variable method. Biometrics **59**, 237–247 (2003)
11. Lin, S.H., Lee, J.C.: Generalized inferences on the common mean of several normal populations. J. Stat. Plan. Inference **134**, 568–582 (2005)
12. Tian, L.: Inferences on the common coefficient of variation. Stat. Med. **24**, 2213–2220 (2005)
13. Tian, L., Wu, J.: Inferences on the common mean of several log-normal populations: the generalized variable approach. Biom. J. **49**, 944–951 (2007)
14. Thangjai, W., Niwitpong, S., Niwitpong, S.: Confidence intervals for the common mean of several normal populations. Stud. Computat. Intell. **692**, 321–331 (2017)

15. Jordan, S.J., Krishnamoorthy, K.: Exact confidence intervals for the common mean of several normal populations. Biometrics **52**, 77–86 (1996)
16. Fung, W.K., Tsang, T.S.: A simulation study comparing tests for the equality of coefficients of variation. Stat. Med. **17**, 2003–2014 (1998)
17. Thangjai, W., Niwitpong, S.: Confidence intervals for the weighted coefficients of variation of two-parameter exponential distributions. Cogent Math. **4**, 1–16 (2017)
18. Weerahandi, S.: Generalized confidence intervals. J. Am. Stat. Assoc. **88**, 899–905 (1993)
19. Ye, R.D., Ma, T.F., Wang, S.G.: Inferences on the common mean of several inverse Gaussian populations. Comput. Stat. Data Anal. **54**, 906–915 (2010)
20. Zou, G.Y., Donner, A.: Construction of confidence limits about effect measures: a general approach. Stat. Med. **27**, 1693–1702 (2008)
21. Zou, G.Y., Taleban, J., Hao, C.Y.: Confidence interval estimation for lognormal data with application to health economics. Comput. Stat. Data Anal. **53**, 3755–3764 (2009)
22. Graybill, F.A., Deal, R.B.: Combining unbiased estimators. Biometrics **15**, 543–550 (1959)

Pattern Classification and Data Analysis

On Some Fuzzy Clustering Algorithms for Time-Series Data

Mizuki Fujita[(✉)] and Yuchi Kanzawa[iD]

Shibaura Institute of Technology, 3-7-5 Toyosu, Koto, Tokyo 135-8548, Japan
{ma21131,kanzawa}@shibaura-it.ac.jp

Abstract. Various fuzzy clustering algorithms have been proposed for vectorial data. However, these methods have not been applied to time-series data. This paper presents three fuzzy clustering algorithms for time-series data based on dynamic time warping (DTW). The first algorithm involves Kullback–Leibler divergence regularization of the DTW k-means objective function. The second algorithm replaces the membership of the DTW k-means objective function with its power. The third algorithm involves q-divergence regularization of the objective function of the first algorithm. Theoretical discussion shows that the third algorithm is a generalization of the first and second algorithms, which is substantiated through numerical experiments.

Keywords: Time-series data · Fuzzy clustering · Dynamic time warping

1 Introduction

Hard c-means (HCM) is the most commonly used type of clustering algorithm [1]. The fuzzy c-means (FCM) [2] approach is an extension of the HCM that allows each object to belong to all or some of the clusters to varying degrees. To distinguish the general FCM method from other proposed, such as entropy-regularized FCM (EFCM) [3], it is referred to as the Bezdek-type FCM (BFCM) in this work. The above mentioned algorithms may misclassify some objects that should be assigned to a large cluster as belonging to a smaller cluster if the cluster sizes are not balanced. To overcome this problem, some approaches introduce variables to control the cluster sizes [4,5]. Such variables have been added to the BFCM and EFCM algorithms to derive the revised BFCM (RBFCM) and revised EFCM (REFCM) [6] algorithms, respectively.

In the aforementioned clustering algorithms, the dissimilariies between the objects and cluster centers are measured as the inner-product-induced squared distances. This measure cannot be used for time-series data because they vary over time. Dynamic time warping (DTW) is a representative dissimilarity with respect to time-series data. Hence, a clustering algorithm using the DTW is proposed [8] herein and referred to as the DTW k-means algorithm.

© Springer Nature Switzerland AG 2022
K. Honda et al. (Eds.): IUKM 2022, LNAI 13199, pp. 169–181, 2022.
https://doi.org/10.1007/978-3-030-98018-4_14

The accuracies that can be achieved with fuzzy clustered results are better than those using hard clustering. Various kinds of fuzzy clustering algorithms have been proposed in literature for vectorial data [2,3]. However, this is not true for time-series data, which is the main motivation for this study.

In this work, we propose three fuzzy clustering algorithms for time-series data. The first algorithm involves the Kullback–Leibler (KL) divergence regularization of the DTW k-means objective function, which is referred to as the KL-divergence-regularized fuzzy DTW c-means (KLFDTWCM); this approach is similar to the REFCM obtained by KL divergence regularization of the HCM objective function. In the second algorithm, the membership of the DTW k-means objective function is replaced with its power, which is referred to as the Bezdek-type fuzzy DTW c-means (BFDTWCM); this method is similar to the RBFCM, where the membership of the HCM objective function is replaced with its power. The third algorithm is obtained by q-divergence regularization of the objective function of the first algorithm (QFDTWCM). The theoretical results indicate that the QFDTWCM approach reduces to the BFDTWCM under a specific condition and to the KLFDTWCM under a different condition. Numerical experiments were performed using artificial datasets to substantiate these observations.

The remainder of this paper is organized as follows. Section 2 introduces the notations used herein and the background regarding some conventional algorithms. Section 3 describes the three proposed algorithms. Section 4 presents the procedures and results of the numerical experiments demonstrating the properties of the proposed algorithms. Finally, Sect. 5 presents the conclusions of this work.

2 Preliminaries

2.1 Divergence

For two probability distributions P and Q, the KL divergence of Q from P, $D_{\mathsf{KL}}(P\|Q)$ is defined to be

$$D_{\mathsf{KL}}(P\|Q) = \sum_{k} P(k) \ln \left(\frac{P(k)}{Q(k)} \right). \tag{1}$$

KL divergence has been used to achieve fuzzy clustering [3] of vectorial data. KL divergence has been extended by using q-logarithmic function

$$\ln_q(x) = \frac{1}{1-q}(x^{1-q} - 1) \quad \text{(for } x > 0\text{)} \tag{2}$$

as

$$D_q(P\|Q) = \frac{1}{1-q} \left(\sum_{k} P(k)^q Q(k)^{1-q} - 1 \right), \tag{3}$$

referred to as q-divergence [7]. In the limit $q \to 1$, the KL-divergence is recovered. q-divergence has been implicitly used to derive fuzzy clustering only for vectorial data [6] although that is not indicated in the literature.

2.2   Clustering for Vectorial Data

Let $X = \{x_k \in \mathbb{R}^D \mid k \in \{1, \ldots, N\}\}$ be a dataset of D-dimensional points. The set of cluster centers is denoted by $v = \{v_i \in \mathbb{R}^D \mid i \in \{1, \ldots, C\}\}$. The membership of x_k with respect to the i-th cluster is denoted by $u_{i,k}$ ($i \in \{1, \ldots, C\}k \in \{1, \ldots, N\}$) and has the following constraint:

$$\sum_{i=1}^{C} u_{i,k} = 1. \tag{4}$$

The variable controlling the i-th cluster size is denoted by α_i, and has the constraint

$$\sum_{i=1}^{C} \alpha_i = 1. \tag{5}$$

The HCM, RBFCM, and REFCM clusters are respectively obtained by solving the following optimization problems:

$$\underset{u,v}{\text{minimize}} \sum_{i=1}^{C} \sum_{k=1}^{N} u_{i,k} \|x_k - v_i\|_2^2, \tag{6}$$

$$\underset{u,v,\alpha}{\text{minimize}} \sum_{i=1}^{C} \sum_{k=1}^{N} (\alpha_i)^{1-m} (u_{i,k})^m \|x_k - v_i\|_2^2, \tag{7}$$

$$\underset{u,v,\alpha}{\text{minimize}} \sum_{i=1}^{C} \sum_{k=1}^{N} u_{i,k} \|x_k - v_i\|_2^2 + \lambda^{-1} \sum_{i=1}^{C} \sum_{k=1}^{N} u_{i,k} \log\left(\frac{u_{i,k}}{\alpha_i}\right). \tag{8}$$

where $m > 1$ and $\lambda > 0$ are the fuzzification parameters. When $m = 1$, the RBFCM is reduced to HCM; the larger the value of m, the fuzzier are the memberships. When $\lambda \to +\infty$, the REFCM is reduced to HCM; the smaller the value of λ, the fuzzier are the memberships.

2.3   Clustering of Time-Series Data: DTW k-Means

Let $X = \{x_k \in \mathbb{R}^D \mid k \in \{1, \ldots, N\}\}$ be a time-series dataset and $x_{k,\ell}$ be its elements at time ℓ. Let $v = \{v_i \in \mathbb{R}^D \mid i \in \{1, \ldots, C\}\}$ be the set of cluster centers set $v_{i,\ell}$ be its elements at time ℓ. Let $\mathsf{DTW}_{i,k}$ be the dissimilarities between the objects x_k and cluster centers v_i as below, with $\mathsf{DTW}_{i,k}$ being defined as follows DTW [8]. Denoting $\Omega_{i,k} \in \{0,1\}^{D \times D}$ as the warping path used to calculate $\mathsf{DTW}_{i,k}$, the membership of x_k with respect to the i-th cluster is given by $u_{i,k}$ ($i \in \{1, \ldots, C\}k \in \{1, \ldots, N\}$). The DTW k-means is obtained by solving the following optimization problem

$$\underset{u,v}{\text{minimize}} \sum_{i=1}^{C} \sum_{k=1}^{N} u_{i,k} \mathsf{DTW}_{i,k}. \tag{9}$$

in accordance with Eq. (4), where

$$\mathrm{DTW}_{i,k} = \sqrt{d(v_{i,D}, x_{k,D})},$$

$$d(v_{i,D}, x_{k,D}) = ||x_{k,D} - v_{i,D}||^2 \tag{10}$$
$$+ \min\{d(v_{i,D-1}, x_{k,D-1}), d(v_{i,D}, x_{k,D-1}), d(v_{i,D-1}, x_{k,D})\}. \tag{11}$$

In addition to DTW, we obtain the warping path that maps the pairs (ℓ, m) for each element in the series to minimize the distance between them. Hence, the warping path is a sequence of pairs (ℓ, m). Here, we consider matrices $\{\Omega_{i,k} \in \{0,1\}^{D \times D}\}_{(i,k)=(1,1)}^{(C,N)}$ whose (ℓ, m)-th element is one if (ℓ, m) is an element of the corresponding warping path and zero otherwise then, we have the cluster centers

$$v_i = \left(\sum_{x_k \in G_i}^{N} \Omega_{i,k} x_k \right) \oslash \left(\sum_{x_k \in G_i} \Omega_{i,k} \mathbf{1} \right), \tag{12}$$

where $\mathbf{1}$ is the D-dimensional vector with all elements equal to one, and $\oslash$ describes element-wise division. The DTW k-means algorithm can be summarized as follows.

Algorithm 1 (DTW k-means). [8]

STEP 1. Set the number of clusters C and initial membership $\{u_{i,k}\}_{(i,k)=(1,1)}^{(C,N)}$.
STEP 2. Calculate $\{v_i\}_{i=1}^{C}$ as

$$v_i = \frac{\sum_{k=1}^{N} u_{i,k} x_k}{\sum_{k=1}^{N} u_{i,k}}. \tag{13}$$

STEP 3. Calculate $\{\mathrm{DTW}_{i,k}\}_{(i,k)=(1,1)}^{(C,N)}$ and update $\{v_i\}_{i=1}^{C}$ as
    (a) Calculate $\mathrm{DTW}_{i,k}$ from Eq. (11).
    (b) Update v_i from Eq. (25).
    (c) Check the limiting criterion for v_i. If the criterion is not satisfied, go to Step (a).
STEP 4. Update $\{u_{i,k}\}_{(i,k)=(1,1)}^{(C,N)}$ as

$$u_{i,k} = \begin{cases} 1 & (i = \arg\min_{1 \le j \le C}\{\mathrm{DTW}_{j,k}\}), \\ 0 & (\text{otherwise}). \end{cases} \tag{14}$$

STEP 5. Check the limiting criterion for (u, v). If the criterion is not satisfied, go to Step 3

3   Proposed Algorithms

3.1   Concept

In this work, we propose three fuzzy clustering algorithms for time-series data.

The first algorithm is similar to the REFCM obtained by KL divergence regularization the DTW k-means objective function, which is referred to as KLFDTWCM. The optimization problem for this is given by

$$\underset{u,v,\alpha}{\text{minimize}} \sum_{i=1}^{C}\sum_{k=1}^{N} u_{i,k}\mathsf{DTW}_{i,k} + \lambda^{-1}\sum_{i=1}^{C}\sum_{k=1}^{N} u_{i,k}\ln\left(\frac{u_{i,k}}{\alpha_i}\right) \qquad (15)$$

subject to Eqs. (4) and (5).

The second algorithm is similar to the RBFCM obtained by replacing the membership of the HCM objective function with its power, which is referred to as BFDTWCM. The optimization problem is then given by

$$\underset{u,v,\alpha}{\text{minimize}} \sum_{i=1}^{C}\sum_{k=1}^{N} (\alpha_i)^{1-m}(u_{i,k})^m\mathsf{DTW}_{i,k} \qquad (16)$$

subject to Eqs. (4) and (5).

The third algorithm is obtained by q-divergence regularization of the BFDTWCM, which is referred to as QFDTWCM. The optimization problem in this case is given by

$$\underset{u,v,\alpha}{\text{minimize}} \sum_{i=1}^{C}\sum_{k=1}^{N} (\alpha_i)^{1-m}(u_{i,k})^m\mathsf{DTW}_{i,k} + \frac{\lambda^{-1}}{m-1}\sum_{i=1}^{C}\sum_{k=1}^{N}(\alpha_i)^{1-m}(u_{i,k})^m \qquad (17)$$

subject to Eqs. (4) and (5). This optimization problem relates the optimization problems for BFDTWCM and KLFDTWCM because Eq. (17) with $\lambda \to +\infty$ reduces to the BFDTWCM method and Eq. (17) with $m \to 1$ reduces to the KLFDTWCM approach. In the next subsection, we present derivation of the update equations for u, v, and α based on of the minimization problem in Eqs. (15), (16), and (17).

3.2   KLFDTWCM, BFDTWCM and QFDTWCM

The KLFDTWCM is obtained by solving the optimization problem in Eqs. (15), (4) and (5), where the Lagrangian $L(u, v, \alpha)$ is defined as

$$L(u,v,\alpha) = \sum_{i=1}^{C}\sum_{k=1}^{N} u_{i,k}\mathsf{DTW}_{i,k} + \lambda^{-1}\sum_{i=1}^{C}\sum_{k=1}^{N} u_{i,k}\ln\left(\frac{u_{i,k}}{\alpha_i}\right)$$
$$+ \sum_{k=1}^{N}\gamma_k\left(1-\sum_{i=1}^{C}u_{i,k}\right) + \beta\left(1-\sum_{i=1}^{C}\alpha_i\right) \qquad (18)$$

using Lagrangian multipliers $(\gamma_1, \cdots, \gamma_{N+1})$. The necessary conditions for optimality are given as

$$\frac{\partial L(u, v, \alpha)}{\partial u_{i,k}} = 0, \tag{19}$$

$$\frac{\partial L(u, v, \alpha)}{\partial \alpha_i} = 0, \tag{20}$$

$$\frac{\partial L(u, v, \alpha)}{\partial \gamma_k} = 0, \tag{21}$$

$$\frac{\partial L(u, v, \alpha)}{\partial \beta} = 0. \tag{22}$$

The optimal membership is obtained from Eqs. (19) and (21) in a manner similar to that of the REFCM as

$$u_{i,k} = \left[\sum_{j=1}^{C} \frac{\alpha_j}{\alpha_i} \exp(-\lambda(\mathrm{DTW}_{j,k} - \mathrm{DTW}_{i,k})) \right]^{-1}. \tag{23}$$

The optimal variable for controlling the cluster sizes is obtained from Eqs. (20) and (22) in a manner similar to that of the REFCM as

$$\alpha_i = \frac{\sum_{k=1}^{N} u_{i,k}}{N}. \tag{24}$$

Recall that in the DTW k-means approach, the cluster centers v_i are calculated using $\Omega_{i,k}$ and x_k belonging to cluster $\#i$, as shown in Eq. (12), which can be equivalently written as

$$v_i = \left(\sum_{k=1}^{N} u_{i,k} \Omega_{i,k} x_k \right) \oslash \left(\sum_{k=1}^{N} u_{i,k} \Omega_{i,k} \mathbf{1} \right). \tag{25}$$

This form can be regarded as the $u_{i,k}$-weighted mean of $\Omega_{i,k} x_k$. Similarly, the cluster centers for KLFDTWCM are calculated using Eq. (25). KLFDTWCM can be described as follows:

Algorithm 2 (KLFDTWCM).

STEP 1. Set the number of clusters C, fuzzification parameter $\lambda > 0$, and initial membership $\{u_{i,k}\}_{(i,k)=(1,1)}^{(C,N)}$.

STEP 2. Calculate v_i from Eq. (13).

STEP 3. Calculate $\{\mathrm{DTW}_{i,k}\}_{(i,k)=(1,1)}^{(C,N)}$ and update $\{v_i\}_{i=1}^{C}$ as

(a) Calculate $\mathrm{DTW}_{i,k}$ from Eq. (11).

(b) Update v_i from Eq. (25).

(c) Check the limiting criterion for v_i. If the criterion is not satisfied, go to Step (a).

STEP 4. Update u from Eq. (23)

STEP 5. Calculate α from Eq. (24)

STEP 6. Check the limiting criterion for (u, v, α). If the criterion is not satisfied, go to Step 3.

The BFDTWCM is obtained by solving the optimization problem in Eqs. (16), (4), and (5). Similar to the derivation of the KLFDTWCM, the optimal membership u, variable for controlling the cluster sizes α, and cluster centers v are obtained as

$$u_{i,k} = \frac{1}{\sum_{j=1}^{C} \frac{\alpha_j}{\alpha_i} \left(\frac{\mathsf{DTW}_{j,k}}{\mathsf{DTW}_{i,k}}\right)^{1/(1-m)}}, \tag{26}$$

$$\alpha_i = \frac{1}{\sum_{j=1}^{C} \left(\frac{\sum_{k=1}^{N}(u_{j,k})^m \mathsf{DTW}_{j,k}}{\sum_{k=1}^{N}(u_{i,k})^m \mathsf{DTW}_{i,k}}\right)^{1/m}}, \tag{27}$$

$$v_i = \left(\sum_{k=1}^{N}(u_{i,k})^m \Omega_{i,k} x_k\right) \oslash \left(\sum_{k=1}^{N}(u_{i,k})^m \Omega_{i,k} \mathbf{1}\right), \tag{28}$$

respectively. The BFDTWCM can be described as follows:

Algorithm 3 (BFDTWCM).

STEP 1. Set the number of clusters C, fuzzification parameter $m > 1$, and initial membership $\{u_{i,k}\}_{(i,k)=(1,1)}^{(C,N)}$.

STEP 2. Calculate v_i from Eq. (13).

STEP 3. Calculate $\{\mathsf{DTW}_{i,k}\}_{(i,k)=(1,1)}^{(C,N)}$ and update $\{v_i\}_{i=1}^{C}$ as
   (a) Calculate $\mathsf{DTW}_{i,k}$ from Eq. (11).
   (b) Update v_i from Eq. (28).
   (c) Check the limiting criterion for v_i. If the criterion is not satisfied, go to Step. (a).

STEP 4. Update u from Eq. (26)

STEP 5. Calculate α from Eq. (27)

STEP 6. Check the limiting criterion for (u, v, α). If the criterion is not satisfied, go to Step. 3.

The QFDTWCM is obtained by solving the optimization problem in Eqs. (17), (4), and (5). Similar to the derivations of BFDTWCM and KLFDTWCM, the optimal membership u and variable for controlling the cluster sizes α are obtained as

$$u_{i,k} = \frac{1}{\sum_{j=1}^{C} \frac{\alpha_j}{\alpha_i} \left(\frac{1-\lambda(1-m)\mathsf{DTW}_{j,k}}{1-\lambda(1-m)\mathsf{DTW}_{i,k}}\right)^{1/(1-m)}}, \tag{29}$$

$$\alpha_i = \frac{1}{\sum_{j=1}^{C} \left(\frac{\sum_{k=1}^{N}(u_{j,k})^m(1-\lambda(1-m)\mathsf{DTW}_{j,k})}{\sum_{k=1}^{N}(u_{i,k})^m(1-\lambda(1-m)\mathsf{DTW}_{i,k})}\right)^{1/m}}, \tag{30}$$

respectively. The optimal cluster centers are defined by Eq. (28). The QFDTWCM can be described as follows:

Algorithm 4 (QFDTWCM).

STEP 1. Set the number of clusters C, fuzzification parameter $m > 1, \lambda > 0$, and
initial membership $\{u_{i,k}\}_{(i,k)=(1,1)}^{(C,N)}$.

STEP 2. Calculate $\{v_i\}_{i=1}^{C}$ from Eq. (13).

STEP 3. Calculate $\{DTW_{i,k}\}_{(i,k)=(1,1)}^{(C,N)}$ and update $\{v_i\}_{i=1}^{C}$ as

(a) Calculate $DTW_{i,k}$ from Eq. (11).

(b) Update v_i from Eq. (28).

(c) Check the limiting criterion for v_i. If the criterion is not satisfied, go to Step (a).

STEP 4. Update u from Eq. (29).

STEP 5. Calculate α from Eq. (30).

STEP 6. Check the limiting criterion for (u, v, α). If the criterion is not satisfied, go to Step 4.

In the remainder of this section, we show that the QFDTWCM with $m - 1 \to +0$ reduces to BFDTWCM and QFDTWCM with $\lambda \to +\infty$ reduces to KLFDTWCM.

The third step in the QFDTWCM approach is exactly equal to that of the BFDTWCM because Eq. (28) is identical to Eq. (28). In the fourth step of the QFDTWCM, the u value in Eq. (29) reduces to that in Eq. (26) of the BFDTWCM as

$$\frac{1}{\sum_{j=1}^{C} \frac{\alpha_j}{\alpha_i} \left(\frac{1-\lambda(1-m)\mathsf{DTW}_{j,k}}{1-\lambda(1-m)\mathsf{DTW}_{i,k}}\right)^{1/(1-m)}}$$

$$= \frac{\alpha_i \left(1/\lambda - (1-m)\mathsf{DTW}_{i,k}\right)^{1/(1-m)}}{\sum_{j=1}^{C} \alpha_j \left(1/\lambda - (1-m)\mathsf{DTW}_{j,k}\right)^{1/(1-m)}}$$

$$\to \frac{\alpha_i \left(-(1-m)\mathsf{DTW}_{i,k}\right)^{1/(1-m)}}{\sum_{j=1}^{C} \alpha_j \left(-(1-m)\mathsf{DTW}_{j,k}\right)^{1/(1-m)}}$$

(with $\lambda \to +\infty$)

$$= \frac{(m-1)\alpha_i \left(\mathsf{DTW}_{i,k}\right)^{1/(1-m)}}{(m-1)\sum_{j=1}^{C} \alpha_j \left(\mathsf{DTW}_{j,k}\right)^{1/(1-m)}}$$

$$= \frac{\alpha_i \left(\mathsf{DTW}_{i,k}\right)^{1/(1-m)}}{\sum_{j=1}^{C} \alpha_j \left(\mathsf{DTW}_{j,k}\right)^{1/(1-m)}}$$

$$= \frac{1}{\sum_{j=1}^{C} \frac{\alpha_j}{\alpha_i} \left(\frac{\mathsf{DTW}_{j,k}}{\mathsf{DTW}_{i,k}}\right)^{1/(1-m)}}. \tag{31}$$

In the fifth step of the QFDTWCM, the α value in Eq. (30) is reduces to that in Eq. (27) of the BFDTWCM as

$$\frac{1}{\sum_{j=1}^{C}\left(\frac{\sum_{k=1}^{N}(u_{j,k})^m(1-\lambda(1-m)\mathrm{DTW}_{j,k})}{\sum_{k=1}^{N}(u_{i,k})^m(1-\lambda(1-m)\mathrm{DTW}_{i,k})}\right)^{1/m}}$$

$$=\frac{\left(\sum_{k=1}^{N}(u_{i,k})^m(1/\lambda-(1-m)\mathrm{DTW}_{i,k})\right)^{1/m}}{\sum_{j=1}^{C}\left(\sum_{k=1}^{N}(u_{j,k})^m(1/\lambda-(1-m)\mathrm{DTW}_{j,k})\right)^{1/m}}$$

$$\rightarrow\frac{\left(\sum_{k=1}^{N}(u_{i,k})^m(-(1-m)\mathrm{DTW}_{i,k})\right)^{1/m}}{\sum_{j=1}^{C}\left(\sum_{k=1}^{N}(u_{j,k})^m(-(1-m)\mathrm{DTW}_{j,k})\right)^{1/m}}$$

(with $\lambda\rightarrow+\infty$)

$$=\frac{(m-1)^{1/m}\left(\sum_{k=1}^{N}(u_{i,k})^m(\mathrm{DTW}_{i,k})\right)^{1/m}}{(m-1)^{1/m}\sum_{j=1}^{C}\left(\sum_{k=1}^{N}(u_{j,k})^m(\mathrm{DTW}_{j,k})\right)^{1/m}}$$

$$=\frac{\left(\sum_{k=1}^{N}(u_{i,k})^m(\mathrm{DTW}_{i,k})\right)^{1/m}}{\sum_{j=1}^{C}\left(\sum_{k=1}^{N}(u_{j,k})^m(\mathrm{DTW}_{j,k})\right)^{1/m}}$$

$$=\frac{1}{\sum_{j=1}^{C}\left(\frac{\sum_{k=1}^{N}(u_{j,k})^m\mathrm{DTW}_{j,k}}{\sum_{k=1}^{N}(u_{i,k})^m\mathrm{DTW}_{i,k}}\right)^{1/m}}. \tag{32}$$

From the above discussion, we can conclude that the QFDTWCM with $\lambda\rightarrow+\infty$ reduces to the BFDTWCM.

The third step of the QFDTWCM with $m=1$ is obviously equal to the third step of the KLFDTWCM because Eq. (28) with $m=1$ is identical to Eq. (11). In the fourth step of the QFDTWCM, the u value in Eq. (29) reduces to that in Eq. (23) of the KLFDTWCM as

$$(1-\lambda(1-m)\mathrm{DTW}_{i,k})^{1/(1-m)}$$
$$\rightarrow\exp(-\lambda(\mathrm{DTW}_{i,k}))\quad(\text{with }m=1). \tag{33}$$

The fifth step of the QFDTWCM reduces to that of the KLFDTWCM because

$$=\frac{1}{\sum_{j=1}^{C}\left(\frac{\sum_{k=1}^{N}(u_{j,k})^m(1-\lambda(1-m)\mathrm{DTW}_{j,k})}{\sum_{k=1}^{N}(u_{i,k})^m(1-\lambda(1-m)\mathrm{DTW}_{i,k})}\right)^{1/m}}$$

$$\rightarrow\frac{1}{\sum_{j=1}^{C}\sum_{k=1}^{N}\frac{u_{j,k}}{u_{i,k}}}\quad(\text{with }m=1)$$

$$=\frac{\sum_{k=1}^{N}u_{i,k}}{\sum_{j=1}^{C}\sum_{k=1}^{N}u_{j,k}}$$

$$=\frac{\sum_{k=1}^{N}u_{i,k}}{N}. \tag{34}$$

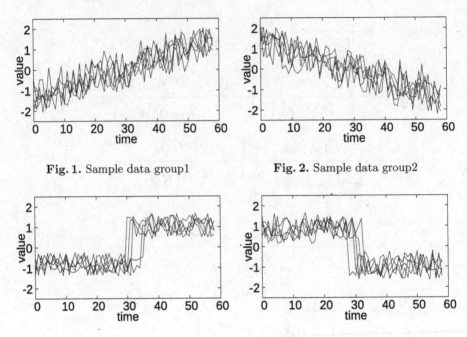

Fig. 1. Sample data group1

Fig. 2. Sample data group2

Fig. 3. Sample data group3

Fig. 4. Sample data group4

From the above discussion, we can conclude that the QFDTWCM with $m-1 \to 0$ reduces to the KLFDTWCM.

As shown herein, the proposed QFDTWCM includes both the BFDTWCM and KLFDTWCM. Thus, the QFDTWCM is a generalization of the BFDTWCM as well as KLFDTWCM.

4   Numerical Experiments

This section presents some numerical examples based on one artificial dataset. The example compares the characteristic features of the proposed clustering algorithm (Algorithm 4) with those of other algorithms (Algorithms 2 and 3) for an artificial dataset, as shown in Figs. 1, 2, 3 and 4 for four clusters ($C = 4$), with each clusters containing five objects ($N = 4 \times 5 = 20$).

The initialization step assigns the initial memberships according to the actual class labels. All three proposed methods with various fuzzification parameter values were able to classify the data adequately, and the obtained membership values are shown in Tables 1, 2, 3, 4, 5, 6, 7, 8 and 9. Tables 1 and 2 show that for the BFDTWCM, when the fuzzification parameter m is larger, the membership values are fuzzier. Tables 3 and 4 show that for the KLFDTWCM, when the fuzzification parameter λ is smaller, the membership values are fuzzier. Tables 5 and 6

Table 1. Sample data memberships of the BFDTWCM, $m = 1.001$

Cluster	Group			
	1	2	3	4
1	1.00	0.00	0.00	0.00
2	0.00	1.00	0.00	0.00
3	0.00	0.00	1.00	0.00
4	0.00	0.00	0.00	1.00

Table 2. Sample data memberships of the BFDTWCM, $m = 1.35$

Cluster	Group			
	1	2	3	4
1	0.77	0.01	0.10	0.00
2	0.01	0.66	0.00	0.11
3	0.21	0.01	0.89	0.00
4	0.01	0.32	0.00	0.89

Table 3. Sample data memberships of the KLFDTWCM, $\lambda = 1.5$

Cluster	Group			
	1	2	3	4
1	0.84	0.00	0.06	0.00
2	0.00	0.70	0.00	0.06
3	0.15	0.00	0.94	0.00
4	0.00	0.30	0.00	0.94

Table 4. Sample data memberships of the KLFDTWCM, $\lambda = 100$

Cluster	Group			
	1	2	3	4
1	1.00	0.00	0.00	0.00
2	0.00	1.00	0.00	0.00
3	0.00	0.00	1.00	0.00
4	0.00	0.00	0.00	1.00

Table 5. Sample data memberships of the QFDTWCM, $(m, \lambda) = (1.2, 3)$

Cluster	Group			
	1	2	3	4
1	0.78	0.00	0.10	0.00
2	0.00	0.61	0.00	0.09
3	0.21	0.00	0.90	0.00
4	0.00	0.39	0.00	0.91

Table 6. Sample data memberships of the QFDTWCM, $(m, \lambda) = (1.001, 3)$

Cluster	Group			
	1	2	3	4
1	0.99	0.00	0.00	0.00
2	0.00	0.98	0.00	0.01
3	0.01	0.00	1.00	0.00
4	0.00	0.02	0.00	0.99

show that for the QFDTWCM, when the fuzzification parameter m is larger, the membership values are fuzzier. Tables 5 and 7 show that for the QFDTWCM, when the fuzzification parameter λ is smaller, the membership values are fuzzier. Furthermore, Tables 6 and 8 show that the QFDTWCM with large values of λ produces results similar to those of the KLFDTWCM, and Tables 7 and 9 show that the QFDTWCM with smaller values of m produces results similar to those of the BFDTWCM. These results indicate that the QFDTWCM combines the features of both BFDTWCM and KLFDTWCM.

Table 7. Sample data memberships of the QFDTWCM, $(m, \lambda) = (1.2, 100)$

Cluster	Group			
	1	2	3	4
1	0.95	0.00	0.03	0.00
2	0.00	0.90	0.00	0.04
3	0.05	0.00	0.97	0.00
4	0.00	0.10	0.00	0.96

Table 8. Sample data memberships of the KLFDTWCM, $\lambda = 3$

Cluster	Group			
	1	2	3	4
1	0.99	0.00	0.00	0.00
2	0.00	0.98	0.00	0.01
3	0.01	0.00	1.00	0.00
4	0.00	0.02	0.00	0.99

Table 9. Sample data memberships of the BFDTWCM, $m = 1.2$

Cluster	Group			
	1	2	3	4
1	0.95	0.00	0.03	0.00
2	0.00	0.90	0.00	0.04
3	0.05	0.00	0.97	0.00
4	0.00	0.10	0.00	0.96

5   Conclusion

This work, propose three fuzzy clustering algorithms for classifying time-series data. The theoretical results indicate that the QFDTWCM approach reduces to the BFDTWCM as $m - 1 \rightarrow +0$ and to the KLFDTWCM as $\lambda \rightarrow +\infty$. Numerical experiments were performed on an artificial dataset to substantiate these properties.

In the future work, these proposed algorithms will be applied to real datasets.

References

1. MacQueen, J.B.: Some methods for classification and analysis of multivariate observations. In: Proceedings of the 5th Berkeley Symposium on Mathematical Statistics and Probability, pp. 281–297 (1967)
2. Bezdek, J.: Pattern Recognition with Fuzzy Objective Function Algorithms. Plenum Press, New York (1981)
3. Miyamoto, S., Mukaidono, M.: Fuzzy c-means as a regularization and maximum entropy approach. In: Proceedings of the 7th International Fuzzy Systems Association World Congress (IFSA 1997), vol. 2, pp. 86–92 (1997)
4. Miyamoto, S., Kurosawa, N.: Controlling cluster volume sizes in fuzzy c-means clustering. In: Proceedings of the SCIS&ISIS2004, pp. 1–4 (2004)
5. Ichihashi, H., Honda, K., Tani, N.: Gaussian mixture PDF approximation and fuzzy c-means clustering with entropy regularization. In: Proceedings of the 4th Asian Fuzzy System Symposium, pp. 217–221 (2000)

6. Miyamoto, S., Ichihashi, H., Honda, K.: Algorithms for Fuzzy Clustering. Springer, Heidelberg (2008)
7. Chernoff, H.: A measure of asymptotic efficiency for tests of a hypothesis based on a sum of observations. Ann. Math. Statist. **23**, 493–507 (1952)
8. Petitjean, F., Ketterlin, A., Gancarski, P.: A global averaging method for dynamic time warping, with applications to clustering. Pattern Recogn. **44**, 678–693 (2011)

On an Multi-directional Searching Algorithm for Two Fuzzy Clustering Methods for Categorical Multivariate Data

Kazune Suzuki[✉] and Yuchi Kanzawa

Shibaura Institute of Technology, 3-7-5 Toyosu, Koto, Tokyo 135-8548, Japan
ma21070@shibaura-it.ac.jp, kanzawa@sic.shibaura-it.ac.jp

Abstract. Clustering for categorical multivariate data is an important task for summarizing co-occurrence information that consists of mutual affinity among objects and items. This work focus on two fuzzy clustering methods for categorical multivariate data. One of the serious limitations for these methods is the local optimality problem. In this work, an algorithm is proposed to address this issue. The proposed algorithm incorporates multiple token search generated from the eigen decomposition of the Hessian of the objective function. Numerical experiments using an artificial dataset shows that the proposed algorithm is valid.

Keywords: Fuzzy clustering · Local optimality problem · Multiple token search

1 Introduction

The hard c-means (HCM) or k-means clustering algorithm [1] partitions objects into groups. This method is called "hard clustering" because each object belongs to only one cluster, whereas Gaussian mixture models and fuzzy clustering are called "soft clustering" because each object belongs to all or some clusters to varying degrees.

Clustering for categorical multivariate data is a method for summarizing co-occurrence information that consists of mutual affinity among objects and items. A multinomial mixture model (MMM) [2] is a probabilistic model for clustering tasks for categorical multivariate data, where each component distribution is defined by multinomial distribution. Honda et al. [3] proposed the fuzzy clustering for categorical multivariate data induced by MMM (FCCMM). The FCCMM method is a fuzzy counterpart to MMMs, where the degree of fuzziness can be controlled by two fuzzification parameters. Kondo et al. [4] extended FCCMM by introducing q-divergence instead of Kullback-Leibler (KL) divergence in FCCMM. Furthermore, Kondo et al. [4] showed that QFCCMM outperforms FCCMM in terms of clustering accuracy.

© Springer Nature Switzerland AG 2022
K. Honda et al. (Eds.): IUKM 2022, LNAI 13199, pp. 182–190, 2022.
https://doi.org/10.1007/978-3-030-98018-4_15

One of the most serious limitations for FCCMM and QFCCMM is the local optimality problem. The problem makes the accuracy of their algorithms dependent on its starting points. Thus, obtaining good starting points has been long addressed. One easily idea to avoid locally optimal solutions is running their algorithms multiple times with differently initial setting, and selecting the result where the optimal objective function value is achieved. However, it is unknown how many times should their algorithms run to obtain the globally optimal solution. Arthur and Vassilvitskii [5] proposed k-means++, which is an algorithm for choosing the initial setting for k-means or HCM, This algorithm not only yields considerable improvement in the clustering accuracy of k-means, but also provides a probabilistic upper bound of error. However, this algorithm cannot be applied directly to the other clustering algorithm such as fuzzy clustering algorithms, nor provides any upper bound of error for those than k-means. Ishikawa and Nakano [6] proposed the mes-EM algorithm for the Gaussian mixture models (GMM) incorporating a multiple token search into the EM algorithm for GMM, employing the primitive initial point (PIP) as its initial point, where the search tokens are generated along the directions spanned by the eigen vectors with negative eigen values of the Hessian of the objective function. This idea can be applied to fuzzy clustering algorithms for categorical multivariate data including FCCMM and QFCCMM, which has a potential to solve the local optimality problem of FCCMM and QFCCMM.

In this study, we propose an algorithm to address the local optimality problem of FCCMM and QFCCMM, by modifying the idea of the mes-EM algorithm. The first modification is considering equality-constraints. The idea of the mes-EM algorithm, incorporating a multiple token generated along the directions spanned by the eigen vectors with negative eigen values of the Hessian of the objective function, cannot be valid as it is for FCCMM or QFCCMM. It is because the FCCMM and QFCCMM optimization problems must consider some equality-constraints for variables. If we apply the idea of the mes-EM algorithm directly to FCCMM or QFCCMM, the generating tokens often violate such the constraint. Then, we generate tokens from the intersection of the space spanned by the eigenvectors with negative eigen values of the Hessian of the objective function and the null space of the constraints. The other modification is concerning the length of tokens. Although the generated tokens show the direction to which the objective function improves, we cannot its length at which the objective function improves. If we easily determine the length of tokens, such the tokens may not only make the objective function value worsen but also violate the inequality-constraints. Then, we reduce the length of tokens if it violates the inequality-constraints or it make the objective function value worsen.

The remainder of this paper is organized as follows. Section 2 introduces the notations used and some conventional algorithms. Section 3 describes the proposed algorithm. Section 4 presents the results of numerical experiments conducted to demonstrate the performance of the proposed algorithm. Finally, Sect. 5 concludes the paper.

2   Preliminaries

2.1 Two Fuzzy Clustering Algorithms for Categorical Multivariate Data

Let $X = \{x_k \in \mathbb{R}^M | k \in \{1, ..., N\}\}$ be a categorical multivariate dataset of M dimensional points. The membership of x_k that belongs to the i-th cluster is denoted by $u_{i,k}$ ($i \in \{1, ..., C\}, k \in \{1, ..., N\}$) and the set of $u_{i,k}$ is denoted by u, which obeys the following constraint:

$$\sum_{i=1}^{C} u_{i,k} = 1, u_{i,k} \in [0,1] \tag{1}$$

The variable controlling the i-th cluster size is denoted by α_i. The i-th element of vector α is denoted by α_i, and α obeys the following constraint:

$$\sum_{i=1}^{C} \alpha_i = 1, \alpha_i \in (0,1) \tag{2}$$

The cluster center set is denoted by $v = \{v_i | v_i \in \mathbb{R}^M, i \in \{1, ..., C\}\}$. The ℓ-th item typicality for i-th cluster is denoted by $v_{i,\ell}$, and v obeys the following constraint:

$$\sum_{\ell=1}^{M} v_{i,\ell} = 1, \quad v_{i,\ell} \in [0,1] \tag{3}$$

The methods FCCMM and QFCCMM are derived by solving the optimization problems,

$$\underset{u,v,\alpha}{\text{minimize}} J_{\mathsf{FCCMM}}(u, v, \alpha), \tag{4}$$

$$\underset{u,v,\alpha}{\text{minimize}} J_{\mathsf{QFCCMM}}(u, v, \alpha), \tag{5}$$

subject to Eqs. (1), (2), and (3), where

$$J_{\mathsf{FCCMM}}(u, v, \alpha) = \sum_{i=1}^{C} \sum_{k=1}^{N} u_{i,k} d_{i,k} + \lambda^{-1} \sum_{i=1}^{C} \sum_{k=1}^{N} u_{i,k} \log\left(\frac{u_{i,k}}{\alpha_i}\right), \tag{6}$$

$$J_{\mathsf{QFCCMM}}(u, v, \alpha) = \sum_{i=1}^{C} \sum_{k=1}^{N} (u_{i,k})^m (\alpha_i)^{1-m} d_{i,k} + \frac{\lambda^{-1}}{m-1} \sum_{i=1}^{C} \sum_{k=1}^{N} (u_{i,k})^m (\alpha_i)^{1-m}, \tag{7}$$

$$d_{i,k} = -\frac{1}{t} \sum_{q=1}^{Q} x_{k,q}\left((v_{i,q})^t - 1\right), \tag{8}$$

and $m > 1$, $\lambda > 0$ and $t < 1$ are fuzzification parameters. The FCCMM and QFCCMM algorithms are summarized as follows.

Algorithm 1 (FCCMM, QFCCMM).

STEP 1. Set the number of clusters as C. Fix $\lambda > 0$ and $t > 0$ for FCCMM, and $m > 1$, $\lambda > 0$ and $t < 1$ for QFCCMM. Assume initial item typicality as v and initial variable controlling cluster sizes as α.

STEP 2. Update u as

$$u_{i,k} = \frac{\alpha_i \exp(-\lambda d_{i,k})}{\sum_{j=1}^{C} \alpha_j \exp(-\lambda d_{j,k})} \qquad (9)$$

for FCCMM, and

$$u_{i,k} = \frac{\alpha_i \left(1 - \lambda \left(1 - m\right) d_{i,k}\right)^{\frac{1}{1-m}}}{\sum_{j=1}^{C} \alpha_j \left(1 - \lambda \left(1 - m\right) d_{j,k}\right)^{\frac{1}{1-m}}} \qquad (10)$$

for QFCCMM.

STEP 3. Update α as

$$\alpha_i = \frac{\sum_{k=1}^{N} u_{i,k}}{N} \qquad (11)$$

for FCCMM, and

$$\alpha_i = \frac{1}{\sum_{j=1}^{C} \left(\frac{\sum_{k=1}^{N} (u_{i,k})^m (1-\lambda(1-m)d_{i,k})}{\sum_{k=1}^{N} (u_{j,k})^m (1-\lambda(1-m)d_{j,k})} \right)^{\frac{1}{m}}} \qquad (12)$$

for QFCCMM.

STEP 4. Update v as

$$v_{i,\ell} = \frac{\left(\sum_{k=1}^{N} u_{i,k} x_{k,\ell}\right)^{1/(1-t)}}{\sum_{r=1}^{M} \left(\sum_{k=1}^{N} u_{i,k} x_{k,r}\right)^{1/(1-t)}} \qquad (13)$$

for FCCMM, and

$$v_{i,\ell} = \frac{\left(\sum_{k=1}^{N} (u_{i,k})^m x_{k,\ell}\right)^{1/(1-t)}}{\sum_{r=1}^{M} \left(\sum_{k=1}^{N} (u_{i,k})^m x_{k,r}\right)^{1/(1-t)}} \qquad (14)$$

for QFCCMM.

STEP 5. Check the limiting criterion for (u, v, α). If the criterion is not satisfied, go to STEP 2.

2.2   Multi-directional in Eigen Space-EM Algorithm for GMM

The mes-EM algorithm was proposed to improve the solution quality of the EM algorithm. The mes-EM algorithm starts from the primitive initial point (PIP), which is the solution for extreme values of inverse temperature in the deterministic annealing [7] context. Let the Hessian of the target function to be minimized have negative eigen values at the PIP. Let $\mathcal{W} = \{w_r, -w_r\}_{r=1}^{R}$ be the orthonormal set of the corresponding eigen vector. Search tokens are generated along the directions

$$\mathcal{W}' = \left\{\sum_{r=1}^{R}(\pm w_r)\right\} = \{(+w_1) + \ldots(+w_R), \ldots, (-w_1) + \cdots + (-w_R)\} \quad (15)$$

in addition to the orthonormal set $\mathcal{W}$. The mes-EM algorithm is the method of running the EM algorithm $2R + 2^R$ times starting from the same PIP with $\mathcal{W} \cup \mathcal{W}'$ as their search directions, and is described below.

Algorithm 2 (mes-EM).

STEP 1.   Calculate all eigen values of the Hessian of the target function at the PIP.
STEP 2.   Generate search directions $\mathcal{W} \cup \mathcal{W}'$ by using the negative eigen values.

3   Proposed Methods

In this section, we propose an algorithm to address the local optimality problem of FCCMM and QFCCMM, by modifying the idea of the mes-EM algorithm.

Consider the FCCMM objective function given in Eq. (4) as the function of $s = (v, \alpha) \in \mathbb{R}^{(C+1)M}$, i.e., $J_{\mathsf{FCCMM}}(s) = J_{\mathsf{FCCMM}}(v, \alpha)$, where, u is considered as the function of (v, α) given as Eq. (9). The PIP for the mes-EM algorithm is the solution for extreme values of inverse temperature in the deterministic annealing context, where as the PIP for FCCMM is the solution of their optimization problem with $\lambda \to 0$, given by $s^{(0)} = (v^{(0)}, \alpha^{(0)})$ where

$$v_{i,\ell}^{(0)} = \frac{\sum_{k=1}^{N} x_{k,\ell}}{\sum_{r=1}^{M} \sum_{k=1}^{N} x_{k,r}}, \quad (16)$$

$$\alpha_i^{(0)} = \frac{1}{C}. \quad (17)$$

The proposed algorithm starts from the PIP.

Let the Hessian of the objective function given by Eq. (4) have negative eigen values at the PIP. Let $\mathcal{W} = \{w_r, -w_r\}_{r=1}^{R}$ be the orthonormal set of the corresponding eigen vector. In the mes-EM algorithm, multiple tokens are generated using the direction in the space spanned by the corresponding eigen vectors to the negative eigen values of the Hessian of the target function, whereas for FCCMM, the generated token $s^{(0)} + \Delta s$, where Δs is in the space spanned by

$\mathcal{W}$, is not always valid. It is because we must consider the equality-constraints given by Eqs. (2) and (3) for (v, α). These constraints are equivalently written as

$$As = 1_{C+1}, \tag{18}$$

$$A = \begin{pmatrix} 1_M^\mathsf{T} & 0_M^\mathsf{T} & \cdots, & 0_M^\mathsf{T} & 0_C^\mathsf{T} \\ 0_M^\mathsf{T} & 1_M^\mathsf{T} & \cdots, & 0_M^\mathsf{T} & 0_C^\mathsf{T} \\ 0_M^\mathsf{T} & 0_M^\mathsf{T} & \cdots, & 1_M^\mathsf{T} & 0_C^\mathsf{T} \\ 0_M^\mathsf{T} & 0_M^\mathsf{T} & \cdots, & 0_M^\mathsf{T} & 1_C^\mathsf{T} \end{pmatrix}, \tag{19}$$

where 1_{C+1}, 1_M, and 1_C are the vector whose all the elements are ones with the dimension of $C+1$, M, and C, respectively, and 0_M, and 0_C are the vector whose all the elements are zeros with the dimension of M and C, respectively. If we have $A\Delta s \neq 0$, then the generated token $s + \Delta s$ violates the equality-constraint as

$$A(s + \Delta s) = As + A\Delta s \neq 1_{C+1}. \tag{20}$$

Then, we generate tokens $s^{(0)} + \Delta s$ where Δs is in the intersection of $\mathsf{span}(\mathcal{W})$ and the null space of A, i.e., $\mathsf{null}(A)$. Such the intersection can be obtained as the righthand singular vectors of AW where $W = (w_1, \ldots, w_R)$.

Although Δs show the direction to which the objective function improves with keeping the equality-constraints given by Eq. (18), or equivalently Eqs. (2) and (3), we cannot know its length at which the objective function value improves. If we easily determine the length of Δs, such the tokens $s^0 + \Delta s$ may not only make the objective function value $J_{\mathsf{FCCMM}}(s^{(0)} + \Delta s)$ worsen but also violate the inequality-constraints $v_{i,\ell} \in [0, 1]$ and $\alpha_i \in (0, 1)$. Then, we reduce the length of tokens if it violates the inequality-constraints or it make the objective function value worsen.

The above discussion is not only for FCCMM but also for QFCCMM, and is summarized into the following algorithm:

Algorithm 3.

STEP 1. Let $\mathcal{S}$, $\mathcal{S}^*$, and $\Delta\mathcal{S}$ be empty sets, add $s^{(0)}$ given by Eqs. (16) and (17) to $\mathcal{S}$.

STEP 2. If $\mathcal{S}$ is empty, output the element of $\mathcal{S}^*$ such that its objective function value is the minimum, and terminate this algorithm. Otherwise, pop s from $\mathcal{S}$, and run Algorithm 1 using the initial setting s, resulting into $\hat{s}$.

STEP 3. Calculate all eigen pairs of $\nabla^2 J_{\mathsf{FCCMM}}(\hat{s})$ for FCCMM or $\nabla^2 J_{\mathsf{QFCCMM}}(\hat{s})$ for QFCCMM. If all the eigen values are positive, $\hat{s}$ is a locally or globally optimal solution. Then, add $\hat{s}$ to $\mathcal{S}^*$, and return to STEP 2. If all the eigen values are negative, $\hat{s}$ is not a locally or globally optimal solution. Then, ignore $\hat{s}$, and return to STEP 2. If at least one eigen value is negative, $\hat{s}$ is a saddle point. Let the corresponding eigen vectors be $\mathcal{W} = \{w_r \in \mathbb{R}^{(C+1)M}\}_{r=1}^R$.

STEP 4. Obtain the orthonormal basis vectors

$$\breve{W} = \{\breve{w}_r, -\breve{w}_r\}_{r=1}^{\breve{R}} \tag{21}$$

of $\mathsf{span}(\mathcal{W}) \cap \mathsf{null}(A)$ and their combinations

$$\breve{W}' = \left\{ \sum_{r=1}^{\breve{R}} (\pm \breve{w}_r) \right\} = \{(+\breve{w}_1) + \ldots (+\breve{w}_{\breve{R}}), \ldots, (-\breve{w}_1) + \cdots + (-\breve{w}_{\breve{R}})\}. \tag{22}$$

Add all the elements of $\breve{W} \cup \breve{W}'$ to $\Delta\mathcal{S}$.

STEP 5. If $\Delta\mathcal{S}$ is empty, return to STEP 2. Otherwise, pop Δs from $\Delta\mathcal{S}$.

STEP 6. Find $0 < \beta \le 1$ such as $0 < \hat{s} + \beta\Delta s < 1$ and $J_{\mathsf{FCCMM}}(\hat{s} + \beta\Delta s) < J_{\mathsf{FCCMM}}(\hat{s})$ for FCCMM or $J_{\mathsf{QFCCMM}}(\hat{s} + \beta\Delta s) < J_{\mathsf{QFCCMM}}(\hat{s})$ for QFC-CMM, add $\hat{s} + \beta\Delta s$ to $\mathcal{S}$, and return to STEP 5. If there does not exist such the value β, ignore Δs and return to STEP 5.

4   Numerical Experiments

This section provides numerical experiments to illustrate Algorithm 3 based on one artificial dataset as shown in Fig. 1. with four clusters ($C = 4$) in the two dimensional unit-simplex. First cluster is composed of 100 objects generated from multinomial distribution with $v_1 = (0.1, 0.1, 0.8)$. Second cluster is composed of 200 objects generated from multinomial distribution with $v_2 = (0.8, 0.1, 0.1)$. Third cluster is composed of 400 objects generated from multinomial distribution with $v_3 = (0.1, 0.8, 0.1)$. Fourth cluster is composed of 400 objects generated from multinomial distribution with $v_4 = (\frac{1}{3}, \frac{1}{3}, \frac{1}{3})$.

The fuzzification parameter λ and t for FCCMM was set as $\lambda \in \{10, 40\}$ and $t = 0.5$. The fuzzification parameter m, λ and t for QFCCMM was set as $m = \{1.0001, 1.2\}, \lambda = 40$ and $t = 0.5$.

For FCCMM with $(\lambda, t) = (10, 0.5)$, after the only output of Algorithm 1 from the PIP was judged as a saddle point at STEP 3. of Algorithm 3, 76 tokens were generated through STEP 4. and STEP 6. of Algorithm 3, all the outputs of Algorithm 1 from these tokens were judged as locally or globally optimal solutions at STEP 3. of Algorithm 3, and Algorithm 3 terminated. Among 76 locally or globally optimal solutions,

10 points are strictly local optimum with ARI = 0.82, and 66 points achieve the minimum objective function value with ARI = 1.0. This result is summarized in Table 1 along with the other cases. These results show that the proposed algorithm produce the globally optimal solution through multiple tokens generated from the PIP. However, many generated tokens are the same convergence point. For example, in the case with FCCMM with $(\lambda, t) = (10, 0.5)$, among 66 solutions with minimal objective function value, here exists only 1 distinct one which means that the algorithm has redundancy. More efficient generating tokens is a future work.

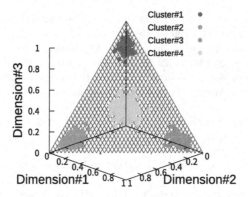

Fig. 1. Artificial dataset

Table 1. Number of saddle points, tokens, and locally/globally optimal solutions along with actual number obtained from Algorithm 3.

Method	Fuzzification parameter			Number of saddle points	Number of tokens	Number of strictly local optimum	Number of solutions with minimal objective function value (Actual number)	ARI of the solution minimal objective function value
	m	λ	t					
EFCCMM		10	0.5	1	76	10	66(1)	1.0
		30	0.5	1	76	5	71(1)	1.0
		40	0.5	1	76	6	70(1)	1.0
QFCCMM	1.2	40	0.5	1	76	10	66(1)	1.0
	1.2	30	0.5	1	76	20	56(1)	1.0
	1.0001	40	0.5	1	76	6	70(1)	1.0
	1.0001	30	0.5	1	76	8	68(1)	1.0

5 Conclusion

In this work, we proposed an algorithm to address the local optimality problem of FCCMM and QFCCMM. Numerical experiments using an artificial dataset shows that the proposed algorithm is valid, though it has a redundancy.

In the future, through improving the proposed algorithm efficiently, the proposed algorithms will be applied to a large number of real datasets. Furthermore, the technique generating multiple tokens will be applied to clustering algorithms for the other types of data, such as spherical data, e.g., in [8].

References

1. MacQueen, J.B.: Some methods for classification and analysis of multivariate observations. In: Proceedings of the 5th Berkeley Symposium on Mathematical Statistics and Probability, pp. 281–297 (1967)
2. Rigouste, L., Cappé, O., Yvon, F.: Inference and evaluation of the multinomial mixture model for text clustering. Inf. Process. Manag. **43**(5), 1260–1280 (2007)

3. Honda, K., Oshio, S., Notsu, A.: Fuzzy co-clustering induced by multinomial mixture models. JACIII **19**(6), 717–726 (2015)
4. Kondo, T., Kanzawa, Y.: Fuzzy clustering methods for categorical multivariate data based on q-divergence. JACIII **22**(4), 524–536 (2018)
5. Arthur, D., Vassilvitskii, S.: k-means++: the advantages of careful seeding. In: Proceedings of the 8th Annual ACM-SIAM Symposium on Discrete Algorithms, pp. 1027–1035 (2007)
6. Ishikawa, Y., Nakano, R.: Landscape of a likelihood surface for a gaussian mixture and its use for the EM algorithm. In: Proceedings of the IJCNN2006, pp. 2413–2419 (2006)
7. Ueda, N., Nakano, R.: Deterministic annealing EM algorithm. Neural Netw. **11**(2), 271–282 (1998)
8. Higashi, M., Kondo, T., Kanzawa, Y.: Fuzzy clustering method for spherical data based on q-divergence. JACIII **23**(3), 561–570 (2019)

On Some Fuzzy Clustering Algorithms with Cluster-Wise Covariance

Toshiki Ishii$^{(\boxtimes)}$ and Yuchi Kanzawa (iD)

Shibaura Institute of Technology, 3-7-5 Toyosu, Koto, Tokyo 135-8548, Japan
{ma21010,kanzawa}@shibaura-it.ac.jp

Abstract. In many fuzzy clustering algorithms, the KL-divergence-regularized method based on the Gaussian mixture model, fuzzy classification maximum likelihood, and a fuzzy mixture of Student's-t distributions have been proposed for cluster-wise anisotropic data, whereas more other types of fuzzification technique have been applied to fuzzy clustering for cluster-wise isotropic data. In this study, some fuzzy clustering algorithms are proposed based on the combinations between four types of fuzzification—namely, the Bezdek-type fuzzification, KL-divergence regularization, fuzzy classification maximum likelihood, and q-divergence-basis—and two types of mixture model—namely, the Gaussian mixture model and t-mixture model. Numerical experiments are conducted to demonstrate the features of the proposed methods.

Keywords: Fuzzy clustering · Cluster-wise anisotropic data · t distribution · q-divergence

1 Introduction

Clustering is a technique for partitioning a set of objects into subsets, where objects in the same cluster are more similar to other objects in other clusters. Fuzzy c-Means (FCM) [1] is the most popular fuzzy clustering algorithm. To differentiate this algorithm from other alternatives that have been proposed, such as q-divergence-based FCM (QFCM), this algorithm is referred to as the Bezdek-type FCM (BFCM) in this paper. Furthermore, a variable controlling the cluster size was introduced into the BFCM, which is referred to as modified BFCM (mBFCM) [2]. The QFCM objective function is obtained by introducing a fuzzification parameter to the q-divergence that appears at the lower bound of the q-log-likelihood of Gaussian mixture model (GMM) with the identity covariance. Thus, the fuzzification is referred to as Q-type fuzzification in this paper. It is noteworthy that the mBFCM objective function can be regarded as diverging the fuzzification parameter of the QFCM objective function. Thus, the fuzzification is referred to as B-type fuzzification in this paper.

One disadvantage of the above mentioned algorithms is that they tend to produce isotropic spherical clusters. Consequently, if the cluster shapes are

© Springer Nature Switzerland AG 2022
K. Honda et al. (Eds.): IUKM 2022, LNAI 13199, pp. 191–203, 2022.
https://doi.org/10.1007/978-3-030-98018-4_16

anisotropic, some objects that should be assigned to a cluster may be misclassified into another cluster. To overcome this issue, some approaches introduce the Mahalanobis distance between objects to capture covariance structures of clusters [3–6]. Yang [3] proposed the fuzzy classification maximum likelihood (FCML) by replacing the membership in the Classification Maximum Likelihood (CML) of GMM with its power. This fuzzification is referred to as F-type fuzzification in this paper, and the method in [3] is referred to as F-type fuzzified GMM (FFGMM). Ichihashi et al. [4] proposed the Kullback-Leibler (KL) divergence-regularized FCM (KFCM) by introducing a fuzzification parameter to the KL divergence appearing at the lower bound of the log-likelihood of GMM. This fuzzification is referred to as KL-type fuzzification in this paper, and this method is referred to as KL-type fuzzified GMM (KLFGMM). Furthermore, F-type fuzzification was applied to the Student's-t mixture models (TMM) [5,6], where the t-distribution is considered as a Gaussian scale mixture model and is evaluated using the expectation of the latent scale based on the EM algorithm. This method is referred to as F-type fuzzified TMM using the EM algorithm (FFTMM-EM). However, the above types of fuzzification are not always applied to GMM or TMM. There is a potential to increase clustering accuracy by combining fuzzification and the base distribution of mixture models.

In this study, we propose seven fuzzy clustering algorithms for classifying data using the anisotropic covariance structures of clusters. The first two proposed methods are referred to as the KL-type fuzzified TMM using the EM algorithm (KLFTMM-EM) and the KL-type fuzzified TMM using the MM algorithm (KLFTMM-MM), whose objective functions are constructed by KL-divergence-regularization of a lower bound of TMM log-likelihood. This is similar to the fact that the KLFGMM objective function is constructed by KL-divergence-regularization of a lower bound of the GMM log-likelihood. Here, for KLFTMM-EM, the t-distribution is considered as the Gaussian scale mixture model and is evaluated using the expectation of the latent scale, whereas for KLFTMM-MM, the t-distribution is evaluated using the hyperplane supporting the logarithmic function in the framework of the MM algorithm. The third proposed method is referred to as the F-type fuzzified TMM using the MM algorithm (FFTMM-MM), which is an alternative of FFTMM-EM. The FFTMM-EM [5,6] objective function is a realization of the FCML framework with the t-distribution as the component, where the t-distribution is considered as the Gaussian scale mixture model and is evaluated using the expectation of the latent scale. In the FFTMM-MM, however, the t-distribution is evaluated using the hyperplane supporting the logarithmic function in the framework of the MM algorithm. The fourth and fifth proposed methods are referred to as Q-type fuzzified GMM (QFGMM) and Q-type fuzzified TMM (QFTMM), respectively, whose objective functions are constructed by q-divergence-regularizing a lower bound of the q-log-likelihood for mixture models, where the component distributions are the Gaussian and t-distribution for QFGMM and QFTMM, respectively. This is similar to the fact that the QFCM objective function can be regarded as q-divergence-regularizing a lower bound of the q-log-likelihood for GMM with the identity covariances. The sixth and seventh proposed methods are referred to as B-type fuzzified GMM (BFGMM) and B-type fuzzified TMM (BFTMM), whose objective functions are constructed from the QFGMM and QFTMM objective functions as λ approaches infinity. This is similar to the fact that the mBFCM objective function can be regarded as diverging the fuzzification parameter λ of the QFCM objective function to infinity. The properties of the proposed methods were analyzed using an artificial dataset along with conventional methods.

The rest of this paper is organized as follows. In Sect. 2, we introduce the notations and conventional methods. The proposed methods are presented in Sect. 3, and the numerical experiments are described in Sect. 4. Our concluding remarks are made in Sect. 5.

2 Preliminaries

Let $X = \{x_k \in \mathbb{R}^M | k \in \{1, ..., N\}\}$ be a dataset of M dimensional points. The membership of x_k to the i-th cluster is denoted by $u_{i,k} (i \in \{1, ..., C\}, k \in \{1, ..., N\})$, and the set of $u_{i,k}$ is denoted by u, which satisfies the following constraint:

$$\sum_{i=1}^{C} u_{i,k} = 1, u_{i,k} \in [0, 1]. \tag{1}$$

The cluster center set is denoted by $v = \{v_i \mid v_i \in \mathbb{R}^M, i \in \{1, ..., C\}\}$. The variable controlling the i-th cluster size is denoted by π_i. The i-th element of vector π is denoted by π_i, and π satisfies the following constraint:

$$\sum_{i=1}^{C} \pi_i = 1, \pi_i \in (0, 1). \tag{2}$$

The methods mBFCM and QFCM are derived by solving the optimization problems,

$$\underset{u,v,\pi}{\text{minimize}} \sum_{i=1}^{C} \sum_{k=1}^{N} (\pi_i)^{1-m} (u_{i,k})^m \|x_k - v_i\|_2^2, \tag{3}$$

$$\underset{u,v,\pi}{\text{minimize}} \sum_{i=1}^{C} \sum_{k=1}^{N} (\pi_i)^{1-m} (u_{i,k})^m \|x_k - v_i\|_2^2 + \frac{\lambda^{-1}}{m-1} \sum_{i=1}^{C} \sum_{k=1}^{N} (\pi_i)^{1-m} (u_{i,k})^m, \tag{4}$$

respectively, subject to Eqs. (1) and (2), where $m > 1$ and $\lambda > 0$ are fuzzification parameters. The QFCM objective function is a Q-type fuzzification, which is obtained by introducing a fuzzification parameter to the q-divergence that appears at the lower bound of the q-log-likelihood of GMM with the identity covariances. The mBFCM objective function is a B-type fuzzification, which is obtained by diverging the fuzzification parameter of the QFCM objective function. It is noteworthy that mBFCM and QFCM cannot capture the covariance structure of clusters because their basis is GMM with identity covariances. The algorithms of mBFCM and QFCM are omitted.

The cluster-wise covariance is denoted by $A = \{A_i \mid A_i \in \mathbb{R}^{M \times M}, i \in \{1, ..., C\}\}$, and the degree of freedom is denoted by $\gamma = \{\gamma_i \mid \gamma_i \in \mathbb{R}, i \in \{1, ..., C\}\}$. The KLFGMM (KFCM) [4], FFGMM (FCML) [3], and FFTMM-EM (FSMM) [6] methods were derived by solving the optimization problems

$$\underset{u,v,\pi,A}{\text{minimize}} \sum_{i=1}^{C} \sum_{k=1}^{N} u_{i,k} d_{i,k}^{(1)} + \lambda^{-1} \sum_{i=1}^{C} \sum_{k=1}^{N} u_{i,k} \ln\left(\frac{u_{i,k}}{\pi_i}\right), \tag{5}$$

$$\underset{u,v,\pi,A}{\text{minimize}} \sum_{i=1}^{C} \sum_{k=1}^{N} (u_{i,k})^m d_{i,k}^{(1)} + \lambda^{-1} \sum_{i=1}^{C} \sum_{k=1}^{N} (u_{i,k})^m \ln\left(\frac{1}{\pi_i}\right), \tag{6}$$

$$\underset{u,v,\pi,A,\gamma}{\text{minimize}} \sum_{i=1}^{C} \sum_{k=1}^{N} (u_{i,k})^m d_{i,k}^{(2)} + \lambda^{-1} \sum_{i=1}^{C} \sum_{k=1}^{N} (u_{i,k})^m \ln\left(\frac{1}{\pi_i}\right), \tag{7}$$

respectively, where $\delta_{i,k}$, $d_{i,k}^{(1)}$, $d_{i,k}^{(2)}$, $\hat{\mu}_{i,k}$, and $\breve{\mu}_{i,k}$ are

$$\delta_{i,k} = (x_k - v_i)^\mathsf{T} A_i^{-1}(x_k - v_i), \tag{8}$$

$$d_{i,k}^{(1)} = \delta_{i,k} + \ln|A_i|, \tag{9}$$

$$d_{i,k}^{(2)} = 2\ln\Gamma\left(\frac{\gamma_i}{2}\right) - \gamma_i \ln\left(\frac{\gamma_i}{2}\right) - \gamma_i\left(\breve{\mu}_{i,k} - \hat{\mu}_{i,k}\right) + 2\breve{\mu}_{i,k}$$
$$\qquad + \hat{\mu}_{i,k}\delta_{i,k} + M\ln(2\pi) + \ln(|A_i|) - M\breve{\mu}_{i,k}, \tag{10}$$

$$\hat{\mu}_{i,k} = \frac{\gamma_i + M}{\gamma_i + \delta_{i,k}}, \tag{11}$$

$$\breve{\mu}_{i,k} = \ln(\hat{\mu}_{i,k}) - \ln\left(\frac{\gamma_i + M}{2}\right) + F\left(\frac{\gamma_i + M}{2}\right), \tag{12}$$

and $\Gamma(x)$ and $F(x)$ are the Gamma and digamma function. The KLFGMM objective function is a KL-type fuzzification, which is obtained by introducing a fuzzification parameter to the KL divergence appearing at the lower bound of the log-likelihood of the Gaussian mixture models. The FFGMM objective function is an F-type fuzzification, which is obtained by replacing the membership in the CML of the GMM with its power. The FFTMM-EM objective function is also an F-type fuzzification, which is obtained by replacing the membership in the CML of the TMM with its power. KLFGMM and FFGMM are based on GMM. FFTMM-EM is based on TMM. The KLFGMM, FFGMM, and FFTMM-EM algorithms are summarized as follows.

Algorithm 1 (KLFGMM, FFGMM, FFTMM-EM)

1. Set the number of clusters C. Set the fuzzification parameter m for FFGMM and FFTMM-EM. Set the fuzzification parameter λ for KLFGMM. Set the initial cluster center v, initial variable controlling cluster size π, initial cluster-wise covariate A, and initial degree of freedom γ.
2. Calculate $\hat{\mu}_{i,k}$ using Eq. (11), and $\breve{\mu}_{i,k}$, as shown in Eq. (12) for FFTMM-EM
3. Calculate d using $d_{i,k} = d_{i,k}^{(1)}$ for KLFGMM and FFGMM, and $d_{i,k} = d_{i,k}^{(2)}$ for FFTMM-EM.
4. Calculate u as

$$u_{i,k} = \frac{\pi_i \exp\left(-\lambda d_{i,k}\right)}{\sum_{j=1}^{C} \pi_j \exp\left(-\lambda d_{j,k}\right)} \tag{13}$$

for KLFGMM, and

$$u_{i,k} = \frac{\left(d_{i,k} - \lambda^{-1}\ln\pi_i\right)^{\frac{1}{1-m}}}{\sum_{j=1}^{C}\left(d_{j,k} - \lambda^{-1}\ln\pi_j\right)^{\frac{1}{1-m}}} \tag{14}$$

for FFGMM and FFTMM-EM.
5. Calculate π using

$$\pi_i = \frac{\sum_{k=1}^{N} u_{i,k}}{N} \tag{15}$$

for KLFGMM, and

$$\pi_i = \frac{\sum_{k=1}^{N}(u_{i,k})^m}{\sum_{j=1}^{C}\sum_{k=1}^{N}(u_{j,k})^m} \tag{16}$$

for FFGMM and FFTMM-EM.
6. Calculate v using

$$v_i = \frac{\sum_{k=1}^{N} u_{i,k}x_k}{\sum_{k=1}^{N} u_{i,k}} \tag{17}$$

for KLFGMM-EM,

$$v_i = \frac{\sum_{k=1}^{N}(u_{i,k})^m x_k}{\sum_{k=1}^{N}(u_{i,k})^m} \tag{18}$$

for FFGMM, and

$$v_i = \frac{\sum_{k=1}^{N}(u_{i,k})^m \hat{\mu}_{i,k}x_k}{\sum_{k=1}^{N}(u_{i,k})^m \hat{\mu}_{i,k}} \tag{19}$$

for FFTMM-EM.

7. Obtain γ by solving the following equation

$$-F\left(\frac{\gamma_i}{2}\right) + \ln\left(\frac{\gamma_i}{2}\right) + F\left(\frac{\gamma_i + M}{2}\right) - \ln\left(\frac{\gamma_i + M}{2}\right) + 1$$

$$+ \frac{1}{\sum_{k=1}^{N}(u_{i,k})^m} \sum_{k=1}^{N}(u_{i,k})^m \left(\ln(\hat{\mu}_{i,k}) - \hat{\mu}_{i,k}\right) = 0 \qquad (20)$$

for FFTMM-EM.

8. Calculate A using

$$A_i = \frac{\sum_{k=1}^{N} u_{i,k}(x_k - v_i)(x_k - v_i)^{\mathsf{T}}}{\sum_{k=1}^{N} u_{i,k}} \qquad (21)$$

for KLFGMM,

$$A_i = \frac{\sum_{k=1}^{N}(u_{i,k})^m(x_k - v_i)(x_k - v_i)^{\mathsf{T}}}{\sum_{k=1}^{N}(u_{i,k})^m} \qquad (22)$$

for FFGMM, and

$$A_i = \frac{\sum_{k=1}^{N}(u_{i,k})^m \hat{\mu}_{i,k}(x_k - v_i)(x_k - v_i)^{\mathsf{T}}}{\sum_{k=1}^{N}(u_{i,k})^m \hat{\mu}_{i,k}} \qquad (23)$$

for FFTMM-EM.

9. Check the limiting criterion for (u, v, π, γ, A). If the criterion is not satisfied, then go to 2.

3   Proposed Methods

3.1   Concept

This study proposes clustering methods with cluster-wise covariance based on the GMM or TMM.

The first two proposed methods are referred to as the KL-type fuzzified TMM using the EM algorithm (KLFTMM-EM) and the KL-type fuzzified TMM using the MM algorithm (KLFTMM-MM). Their objective functions are constructed via KL-divergence-regularization of a lower bound of the TMM log-likelihood. This is similar to the construction of the KLFGMM objective function via KL-divergence-regularization of a lower bound of the GMM log-likelihood. Here, for KLFTMM-EM, the t-distribution is considered as the Gaussian scale mixture model and is evaluated using the expectation of the latent scale, whereas for KLFTMM-MM, the t-distribution is evaluated using the hyperplane supporting the logarithmic function in the framework of the MM algorithm. The KLFTMM-EM and KLFTMM-MM objective functions are therefore given as

$$\underset{u,v,\pi,A,\gamma}{\text{minimize}} \sum_{i=1}^{C} \sum_{k=1}^{N} u_{i,k} d_{i,k}^{(2)} + \lambda^{-1} \sum_{i=1}^{C} \sum_{k=1}^{N} u_{i,k} \ln\left(\frac{u_{i,k}}{\pi_i}\right), \tag{24}$$

$$\underset{u,v,\pi,A,\gamma}{\text{minimize}} \sum_{i=1}^{C} \sum_{k=1}^{N} u_{i,k} d_{i,k}^{(3)} + \lambda^{-1} \sum_{i=1}^{C} \sum_{k=1}^{N} u_{i,k} \ln\left(\frac{u_{i,k}}{\pi_i}\right), \tag{25}$$

respectively, where

$$d_{i,k}^{(3)} = -2\ln\Gamma\left(\frac{\gamma_i + M}{2}\right) + 2\ln\Gamma\left(\frac{\gamma_i}{2}\right) + M\ln(\gamma_i\pi) + \ln(|A_i|)$$

$$+ \hat{\mu}_{i,k}(\gamma_i + \delta_{i,k}) - (\gamma_i + M)\left(\ln(\gamma_i) - \ln\left(\frac{(\gamma_i + M)}{\hat{\mu}_{i,k}}\right) + 1\right). \tag{26}$$

The third proposed method is referred to as the F-type fuzzified TMM using the MM algorithm (FFTMM-MM), which is an alternative to FFTMM-EM. The FFTMM-EM objective function is a realization of the FCML framework with the t-distribution as the component, where the t-distribution is considered as the Gaussian scale mixture model and is evaluated using the expectation of the latent scale, whereas, in FFTMM-MM, the t-distribution is evaluated using the hyperplane supporting the logarithmic function in the framework of the MM algorithm. The FFTMM-MM objective function is then given as

$$\underset{u,v,\pi,A,\gamma}{\text{minimize}} \sum_{i=1}^{C} \sum_{k=1}^{N} (u_{i,k})^m d_{i,k}^{(3)} + \lambda^{-1} \sum_{i=1}^{C} \sum_{k=1}^{N} (u_{i,k})^m \ln\left(\frac{1}{\pi_i}\right). \tag{27}$$

The fourth and fifth proposed methods are referred to as the Q-type fuzzified GMM (QFGMM) and the Q-type fuzzified TMM (QFTMM), respectively, which are constructed by q-divergence-regularizing a lower bound of the q-log-likelihood for mixture models, where the component distributions are the Gaussian and t-distribution for QFGMM and QFTMM, respectively. This is similar to the QFCM objective function constructed by q-divergence-regularizing a lower bound of the q-log-likelihood for GMM with identity covariances. The QFGMM and QFTMM objective functions are therefore given as

$$\underset{u,v,\pi,A}{\text{minimize}} \sum_{i=1}^{C} \sum_{k=1}^{N} (\pi_i)^{1-m}(u_{i,k})^m d_{i,k}^{(4)} + \frac{\lambda^{-1}}{m-1} \sum_{i=1}^{C} \sum_{k=1}^{N} (\pi_i)^{1-m}(u_{i,k})^m, \tag{28}$$

$$\underset{u,v,\pi,A,\gamma}{\text{minimize}} \sum_{i=1}^{C} \sum_{k=1}^{N} (\pi_i)^{1-m}(u_{i,k})^m d_{i,k}^{(5)} + \frac{\lambda^{-1}}{m-1} \sum_{i=1}^{C} \sum_{k=1}^{N} (\pi_i)^{1-m}(u_{i,k})^m, \tag{29}$$

respectively, where

$$d_{i,k}^{(4)} = \frac{1}{m-1} \left(\left((2\pi)^{\frac{M}{2}} |A_i|^{\frac{1}{2}} \exp\left(\frac{1}{2} \delta_{i,k} \right) \right)^{m-1} - 1 \right), \tag{30}$$

$$d_{i,k}^{(5)} = \frac{1}{m-1} \left(\left(\frac{\Gamma\left(\frac{\gamma_i}{2}\right) |A_i|^{\frac{1}{2}}}{(\pi\gamma_i)^{-\frac{M}{2}} \Gamma\left(\frac{\gamma_i+M}{2}\right)} \left(1 + \frac{1}{\gamma_i} \delta_{i,k} \right)^{\frac{\gamma_i+M}{2}} \right)^{m-1} - 1 \right). \tag{31}$$

The sixth and seventh proposed methods are referred to as B-type fuzzified GMM (BFGMM) and B-type fuzzified TMM (BFTMM) because their objective functions are constructed from the QFGMM and QFTMM objective functions, respectively, by making λ approach infinity. This is similar to the fact that the mBFCM objective function is constructed by diverging the fuzzification parameter λ of the QFCM objective function to infinity. The BFGMM and BFTMM objective functions are then given as

$$\underset{u,v,\pi,A}{\text{minimize}} \sum_{i=1}^{C} \sum_{k=1}^{N} (\pi_i)^{1-m} (u_{i,k})^m d_{i,k}^{(4)}, \tag{32}$$

$$\underset{u,v,\pi,A,\gamma}{\text{minimize}} \sum_{i=1}^{C} \sum_{k=1}^{N} (\pi_i)^{1-m} (u_{i,k})^m d_{i,k}^{(5)}. \tag{33}$$

3.2   Algorithm

The proposed clustering algorithms are obtained by solving the optimization problems given in Eqs. (24), (25), (27), (28), (29), (32), and (33) subject to the constraints in Eqs. (1) and (2), respectively. The analysis of the necessary conditions of optimality, although the detail is omitted for brevity, is summarized by the following algorithm:

Algorithm 2 (KLFTMM-EM,   KLFTMM-MM,   FFTMM-MM, QFGMM, QFTMM, BFGMM, BFTMM)

1. Set the number of clusters C. Set the fuzzification parameter m for FFTMM-MM, QFGMM, QFTMM, BFGMM, and BFTMM. Set the fuzzification parameter λ for KLFTMM-EM, KLFTMM-MM, QFGMM, and QFTMM. Set the initial cluster center v, initial variable controlling cluster size π, initial cluster-wise covariate A, and initial degree of freedom γ.
2. Calculate $\hat{\mu}_{i,k}$ using Eq. (11), and $\check{\mu}_{i,k}$ as Eq. (12) for KLFTMM-EM, KLFTMM-MM, FFTMM-MM, QFTMM, and BFTMM.
3. Calculate d as $d_{i,k} = d_{i,k}^{(2)}$ for KLFTMM-EM, $d_{i,k} = d_{i,k}^{(3)}$ for KLFTMM-MM and FFTMM-MM, $d_{i,k} = d_{i,k}^{(4)}$ for QFGMM and BFGMM, and $d_{i,k} = d_{i,k}^{(5)}$ QFTMM and BFTMM.

4. Calculate u as

$$u_{i,k} = \frac{\pi_i \exp\left(-\lambda d_{i,k}\right)}{\sum_{j=1}^{C} \pi_j \exp\left(-\lambda d_{j,k}\right)} \tag{34}$$

for KLFTMM-EM and KLFTMM-MM,

$$u_{i,k} = \frac{\left(d_{i,k} - \lambda^{-1} \ln \pi_i\right)^{\frac{1}{1-m}}}{\sum_{j=1}^{C} \left(d_{j,k} - \lambda^{-1} \ln \pi_j\right)^{\frac{1}{1-m}}} \tag{35}$$

for FFTMM-MM,

$$u_{i,k} = \frac{\pi_i \left(1 - \lambda\left(1 - m\right) d_{i,k}\right)^{\frac{1}{1-m}}}{\sum_{j=1}^{C} \pi_j \left(1 - \lambda\left(1 - m\right) d_{j,k}\right)^{\frac{1}{1-m}}} \tag{36}$$

for QFGMM and QFTMM, and

$$u_{i,k} = \frac{\pi_i \left(d_{i,k}\right)^{\frac{1}{1-m}}}{\sum_{j=1}^{C} \pi_j \left(d_{j,k}\right)^{\frac{1}{1-m}}} \tag{37}$$

for BFGMM and BFTMM.
5. Calculate π using

$$\pi_i = \frac{\sum_{k=1}^{N} u_{i,k}}{N} \tag{38}$$

for KLFTMM-EM and KLFTMM-MM,

$$\pi_i = \frac{\sum_{k=1}^{N} \left(u_{i,k}\right)^m}{\sum_{j=1}^{C} \sum_{k=1}^{N} \left(u_{j,k}\right)^m} \tag{39}$$

for FFTMM-MM,

$$\pi_i = \frac{\left(\sum_{k=1}^{N} \left(u_{i,k}\right)^m \left(1 - \lambda\left(1 - m\right) d_{i,k}\right)\right)^{\frac{1}{m}}}{\sum_{j=1}^{C} \left(\sum_{k=1}^{N} \left(u_{j,k}\right)^m \left(1 - \lambda\left(1 - m\right) d_{j,k}\right)\right)^{\frac{1}{m}}} \tag{40}$$

for QFGMM and QFTMM, and

$$\pi_i = \frac{\left(\sum_{k=1}^{N} \left(u_{i,k}\right)^m d_{i,k}\right)^{\frac{1}{m}}}{\sum_{j=1}^{C} \left(\sum_{k=1}^{N} \left(u_{j,k}\right)^m d_{j,k}\right)^{\frac{1}{m}}} \tag{41}$$

for BFGMM and BFTMM.

6. Calculate v using

$$v_i = \frac{\sum_{k=1}^{N} u_{i,k}\hat{\mu}_{i,k}x_k}{\sum_{k=1}^{N} u_{i,k}\hat{\mu}_{i,k}} \tag{42}$$

for KLFTMM-EM and KLFTMM-MM,

$$v_i = \frac{\sum_{k=1}^{N} (u_{i,k})^m \hat{\mu}_{i,k}x_k}{\sum_{k=1}^{N} (u_{i,k})^m \hat{\mu}_{i,k}} \tag{43}$$

for FFTMM-MM,

$$v_i = \frac{\sum_{k=1}^{N} (u_{i,k})^m \exp\left(-\frac{1}{2}\delta_{i,k}(1-m)\right) x_k}{\sum_{k=1}^{N} (u_{i,k})^m \exp\left(-\frac{1}{2}\delta_{i,k}(1-m)\right)} \tag{44}$$

for QFGMM and BFGMM, and

$$v_i = \frac{\sum_{k=1}^{N} (u_{i,k})^m \hat{w}_{i,k}x_k}{\sum_{k=1}^{N} (u_{i,k})^m \hat{w}_{i,k}}, \tag{45}$$

$$\hat{w}_{i,k} = \frac{\gamma_i + M}{(\gamma_i + \delta_{i,k})^{1+\frac{(1-m)(\gamma_i+M)}{2}}} \tag{46}$$

for QFTMM and BFTMM.
7. Obtain γ by solving the equation

$$-F\left(\frac{\gamma_i}{2}\right) + \ln\left(\frac{\gamma_i}{2}\right) + F\left(\frac{\gamma_i+M}{2}\right) - \ln\left(\frac{\gamma_i+M}{2}\right)$$
$$+1 + \frac{1}{\sum_{k=1}^{N} u_{i,k}} \sum_{k=1}^{N} u_{i,k}\left(\ln(\hat{\mu}_{i,k}) - \hat{\mu}_{i,k}\right) = 0 \tag{47}$$

for KLFTMM-EM,

$$F\left(\frac{\gamma_i+M}{2}\right) - F\left(\frac{\gamma_i}{2}\right) + \ln(\gamma_i) - \ln(\gamma_i+M) + 1$$
$$+ \frac{1}{\sum_{k=1}^{N} u_{i,k}} \sum_{k=1}^{N} u_{i,k}\left(\ln(\hat{\mu}_{i,k}) - \hat{\mu}_{i,k}\right) = 0 \tag{48}$$

for KLFTMM-MM,

$$F\left(\frac{\gamma_i+M}{2}\right) - F\left(\frac{\gamma_i}{2}\right) + \ln(\gamma_i) - \ln(\gamma_i+M) + 1$$
$$+ \frac{1}{\sum_{k=1}^{N} (u_{i,k})^m} \sum_{k=1}^{N} (u_{i,k})^m\left(\ln(\hat{\mu}_{i,k}) - \hat{\mu}_{i,k}\right) = 0 \tag{49}$$

for FFTMM-MM, and

$$\ln\left(\frac{\gamma_i}{2}\right) - F\left(\frac{\gamma_i}{2}\right) + 1$$

$$+ \frac{1}{\sum_{k=1}^{N}(u_{i,k})^m f_i(x_k)^{1-m}} \sum_{k=1}^{N}(u_{i,k})^m f_i(x_k)^{1-m}(\check{\mu}_{i,k} - \hat{\mu}_{i,k}) = 0, \quad (50)$$

$$f_i(x_k) = \left(\frac{\Gamma\left(\frac{\gamma_i+M}{2}\right)|A_i|^{-\frac{1}{2}}}{(\pi\gamma_i)^{\frac{M}{2}}\Gamma\left(\frac{\gamma_i}{2}\right)}\left(1 + \frac{1}{\gamma_i}\delta_{i,k}\right)\right) \quad (51)$$

for QFTMM and BFTMM.
8. Calculate A using

$$A_i = \frac{\sum_{k=1}^{N} u_{i,k}\hat{\mu}_{i,k}(x_k - v_i)(x_k - v_i)^{\mathsf{T}}}{\sum_{k=1}^{N} u_{i,k}} \quad (52)$$

for KLFTMM-EM and KLFTMM-MM,

$$A_i = \frac{\sum_{k=1}^{N}(u_{i,k})^m \hat{\mu}_{i,k}(x_k - v_i)(x_k - v_i)^{\mathsf{T}}}{\sum_{k=1}^{N}(u_{i,k})^m} \quad (53)$$

for FFTMM-MM,

$$A_i = \frac{\sum_{k=1}^{N}(u_{i,k})^m \exp\left(-\frac{1}{2}\delta_{i,k}(1-m)\right)(x_k - v_i)(x_k - v_i)^{\mathsf{T}}}{\sum_{k=1}^{N}(u_{i,k})^m \exp\left(-\frac{1}{2}\delta_{i,k}(1-m)\right)} \quad (54)$$

for QFGMM and BFGMM, and

$$A_i = \frac{\sum_{k=1}^{N}(u_{i,k})^m \hat{w}_{i,k}(x_k - v_i)(x_k - v_i)^{\mathsf{T}}}{\sum_{k=1}^{N}(u_{i,k})^m \hat{v}_{i,k}}, \quad (55)$$

$$\hat{v}_{i,k} = \frac{1}{(\gamma_i + \delta_{i,k})^{\frac{(1-m)(\gamma_i+M)}{2}}} \quad (56)$$

for QFTMM and BFTMM.
9. Check the limiting criterion for (u, v, π, γ, A). If the criterion is not satisfied, then go to 2.

4 Numerical Experiments

This section provides an experiment based on an artificial dataset. We consider a random sample consisting of 300 simulated points from a two-component bivariate GMM, to which 150 noise points (outliers) were added from a uniform distribution over the range $[-10, 10]$ for each variable. The means of the mixture component densities are $v_1 = (5.0, 5.0)^{\mathsf{T}}$, $v_2 = (-5.0, -5.0)^{\mathsf{T}}$, and their covariances are $A_1 = \begin{pmatrix} 2.0 & -0.5 \\ -0.5 & 0.5 \end{pmatrix}$, $A_2 = \begin{pmatrix} 1.0 & 0.5 \\ 0.5 & 0.1 \end{pmatrix}$, and their mixing proportions are $\pi_1 = \pi_2 = 0.5$. The experimental results were evaluated using the sum of the absolute deviances between the obtained values from each algorithm and the corresponding correct values for mixing proportions, means of the mixture component densities, and covariances of the mixture component densities; here, the smaller the values, the higher the accuracy. The fuzzification parameters m and λ vary as follows: $m \in \{1+10^{-3}, 1+10^{-2}, 1+10^{-1}, 1.2, 1.25, 1.5, 2.0, 2.5, 5.0, 10.0, 100.0\}$, $\lambda \in \{10^{-3}, 10^{-2}, 10^{-1}, 0.25, 0.5, 1.0, 1.25, 1.5, 2.0, 5.0, 10.0, 100.0\}$. In Table 1, we provide a summary of the obtained results with the highest accuracy for each method. As can be noticed, the methods based on TMM offer enhanced data classification capabilities compared to those based on GMM. It is because t-distribution is robust to outliers. Among the methods based on GMM, conventional methods are more accurate than the proposed methods. All the proposed methods based on TMM are more accurate than conventional methods with FFTMM-MM being the most accurate on all methods. Its theoretical reason will be investigated in future research (Fig. 1).

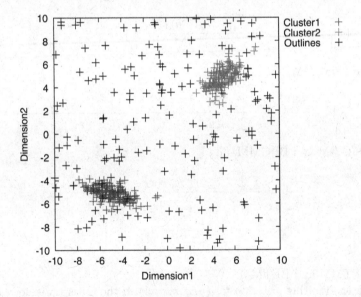

Fig. 1. Artificial dataset

Table 1. Experimental results

		KL-type	F-type	Q-type	B-type
GMM		25.298	25.303	25.346	25.423
TMM	EM	1.423	1.248	1.409	1.619
	MM	1.404	1.238		

5   Conclusion

In this study, we proposed seven fuzzy clustering algorithms for classifying data using the anisotropic covariance structures of clusters. The experimental results showed that the methods based on TMM are more accurate than those based on GMM; moreover, they are less sensitive to outliers. In future research, the relation between each fuzzification type and its clustering accuracy will be investigated. Furthermore, real datasets will be applied to the proposed methods, and the clustering accuracy will be compared with that of the conventional methods.

References

1. Bezdek, J.: Pattern Recognition with Fuzzy Objective Function Algorithms. Plenum Press, New York (1981)
2. Miyamoto, S., Kurosawa, N.: Controlling cluster volume sizes in fuzzy c-means clustering. In: Proceedings of SCIS&ISIS 2004, pp. 1–4 (2004)
3. Yang, M.-S.: On a class of fuzzy classification maximum likelihood procedures. Fuzzy Sets Syst. **57**, 365–375 (1993)
4. Ichihashi, H., Honda, K., Tani, N.: Gaussian mixture PDF approximation and fuzzy c-means clustering with entropy regularization. In: Proceedings of 4th Asian Fuzzy System Symposium, pp. 217–221 (2000)
5. Yang, M.-S., Lin, C.-Y., Tian, Y.-C.: Fuzzy classification maximum likelihood clustering with T-distribution. Appl. Mech. Mater. **598**, 392–397 (2014)
6. Chatzis, S., Varvarigou, T.: Robust fuzzy clustering using mixtures of student's-t distributions. Pattern Recogn. Lett. **29**, 1901–1905 (2008)

Hybrid Rule-Based Classification by Integrating Expert Knowledge and Data

Lianmeng Jiao[✉], Haonan Ma, and Quan Pan

School of Automation, Northwestern Polytechnical University,
Xi'an 710072, People's Republic of China
{jiaolianmeng,quanpan}@nwpu.edu.cn,mhn2020@mail.nwpu.edu.cn

Abstract. The common methods for dealing with classification problems include data-driven models and knowledge-driven models. Recently, some methods were proposed to combine the data-driven model with the knowledge-driven model to construct a hybrid model, which improves the classification performance by complementing each other. However, most of the existing methods just assume that the expert knowledge is known in advance, and do not indicate how to obtain it. To this end, this paper proposes a way to obtain knowledge from experts represented by rules through active learning. Then, a hybrid rule-based classification model is developed by integrating the knowledge-driven rule base and the rule base learned from the training data using genetic algorithm. Experiments based on real datasets demonstrate the superiority of the proposed classification model.

Keywords: Expert knowledge acquisition · Hybrid classification · Active learning

1 Introduction

According to the types of information used in the classification problem, classification models are mainly divided into knowledge-driven ones and data-driven ones. The knowledge-driven classification models rely on experts with certain professional knowledge who understand a specific field. The most common example is the expert system. For example, in [18], the authors develop an expert system for knee problems diagnosis, and in [1], an expert system is designed to provide appropriate solutions for human gingival problems. The knowledge-driven model provides a comprehensible way to understand the classification process. However, because the representation of knowledge in the classification model is fixed, it has poor characterization for the specifics of the data distribution and the dynamics of system behavior. The data-driven classification models learn classifiers from the available training data directly. Common data-driven classification models include support vector machines [13], k nearest neighbors [8], artificial neural networks [21] and fuzzy rule-based systems [20], etc. The

K. Honda et al. (Eds.): IUKM 2022, LNAI 13199, pp. 204–215, 2022.
https://doi.org/10.1007/978-3-030-98018-4_17

training data trends to provide a relatively fine estimates for the real class-conditional distributions, but they may be unreliable in some specific regions of feature space, due to limited training patterns and the potential measurement noise.

In order to inherit the advantages of data-driven and knowledge-driven models, some scholars have proposed to integrate them to solve the classification problem [14]. According to how to combine expert knowledge and data, these hybrid models can be divided into the following three categories:

- The experts first gives an initial model, and then the parameters of the model are optimized by the data. In [5], Bayesian networks are first constructed by experts, and then some parameters in the networks are learned from the data. In [19], the expert knowledge is first transformed into the form of rules, and then the fuzzy membership functions are learned from the data. In [22], the if-then rule base is first constructed by experts, and then the modification of rule parameters is determined by the data. In [9], fuzzy partition is first designed by experts to ensure high interpretability, and then data is used to update system parameters.
- The experts give corresponding suggestions on the models learned from the data. In [23], expert knowledge is transformed into the corresponding constraints, under which the model is learned from the data. In [6], expert knowledge is introduced when data is scarce by being asked to determine whether to revise the boundaries of the data-driven network or not. In [16], a method for human pose classification is proposed first to learn the decision tree from the data and then to prune the decision tree by experts.
- The expert knowledge and data information are transformed into the same model for fusion. In [10], the form of fuzzy rules is taken as a common model to characterize the two types of information. In fuzzy partition level, expert knowledge and data are first fused, and then the second fusion is carried out at the rule level. In [3], the model proposed in [10] was improved on the whole and verified on UCI data. In [2] and [4], an overall summary of the previous work is made and some improvement measures are put forward. In [15], a hybrid intelligent system for medical diagnosis is proposed. First, it learns incremental neural networks from data, and then transforms the network into rules. If expert knowledge is available, it allows expert knowledge to modify the rules, and then maps rules back to the network, and finally merges the rules at the decision-making layer. In [12], the method of evidential reasoning is adopted and the expert rule base and data rule base are integrated under the belief function framework.

In the above papers, some feasible schemes to solve classification problems by combining expert knowledge with data have been proposed. However, in the existing methods, all of them assume that the hybrid classification is carried out under the condition that the expert knowledge already exists, and do not point out how to obtain it. In fact, when dealing with classification problems, people often make decisions based on their subconscious mind, so it is not easy to express and model human knowledge. To solve this problem, an expert knowledge acquisition method based on active learning is proposed in this paper. Through active learning, unlabeled samples that contribute most to the current model are selected, and these samples are given to experts to mark, which are further transformed into rules. In this way, the acquired expert knowledge can well complement the initial training data. On this basis, a **H**ybrid rule-based **C**lassification model by **I**ntegrating **E**xpert knowledge and **D**ata (HCIED) is proposed to improve the classification performance. The learned expert rule base is combined with the rule base learned from training data by genetic algorithm, which selects the most important rules for classification.

The remaining paper will be arranged as follows. Section 2 describes details of the proposed hybrid rule-based classification model by integrating expert knowledge and data. The experiments is carried out in Sect. 3, and the paper is summarized in Sect. 4.

2   HCIED: Hybrid Rule-Based Classification by Integrating Expert Knowledge and Data

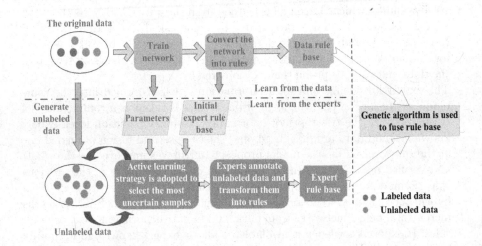

Fig. 1. The structure of the proposed HCIED model

The structure of the proposed HCIED model is shown in Fig. 1. It can be seen that the proposed hybrid classification model is composed of three modules: learning rule base from data (top-left part), learning rule base from experts (bottom-left part), and combining them through genetic algorithm (right part). The process of each module is described in detail as follows. First, learn a network model from the data and record the parameters of the network, and then convert the network into rules. After the conversion is completed, the rules with the highest confidence in each category are selected from the rule base to form the initial expert rule base. At the same time, a group of the unlabeled samples are generated from the initial data, and those samples that minimize the overall information entropy are selected and handed over to experts for annotation. The rule base are updated accordingly with the annotations of experts. When the expert rule base and the data-driven rule base are all learned, genetic algorithm is used to fuse these two rule bases by selecting the antecedent attribute of the rule and selecting the rule itself at the same time.

2.1 Rule Base Learning from the Data

As reviewed in the introduction, there are several proposals for rule base learning from the data in literatures. Here, we adopt the two-step rule extraction strategy in [15] for its well performance. First, a classification network is learned from data, which can be regarded as the process of extracting information from data. Once the network is ready, it is then converted into the form of rules. In the following part, we summarize the main processes of these two steps. .

The classification network learning process is carried out as follows. At the beginning, the first training sample is read in to construct the first node, the mean value of the Gaussian membership function associated with this node is set as the attribute value of the first sample, and the standard deviation of the Gaussian membership function is set as the initial value. The parameters of the first node and the label of the first sample are stored in matrices W_p and W_t, respectively. Then, the next training sample is read in and its distance to the learned node is measured. If the current training sample is considered to belong to a certain node, the parameters of this node are updated, and if the current training sample is different from the existing node, a new node is constructed using it. All training samples perform this process until the end of training.

When all the nodes of the network have been learned, the network will be transformed into the form of rules. On the one hand, the form of rules has a good interpretability. On the other hand, the knowledge of experts is generally presented in the form of rules. The form of the rules can make it easy for experts to understand and facilitate the integration of the rules afterwards.

In network learning, the main parameters include the mean matrix W_p, the label matrix W_t, the standard deviation matrix, and the number of samples contained in each node. In the process of transforming network into rules, the number of linguistic variables is first set, for example, using three linguistic variables: small, medium, and large. Then, the maximum and minimum values of each attribute in the matrix W_p are selected, based on which the fuzzy partition of each attribute are designed (this fuzzy partition will be used again when acquiring expert knowledge). After the fuzzy partition is obtained, each attribute value in the matrix W_p is converted into the corresponding rule antecedent variables. The conclusion of the rule is the value stored in the matrix W_t, and the weight of the rule is calculated by the number of samples contained in the node. In this way, each node in the network will be converted into a rule.

2.2   Knowledge Acquiring from Experts with Active Learning

In many learning tasks, one of the biggest challenges is to obtain enough labeled data for modeling. However, the acquisition of labeled data is expensive and usually requires a lot of human resources. In many fields, unlabeled data is easy to obtain. For these unlabeled samples, their contributions to the classification model are different. The principle of the active learning strategy is to select the sample that contributes the most to the classification model for labeling [11]. In this paper, we design a way to obtain knowledge from experts using active learning strategy. The main idea is that to find out those unlabeled samples that minimize the overall information entropy of the current classification model, and then give this sample to experts for labeling. These labeled samples are used to construct the knowledge rule base, which will be further combined with the rule base learned from the initial training sample. The generated knowledge rule base is considered to compensate for the weakness of the data-driven model effectively. In the following part, we will outline the main steps of the proposed method for acquiring knowledge from experts.

Step 1: Calculate the mean and standard deviation of each attribute for each class of training samples, and generate N_U unlabeled samples (100 as default value) for each class using Gaussian distribution with the corresponding mean and standard deviation.

Step 2: From the data-driven rule base, the rules with the highest confidence in each category are selected to form the initial expert rule base.

Step 3: The expert rule base is used to predict the labels of all unlabeled samples, and the pseudo labels of the current unlabeled samples are obtained as

$$y_i^* = arg \max_y P\left(y \mid x_i, R\right), \quad i = 1, 2, \ldots, N_U, \tag{1}$$

where x_i is an unlabeled samples and R represents the current expert rule base.

Step 4: Each pseudo-labeled sample is used to train a rule, which is then added to the current expert rule base in turn to get a temporary expert rule base R_i as

$$R_i = R \cup r_i, \quad i = 1, 2, \ldots, N_U, \tag{2}$$

where R represents the current expert rule base and r_i represents the rule learned from the pseudo-labeled sample x_i.

Step 5: Each temporary expert rule base R_i is used to measure the information entropy of the remaining pseudo-labeled samples as

$$H\left(Y \mid x_j, R_i\right) = -\sum_{y \in Y} P\left(y \mid x_j, R_i\right) log P\left(y \mid x_j, R_i\right),$$
$$j = 1, 2, \ldots, N_U, j \neq i, \text{ and } i = 1, 2, \ldots, N_U, \tag{3}$$

where Y is the set of class labels.

Step 6: Calculate the sum of the information entropy associated with each temporary expert rule base R_i, and select the sample x_i that minimizes the overall information entropy as

$$i^* = arg \min_i \sum_j H(Y \mid x_j, R_i),$$

$$j = 1, 2, \ldots, N_U, j \neq i, \text{ and } i = 1, 2, \ldots, N_U. \quad (4)$$

Step 7: The selected sample x_{i*} is given to experts to label, and the obtained real label is compared with the pseudo label. If the pseudo label is the same as the real label, the expert rule base will be updated as R_{i*}. If the real label is different from the pseudo label, then learn a rule from the expert-labeled sample and add it to the expert rule base. Besides, it is also needed to search for the most uncertain sample under the updated expert rule base for expert querying, and add the corresponding learned rule to the expert rule base.

Step 8: Execute Step 3 to Step 8 iteratively to get the updated expert rule base until a number of βN_U samples (the dafault value of β is set as 0.2) are labeled by experts.

Step 9: Remove the initial rules from the updated rule base to get the final expert rule base.

Using this expert knowledge extraction strategy, those samples in the attribute space uncovered by training samples have priorities to be labeled by experts. Therefore, it is possible to obtain expert knowledge that is most complementary to the current data-driven model. In next section, we will develop a hybrid model which provides complementary features from both the data-driven rule base and expert rule base.

2.3   Fusion of Rule Bases with Genetic Algorithm

After the knowledge acquiring procedure, an expert rule base different from the data rule base is obtained. During the integration of two rule bases from different sources, there may be rule conflicts and rule redundancy. Genetic algorithm is a random search strategy designed for optimization problems. Thus, genetic algorithm is able to solve problems that may arise during the integration of rule bases. Through binary encoding, the antecedent attribute of the rule and the rule itself are selected at the same time. In this way, the legibility of the rules can be increased, and the size of the rule base can be reduced at the same time. An illustration of the fusion process with genetic algorithm is shown in Example 1.

Example 1. Let $R_h = \begin{cases} 1\,2\,1, 2 \\ 2\,3\,2, 3 \\ 2\,2\,3, 1 \\ 2\,3\,1, 3 \\ 2\,2\,1, 2 \end{cases}$ represents a hybrid rule base containing 5

rules with 3 categories determined by 3 attributes. Using three linguistic labels

$\{1: small, 2: medium, 3: large\}$ on each attribute. It is assumed that the first 3 rules are learned from data, and the last 2 rules are learned from experts. The first three columns represent the antecedent of the rule, and the last column represents the consequent of the rule. The first line of R_h indicates the rule "If Att_1 is small and Att_2 is medium and Att_3 is small, Then class 2". Then this rule base is encoded. Take the first rule as an example again. Suppose its coding form is $\{1, 1, 1, 1\}$, and then the first three "1" represent that all three attributes of the rule are selected, and the last "1" represents that this rule is selected (If this is 0, it means that the rule is not selected). After the rule in R_h are fused by genetic algorithm, the code of the fused rule base is $\left\{\begin{matrix} 0\ 1\ 1\ 1 \\ 0\ 0\ 0\ 0 \\ 0\ 0\ 1\ 1 \\ 1\ 1\ 0\ 1 \\ 0\ 0\ 0\ 0 \end{matrix}\right\}$. Decode it, and the fused rule base will be obtained as $R_h' = \left\{\begin{matrix} 0\ 2\ 1, 2 \\ 0\ 0\ 0, 0 \\ 0\ 0\ 3, 1 \\ 2\ 3\ 0, 3 \\ 0\ 0\ 0, 0 \end{matrix}\right\}$. It can be seen from the rule base that only two rules learned from the data are retained, and one rule learned from experts is retained. In addition, the legibility of the rule base is improved by reducing unimportant attributes.

After the rule base is fused and optimized by genetic algorithm, multiple rule bases with different performances on the training set will be obtained. Note that some rule bases that perform well on the training set may perform poorly on the test set, while some rule bases that perform poorly on the training set may perform well on the test set. Therefore, multiple rule bases are used for fusion at the decision level, in which the weight of each rule base is set as the classification accuracy of each rule base in the training set.

3   Experiments

The experiment in this paper is divided into two parts: first, the model before fusion is compared with the model after fusion; second, the model in this paper is compared with other methods. The experiment used 5-fold cross validation. The main characteristics of the data sets are summarized in Table 1.

3.1   Comparison Before and After Fusion

This section will show a comparison experiment between the fusion model and the model before fusion. The specific parameter settings in the experiment are as follows: the initial standard deviation is set to 5, the population size and number of iterations of the genetic algorithm are set to 200, the crossover probability is 0.8, and the mutation probability is 0.01 (parameter settings refer to [15]).

Table 1. Statistics of the benchmark data sets used in the experiment

Dataset	#instances	#classes	#attributes
Wine	178	3	13
Ionosphere	351	2	34
New-thyroid	215	3	5
Sonar	208	2	60
Car	406	4	6
Iris	150	3	4
Wisconsin	699	2	9

Figure 2 shows the performance of the hybrid model on the training set and the test set after fusing multiple rule bases for different data sets. The x-axis represents the number of fused rule bases. Lines tr_H and te_H represent the performance of the hybrid model on the training set and test set, respectively. Lines tr_D and te_D represent the performance of the data-driven model on the training set and test set. Lines tr_E and te_E indicate the performance of the expert-driven model on the training set and test set.

It can be seen from the results that the performance of the proposed hybrid model performs better than both of the data-driven model and the expert-driven model. In addition when multiple rule bases are fused, the performance of the hybrid model on the test set is basically the same as that on the training set, so the optimal number of fusions can be determined by the performance on the training set. In most cases, the best performance can be obtained by fusing a number of rule bases less than six. It should be noted that since the expert knowledge obtained in this paper is for the sample with the greatest contribution to the current model, the classification effect of these expert rules is not good for the entire sample, and its advantages can only be exerted after fusion.

3.2 Compared with Other Methods

In this section, the classification accuracy of the proposed HCIED is compared with HIS [15] (a representative hybrid classification model), CHI [7] (a classical fuzzy rule-based model) and C4.5 [17] (a classical decision tree model).

In Table 2, the algorithm with the highest classification accuracy for each data set is marked in bold. It can be seen from the table that the model proposed in this paper performs better than other methods in most cases. In the case that the training set does not cover all possible situations, the acquired expert knowledge can well complement the initial training set, and so HCIED performs best in this case. When the coverage of the training set itself is complete, HCIED may be slightly inferior to some data-driven models, for example, it performs poorly on *cars* data sets and *ionosphere* data sets than C4.5. However, in most practical application scenarios, the training set cannot cover all possible situations, so it is a wise choice to introduce expert knowledge.

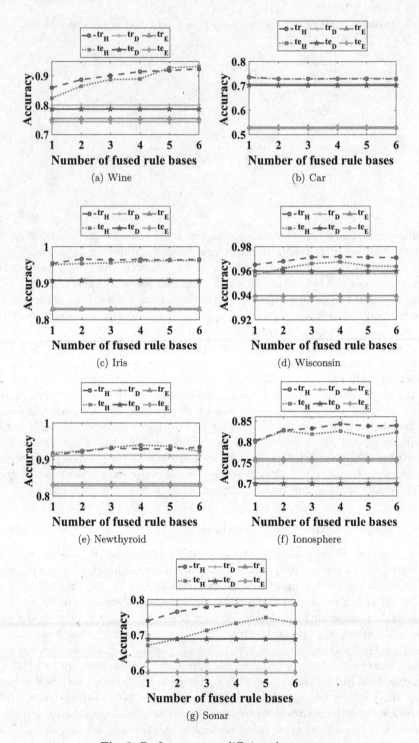

Fig. 2. Performance on different data sets

Table 2. The accuracy of each method on the test set

	Iris	Wine	Ionosphere	Wisconsin	New-thyroid	Sonar	Cars
Data-driven model	90.66	78.68	69.97	95.99	87.90	68.81	70.23
Knowledge-driven model	83.11	74.42	75.41	93.60	82.58	59.33	52.58
HCIED	**96.22**	**92.83**	82.53	**96.42**	**93.64**	**74.91**	73.06
HIS	94.26	88.5	80.98	90.38	91.72	72.65	74.87
CHI	92.27	92.77	66.4	90.2	84.18	74.61	68.97
C4.5	94.25	91.22	**88.72**	94.51	92.09	72.09	**82.15**

4   Conclusions

In this paper, a hybrid rule-based classification model by integrating expert knowledge and data is proposed to overcome the inability to obtain expert knowledge in traditional fusion methods. The main contribution of this paper is to propose a method to acquire expert knowledge, which is not found in other hybrid classification models. Secondly, this paper proposes a method to fuse two rule bases using genetic algorithm, and fuse them at the rule layer and decision layer, respectively. The experiment based on real data sets have shown that the proposed model is competitive compared with the classical methods, especially when the training set cannot completely cover all possible situations. In practical application, the training set usually cannot cover all situations completely, so this also reflects the research value of this paper. For the future work, in order to make the participation of experts more efficient, experts will be directly asked to label the most needed rule premises rather than single data samples, which will greatly reduce the workload of expert labeling and improve the performance of the model.

Acknowledgments. This work was funded by National Natural Science Foundation of China (Grant Nos. 62171386, 61801386 and 61790552), and China Postdoctoral Science Foundation (Grant Nos. 2019M653743 and 2020T130538).

References

1. Abu Ghali, M.J., Mukhaimer, M.N., Abu Yousef, M.K., Abu-Naser, S.S.: Expert system for problems of teeth and gums. Int. J. Eng. Inform. Syst. 1(4), 198–206 (2017)
2. Alonso, J.M., Magdalena, L., Guillaume, S.: Designing highly interpretable fuzzy rule-based systems with integration of expert and induced knowledge. In: Proceedings of the 2008 International Conference on Information Processing & Management of Uncertainty in Knowledge-Based Systems, pp. 682–689. Springer, Málaga (2008)
3. Alonso, J.M.: Interpretable fuzzy systems modeling with cooperation between expert and induced knowledge. Ph.D. thesis, University of Santiago de Compostela, Santiago de Compostela (2007)

4. Alonso, J.M., Magdalena, L., Guillaume, S.: HILK: a new methodology for designing highly interpretable linguistic knowledge bases using the fuzzy logic formalism. Int. J. Intell. Syst. **23**(7), 761–794 (2008)
5. Biçer, I., Sevis, D., Bilgiç, T.: Bayesian credit scoring model with integration of expert knowledge and customer data. In: Proceedings of the 24th Mini EURO Conference on Continuous Optimization and Information Technologies in the Financial Sector, pp. 324–329. Izmir University of Economics, Izmir (2010)
6. Cano, A., Masegosa, A.R., Moral, S.: A method for integrating expert knowledge when learning Bayesian networks from data. IEEE Trans. Syst. Man Cybern. Part B (Cybern.) **41**(5), 1382–1394 (2011)
7. Chi, Z., Yan, H., Pham, T.: Fuzzy Algorithms: With Applications to Image Processing and Pattern Recognition. World Scientific, Singapore (1996)
8. Gou, J., Ma, H., Ou, W., Zeng, S., Rao, Y., Yang, H.: A generalized mean distance-based k-nearest neighbor classifier. Expert Syst. Appl. **115**(1), 356–372 (2019)
9. Guillaume, S., Charnomordic, B.: Fuzzy inference systems: an integrated modeling environment for collaboration between expert knowledge and data using FisPro. Expert Syst. Appl. **39**(10), 8744–8755 (2012)
10. Guillaume, S., Magdalena, L.: Expert guided integration of induced knowledge into a fuzzy knowledge base. Soft. Comput. **10**(9), 773–784 (2006)
11. Guo, Y., Greiner, R.: Optimistic active-learning using mutual information. In: Proceedings of the 20th International Joint Conference on Artificial Intelligence, pp. 823–829. AAAI, Hyderabad (2007)
12. Jiao, L., Denoeux, T., Pan, Q.: A hybrid belief rule-based classification system based on uncertain training data and expert knowledge. IEEE Trans. Syst. Man Cybern: Syst. **46**(12), 1711–1723 (2015)
13. Liu, T.Y., Yang, Y., Wan, H., Zeng, H.J., Chen, Z., Ma, W.Y.: Support vector machines classification with a very large-scale taxonomy. ACM SIGKDD Explor. Newsl. **7**(1), 36–43 (2005)
14. Ma, H., Jiao, L., Pan, Q.: Hybrid classification by integrating expert knowledge and data: literature review. In: Proceedings of the 40th Chinese Control Conference, pp. 3231–3235. IEEE, Shanghai (2021)
15. Meesad, P., Yen, G.G.: Combined numerical and linguistic knowledge representation and its application to medical diagnosis. IEEE Trans. Syst. Man Cybern. Part A (Syst. Hum.) **33**(2), 206–222 (2003)
16. Mirčevska, V., Luštrek, M., Gams, M.: Combining machine learning and expert knowledge for classifying human posture. In: Proceedings of the 18th International Electrotechnical and Computer Science Conference, pp. 183–186. IEEE, Portorož (2009)
17. Quinlan, J.R.: C4.5: Programs for Machine Learning. Elsevier, Amsterdam (2014)
18. Samhan, L.F., Alfarra, A.H., Abu-Naser, S.S.: An expert system for knee problems diagnosis. Int. J. Acad. Inf. Syst. Res. **5**(4), 59–66 (2021)
19. Tang, W., Mao, K., Mak, L.O., Ng, G.W.: Adaptive fuzzy rule-based classification system integrating both expert knowledge and data. In: Proceedings of the 24th International Conference on Tools with Artificial Intelligence, pp. 814–821. IEEE, Athens (2012)
20. Trawiński, K., Cordón, O., Sánchez, L., Quirin, A.: A genetic fuzzy linguistic combination method for fuzzy rule-based multiclassifiers. IEEE Trans. Fuzzy Syst. **21**(5), 950–965 (2013)
21. Yegnanarayana, B.: Artificial Neural Networks. PHI Learning Pvt. Ltd., Delhi (2009)

22. Zhou, Z.J., Hu, C.H., Yang, J.B., Xu, D.L., Chen, M.Y., Zhou, D.H.: A sequential learning algorithm for online constructing belief-rule-based systems. Expert Syst. Appl. **37**(2), 1790–1799 (2010)
23. Zhou, Z.J., Hu, C.H., Yang, J.B., Xu, D.L., Zhou, D.H.: Online updating belief rule based system for pipeline leak detection under expert intervention. Expert Syst. Appl. **36**(4), 7700–7709 (2009)

A Comparative Study on Utilization of Semantic Information in Fuzzy Co-clustering

Yusuke Takahata, Katsuhiro Honda(✉) ⓘ, and Seiki Ubukata ⓘ

Osaka Prefecture University, Sakai, Osaka 599-8531, Japan
{honda,subukata}@cs.osakafu-u.ac.jp

Abstract. Fuzzy co-clustering is a technique for extracting co-clusters of mutually familiar pairs of objects and items from co-occurrence information among them, and has been utilized in document analysis on document-keyword relations and market analysis on purchase preferences of customers with products. Recently, multi-view data clustering attracts much attentions with the goal of revealing the intrinsic features among multi-source data stored over different organizations. In this paper, three-mode document data analysis is considered under multi-view analysis of document-keyword relations in conjunction with semantic information among keywords, where the results of two different approaches are compared. Fuzzy Bag-of-Words (Fuzzy BoW) introduces semantic information among keywords such that co-occurrence degrees are counted supported by fuzzy mapping of semantically similar keywords. On the other hand, three-mode fuzzy co-clustering simultaneously considers the cluster-wise aggregation degree among documents, keywords and semantic similarities. Numerical results with a Japanese novel document demonstrate the different features of these two approaches.

Keywords: Fuzzy co-clustering · Fuzzy Bag-of-Words · Semantic similarity

1 Introduction

Fuzzy co-clustering is a technique for extracting co-clusters of mutually familiar pairs of objects and items from co-occurrence information among them, and has been utilized in document analysis on document-keyword relations and market analysis on purchase preferences of customers with products. Fuzzy clustering for categorical multivariate data (FCCM) [1] and fuzzy co-clustering for document and keywords (Fuzzy CoDoK) [2] performed fuzzy c-means (FCM)-type clustering [3,4] with the aggregation measure of object and item memberships. Fuzzy co-clustering induced by multinomial mixture models (FCCMM) [5] improved the applicability of FCCM by introducing a modified log-likelihood-type measure of multinomial mixture models (MMMs), which is more useful in real applications with a statistical guideline for fuzziness tuning.

© Springer Nature Switzerland AG 2022
K. Honda et al. (Eds.): IUKM 2022, LNAI 13199, pp. 216–225, 2022.
https://doi.org/10.1007/978-3-030-98018-4_18

Recently, multi-view data clustering [6] attracts much attentions with the goal of revealing the intrinsic features among multi-source data stored over different organizations. Besides k-means-type multi-view clustering, several models for multi-view co-clustering has been proposed [7,8], where multiple co-occurrence information matrices are jointly utilized.

In this paper, three-mode document data analysis is considered under multi-view analysis of document-keyword relations in conjunction with semantic information among keywords, where the results of two different approaches are compared. Bag-of-Words (BoW) [9] is a basic approach for transforming unstructured document data into multi-dimensional numerical features by counting appearances of each keyword. However, each document usually uses only a small portion of many keywords and BoW matrices are often very sparse. Fuzzy Bag-of-Words (Fuzzy BoW) [10] improves such sparse BoW matrices by imputing their elements under consideration of semantic similarities among keywords. For example, semantic information among keywords are introduced such that co-occurrence degrees are counted supported by fuzzy mapping of semantically similar keywords. By applying fuzzy co-clustering to fuzzy BoW matrices, we can expect that document-keyword co-clusters are extracted under consideration of the additional semantic information.

Three-mode fuzzy co-clustering is another type of multi-view co-clustering, where co-occurrences among three-mode elements are jointly analyzed by calculating three-types of fuzzy memberships. Besides FCM-type models [11], a probabilistic concept-induced model [12] was demonstrated to be efficient in parameter tuning supported by a statistical guideline. If we regard document × keyword co-occurrence and semantic similarity among keywords as three-mode data of document × keyword × keyword, we can simultaneously consider the cluster-wise aggregation degree among documents, keywords and semantic similarities.

In this paper, the above two approaches are empirically compared under the context of document × keyword relational analysis utilizing semantic information. Numerical results with a Japanese novel document demonstrate the different features of these two approaches. The remaining parts of this paper are organized as follows: Sect. 2 briefly reviews fuzzy co-clustering, fuzzy BoW and three-mode fuzzy co-clustering. The experimental result is presented in Sect. 3 and the summary conclusion is given in Sect. 4.

2   Fuzzy Co-clustering, Fuzzy BoW and Three-Mode Fuzzy Co-clustering

2.1   Fuzzy Co-clustering

Assume that we have $n \times m$ co-occurrence information data matrix $R = \{r_{ij}\}$ among n documents and m keywords, where r_{ij} can be the count of appearance of keyword j in document i based on the BoW concept [9] or its extension to the term frequency-inverse document frequency (TF-IDF) weight [13]. The goal of fuzzy co-clustering is to simultaneously partition documents and keywords into C co-clusters of familiar pairs by estimating two types of fuzzy memberships of u_{ci} for document i and w_{cj} for keyword j in cluster c.

In FCCMM [5], u_{ci} is identified with the probability of i belonging to model c such that $\sum_{c=1}^{C} u_{ci} = 1$ while w_{cj} is the probability of j appearing in each model c such that $\sum_{j=1}^{m} w_{cj} = 1$. By introducing the adjustable weight λ_u for fuzziness degree of document partition, FCCMM extended the conventional MMMs into a fuzzy co-clustering with the following objective function to be maximized:

$$J_{fccmm} = \sum_{c=1}^{C} \sum_{i=1}^{n} \sum_{j=1}^{m} u_{ci} r_{ij} \log w_{cj}$$
$$+ \lambda_u \sum_{c=1}^{C} \sum_{i=1}^{n} u_{ci} \log \frac{\alpha_c}{u_{ci}}, \tag{1}$$

where α_c represents the volume of cluster c such that $\sum_{c=1}^{C} \alpha_c = 1$. When $\lambda_u = 1$, the objective function is reduced to the pseudo-log-likelihood to be maximized in MMMs. Document partition becomes fuzzier than MMMs with $\lambda_u > 1$ while crisper with $\lambda_u < 1$. That is, if a decision maker wants to make cluster boundaries clearer, he/she can choose $\lambda_u < 1$ while $\lambda_u > 1$ can be alternatively selected if he/she prefers smooth boundaries for avoiding noise influences. Starting from random initialization, the FCCMM algorithm iterates two phases of document partitioning u_{ci} and keyword probability estimation w_{cj} until convergence.

Unfortunately, in general document analysis tasks, the BoW matrix R can be very sparse because each document includes only a small portion of all keywords and most elements r_{ij} are remained $r_{ij} = 0$. Then, we often fail to derive satisfactory results by suffering from data sparseness.

In the following parts of this paper, two approaches for enriching the document clustering are considered by utilizing additional semantic information among keywords. That is, it is also assumed that we have additional $m \times m$ semantic information matrix $S = \{s_{jk}\}$ among m keywords, where s_{jk} represents the similarity degree among two keywords j and k. This type of semantic information can be constructed by utilizing such word embedding techniques as word2vec [14], which constructs a neural network model to learn word associations from a text corpus.

2.2 Fuzzy BoW

Fuzzy BoW [10] is a technique for enriching a sparse BoW matrix by utilizing fuzzy mapping of semantically similar keywords in counting the appearance of each keyword.

Assume that $A_{t_i}(w)$ is a mapping function and the frequency of a basis keyword t_i is calculated by summing up $A_{t_i}(w)$ for all words w in a sentence. Usually, the BoW model adopts the frequency of each representative keyword by counting the number of exact word matching by employing the following membership function:

$$A_{t_i}(w) = \begin{cases} 1, \text{ if } w \text{ is } t_i \\ 0, \text{ otherwise} \end{cases} \tag{2}$$

Then, each keyword occurrence can activate only a single keyword frequency and causes a very sparse nature in the BoW matrix.

On the other hand, the FBoW model adopts semantic matching or fuzzy mapping to project the words occurred in documents to the basis keywords. To implement semantic matching, a fuzzy membership function is considered as follows:

$$A_{t_i}(w) = \begin{cases} \cos(W[t_i], W[w]), \text{ if } \cos(W[t_i], W[w]) > \gamma \\ 0, \qquad\qquad\qquad \text{ otherwise} \end{cases} \tag{3}$$

where $W[w]$ denote word embeddings for word w such that they represent mutual semantic similarities among words. γ is a parameter for thresholding the cosine similarity among keywords. In [10], γ was set as $\gamma = 0$ and word2vec [14] was adopted in word embeddings. In this paper, we consider to thresholding the influences of semantically similar words by setting $\gamma > 0$ because $\gamma = 0$ can cause inappropriate influences of unfamiliar words with small $\cos(W[t_i], W[w])$.

Here, the fuzzy membership function of Eq. (3) can be used for constructing the $m \times m$ semantic information matrix $S = \{s_{jk}\}$ such that $s_{jk} = A_{t_j}(w_k)$. Then, the numerical vector representation z of a document under fuzzy BoW model is given by

$$z = xS, \tag{4}$$

where x is a vector composed of the number of occurrence of words, whose dimension is equivalent to the number of the basis keywords.

By applying the fuzzy co-clustering algorithm to Fuzzy BoW matrices, we can expect to extract fuzzy co-clusters of documents and keywords by considering not only the direct appearance of keywords in each document but also the intrinsic semantic connection with potential keywords, which are not appeared in the document.

2.3   Three-Mode Fuzzy Co-clustering

When we have a three-mode co-occurrence information among three elements, their intrinsic co-cluster structures can be extracted by three-mode fuzzy co-clustering [11], where mutual connection among the three elements were assumed to be jointly represented by two different types of co-occurrence information matrices. In this paper, the goal of document clustering is to extract co-clusters of familiar document-keyword pairs by considering not only the direct co-occurrence information of $R = \{r_{ij}\}$ but also the intrinsic semantic information matrix $S = \{s_{jk}\}$ such that each intra-cluster document i has not only direct connection r_{ij} with keyword j but also intrinsic connection $r_{ij} \cdot s_{jk}$ with keyword k.

Three-mode fuzzy co-clustering for categorical multidimensional data based on probabilistic concept (3FCCMP) [12] is a three-mode fuzzy co-clustering model, which improved the parameter tuning cost under the support of a statistical concept. 3FCCMP tries to estimate three types of fuzzy memberships by combining two multinomial distributions. The first two memberships of u_{ci} and w_{cj} are designed to have similar roles to those of FCCMM such that u_{ci} and w_{cj} represent the degree of belongingness of document i and keyword j to cluster c, respectively. On the other hand, the third membership z_{ck} is designed to represent the degree of belongingness of the third element k, which is to be estimated by considering virtual co-occurrence degree among documents and the third element buried in the matrix product $R \times S$. When the dimension of the third element is p, z_{ck} has a similar constraint with w_{cj} as $\sum_{k=1}^{p} z_{ck} = 1$.

In this paper, the intrinsic semantic information matrix $S = \{s_{jk}\}$ is identified with the co-occurrence degree among the second and third elements such that s_{jk} represents the similarity degree among direct keyword j and intrinsic keyword k. Then, the dimension of the third element is the same with that of the second one as $p = m$. The objective function to be maximized is given as:

$$
J_{3fccmp} = \sum_{c=1}^{C} \sum_{i=1}^{n} \sum_{j=1}^{m} \sum_{k=1}^{m} u_{ci} r_{ij} \log \left((w_{cj})^{1/m} (z_{ck})^{s_{jk}} \right)
$$

$$
+ \lambda_u \sum_{c=1}^{C} \sum_{i=1}^{n} u_{ci} \log \frac{\alpha_c}{u_{ci}}, \tag{5}
$$

where α_c represents the volume of cluster c such that $\sum_{c=1}^{C} \alpha_c = 1$. The adjustable weight λ_u has a similar role to that of FCCMM for fuzziness degree of document partition such that a larger λ_u implies a fuzzier object partition.

Starting from random initialization, the 3FCCMP algorithm iterates four phases of document partitioning u_{ci}, cluster volume calculation α_c, direct keyword probability estimation w_{cj} and intrinsic keyword probability estimation z_{ck} until convergence.

3   Numerical Experiment

3.1   Data Set

The data set used in [5, 15] was constructed from the text document of a Japanese novel "Kokoro" written by Soseki Natsume, which can be downloaded from Aozora Bunko (http://www.aozora.gr.jp).[1] The novel is composed of 3 chapters, each of which includes 36, 18, 56 sections, respectively. In this experiment, the goal is to partition the 110 documents drawn from each section ($n = 110$) into C document clusters, where the co-occurrence frequencies with 83 most frequently used substantives and verbs (the number of keywords is $m = 83$) were used for constructing a 110×83 co-occurrence matrix R. Its elements were preprocessed

---

[1] English translation is also available in Eldritch Press (http://www.ibiblio.org/eldritch/).

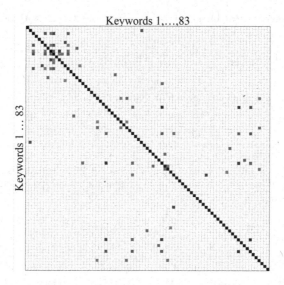

Fig. 1. 83 × 83 semantic information matrix S.

into their normalized TF-IDF weights [13] such that each of column elements has unit-variance and zero-minimum.

Additionally, the second semantic information matrix among the 83 keywords were constructed by utilizing the word embedding vectors estimated by word2vec [14]. In this experiment, the public word embedding vectors of Wikipedia Entity Vectors [16] was used, which were trained with skip-gram algorithm using Japanese Wikipedia texts as the corpus. In order to emphasize the mutual similarity among the 83 keywords, their embedding vectors were preprocessed to have zero-mean. By using the 100 dimensional embedding vectors drawn from `jawiki.word_vectors.100d.txt` [17], the fuzzy membership function of Eq. (3) was adopted for constructing the 83 × 83 semantic information matrix $S = \{s_{jk}\}$ such that $s_{jk} = A_{t_j}(w_k)$ with $\gamma = 0.5$. Figure 1 depicts the matrix S in grayscale presentation such that black and white cells indicates $s_{jk} = 1$ and $s_{jk} = 0$, respectively.

The goal of fuzzy co-clustering is to partition the 110 text documents into C co-clusters for revealing the original chapter structure withholding the chapter information of each document in conjunction with selecting chapter-wise typical keywords from 83 candidates.

3.2   Results of Document Clustering

In this section, the experimental results given by the two multi-view co-clustering approaches, i.e., FCCMM with Fuzzy BoW and 3FCCMP, are compared with the result of FCCMM, which were reported in Ref. [5] under the conventional two-mode fuzzy co-clustering context with co-occurrence matrix R only. The fuzzification penalty was set as $\lambda_u = 1.0$ in all algorithms, which implies that

Table 1. Contingency tables given by FCCMM, FCCMM with Fuzzy BoW and 3FCCMP with $C = 3$ and $\lambda_u = 1.0$.

Cluster		FCCMM			Fuzzy BoW			3FCCMP		
		1	2	3	1	2	3	1	2	3
Chapter	1	25	11	0	30	5	1	31	5	0
	2	0	18	0	0	18	0	0	18	0
	3	17	1	38	14	9	33	14	6	36

Table 2. Extracted keywords by FCCMM and its modification with Fuzzy BoW: cluster-wise typical keywords having top 10 largest memberships, where English translations are given in ( ).

Rank	FCCMM			Fuzzy BoW		
	$c = 1$	$c = 2$	$c = 3$	$c = 1$	$c = 2$	$c = 3$
1	私 (I)	する (do)	私 (I)	これ (this)	できる (can)	できる (can)
2	する (do)	それ (it)	する (do)	ある (be)	それ (it)	それ (it)
3	なる (become)	なる (become)	なる (become)	それ (it)	これ (this)	ある (be)
4	それ (it)	父 (father)	それ (it)	する (do)	する (do)	向う (go)
5	ある (be)	いう (say)	思う (think)	いる (be)	ある (be)	これ (this)
6	思う (think)	思う (think)	ある (be)	なる (become)	いる (be)	する (do)
7	いう (say)	私 (I)	見る (see)	できる (can)	なる (become)	いる (be)
8	いる (be)	聞く (listen)	K (a name)	私 (I)	叔父 (uncle)	自分 (myself)
9	人 (person)	ある (be)	いう (say)	あなた (you)	位置 (location)	なる (become)
10	先生 (master)	いる (be)	お嬢さん (lady)	いう (say)	思う (think)	私 (I)

they correspond to their probabilistic counterparts. By the way, for avoiding overflow in membership calculation, 3FCCMP was implemented with a degraded version of S matrix by dividing each element s_{jk} by 2.

For comparing the characteristics of keyword typicalities among the three algorithms, the clustering results of two multi-view models were selected so as

Table 3. Extracted keywords by 3FCCMP: cluster-wise typical direct keywords and intrinsic keywords having top 10 largest memberships, where English translations are given in ( ).

Rank	Direct keywords			Intrinsic keywords		
	$c = 1$	$c = 2$	$c = 3$	$c = 1$	$c = 2$	$c = 3$
1	私 (I)	する (do)	私 (I)	する (do)	する (do)	する (do)
2	それ (it)	それ (it)	する (do)	これ (this)	これ (this)	これ (this)
3	する (do)	思う (think)	なる (become)	ある (be)	ある (be)	ある (be)
4	なる (become)	父 (father)	それ (it)	なる (become)	なる (become)	なる (become)
5	先生 (master)	私 (I)	ある (be)	いる (be)	いる (be)	いる (be)
6	ある (be)	なる (become)	思う (think)	それ (it)	父 (father)	それ (it)
7	いう (say)	母 (mother)	自分 (I)	私 (I)	叔父 (uncle)	私 (I)
8	思う (think)	いう (say)	いう (say)	あなた (you)	母 (mother)	できる (can)
9	聞く (listen)	ある (be)	K (a name)	来る (come)	兄 (brother)	自分 (myself)
10	人 (person)	いる (be)	見る (see)	できる (can)	それ (it)	あなた (you)

to be most similar to that of FCCMM from multiple results with many random initializations. Then, the contingency tables on chapter vs. cluster matching after maximum membership assignment were given as Table 1. As mentioned in Ref. [5], Chap. 3 is composed of two sub-stories, one of which is inseparable with Chap. 1, and was partially assigned not only to Cluster 3 but also to Cluster 1. Additionally, a part of Chap. 1 was inseparable with Chap. 2.

3.3   Comparison of Typical Keyword Extraction

First, cluster-wise typical keywords having the 10 largest memberships w_{cj} are compared between FCCMM and its modification with Fuzzy BoW in Table 2. Although the two models derived similar document clusters as shown in Table 1, the selected typical keywords were slightly different. The conventional FCCMM extracted some cluster-wise unique keywords as "先生", "父", "K" and "お嬢さん", which were used only in a certain chapter. On the other hand,

FCCMM with Fuzzy BoW extracted only general keywords, which were used in multiple chapters, because Fuzzy BoW emphasized their influences by imputing semantic similarities among them even if they were not appeared in the document. This result implies that Fuzzy BoW may conceal cluster-wise unique keywords from the viewpoint of document cluster characterization.

Next, cluster-wise typical keywords extracted by 3FCCMP are shown in Table 3, where direct keywords drawn by w_{cj} and intrinsic keywords drawn by z_{ck} are compared. In direct keyword extraction, some cluster-wise unique keywords as "先生", "父", "母" and "K" were successfully extracted. Additionally, in intrinsic keyword extraction, such semantically familiar keywords as "叔父" and "兄" were also emphasized supported by "父" and "母" although they were buried in the FCCMM result. Because 3FCCMP is a hybrid of *direct connection analysis* and *intrinsic connection analysis*, it seems to be useful for characterizing each document cluster supported by two-view analysis.

4   Conclusions

In this paper, characterization of document clusters were empirically studied with two multi-view fuzzy co-clustering approaches. The comparative results demonstrated that the three-mode fuzzy co-clustering model of 3FCCMP seems to be useful for characterizing document clusters supported by semantical similarity information among keywords while Fuzzy BoW may conceal cluster-wise unique keywords when FCCMM is implemented with it.

Possible future works include application to other document datasets and comparative study on parameter sensitivity.

Acknowledgment. This work was supported in part by JSPS KAKENHI Grant Number JP18K11474.

References

1. Oh, C.-H., Honda, K., Ichihashi, H.: Fuzzy clustering for categorical multivariate data. In: Joint 9th IFSA World Congress and 20th NAFIPS International Conference, pp. 2154–2159 (2001)
2. Kummamuru, K., Dhawale, A., Krishnapuram, R.: Fuzzy co-clustering of documents and keywords. In: Proceedings of the 12th IEEE International Conference on Fuzzy Systems, vol. 2, pp. 772–777 (2003)
3. Bezdek, J.C.: Pattern Recognition with Fuzzy Objective Function Algorithms. Plenum Press, New York (1981)
4. Miyamoto, S., Ichihashi, H., Honda, K.: Algorithms for Fuzzy Clustering. Springer, Heidelberg (2008)
5. Honda, K., Oshio, S., Notsu, A.: Fuzzy co-clustering induced by multinomial mixture models. J. Adv. Comput. Intell. Intell. Inform. 19(6), 717–726 (2015)
6. Yang, Y., Wang, H.: Multi-view clustering: a survey. Big Data Min. Anal. 1, 83–107 (2018)

7. Bisson, G., Grimal, C.: Co-clustering of multi-view datasets: a parallelizable approach. In: Proceedings of the 2012 IEEE 12th International Conference on Data Mining, pp. 828–833 (2012)
8. Nishida, Y., Honda, K.: Visualization of potential technical solutions by self-organizing maps and co-cluster extraction. J. Adv. Comput. Intell. Intell. Inform. **24**(1), 65–72 (2020)
9. Lan, M., Tan, C.L., Su, J., Lu, Y.: Supervised and traditional term weighting methods for automatic text categorization. IEEE Trans. Pattern Anal. Mach. Intell. **31**(4), 721–735 (2009)
10. Zhao, L., Mao, K.: Fuzzy bag-of-words model for document representation. IEEE Trans. Fuzzy Syst. **26**(2), 794–804 (2018)
11. Chen, T.-C.T., Honda, K.: Fuzzy Collaborative Forecasting and Clustering, SpringerBriefs in Applied Sciences and Technology. Springer, Heidelberg (2019)
12. Honda, K., Hayashi, I., Ubukata, S., Notsu, A.: Three-mode fuzzy co-clustering based on probabilistic concept and comparison with FCM-type algorithms. J. Adv. Comput. Intell. Intell. Inform. **25**(4), 478–488 (2021)
13. Salton, G., Buckley, C.: Term-weighting approaches in automatic text retrieval. Inf. Process. Manag. **24**(5), 513–523 (1988)
14. Mikolov, T., Chen, K., Corrado, G., Dean, J.: Efficient estimation of word representations in vector space. In: International Conference on Learning Representations (2013). https://arxiv.org/pdf/1301.3781.pdf
15. Honda, K., Notsu, A., Ichihashi, H.: Fuzzy PCA-guided robust k-means clustering. IEEE Trans. Fuzzy Syst. **13**(4), 508–516 (2005)
16. Suzuki, M., Matsuda, K., Sekine, S., Okazaki, N., Inui, K.: A joint neural model for fine-grained named entity classification of Wikipedia articles. IEICE Trans. Inf. Syst. **E101-D**(1), 73–81 (2018)
17. Wikipedia Entity Vectors Homepage. https://github.com/singletongue/WikiEntVec. Accessed 28 Oct 2021

On the Effects of Data Protection on Multi-database Data-Driven Models

Lili Jiang[iD] and Vicenç Torra[(✉)][iD]

Department of Computing Sciences, Umeå University, Umeå, Sweden
ljiang@cs.umu.se, vtorra@ieee.org

Abstract. This paper analyses the effects of masking mechanism for privacy preservation in data-driven models (regression) with respect to database integration. Especially two data masking methods (microaggregation and rank swapping) are applied on two public datasets to evaluate the linear regression model in terms of privacy protection and prediction performance. Our preliminary experimental results show that both methods achieve a good trade-off of privacy protection and information loss. We also show that for some experiments although data integration produces some incorrect links, the linear regression model is still comparable, with respect to prediction error, to the one inferred from the original data.

Keywords: Data protection · Masking methods · Reidentification · Microaggregation · Rank swapping · Multidatabase integration

1 Introduction

Data protection mechanisms for databases are usually implemented by means of applying a distortion into the database. Masking methods are the mechanisms to produce such distortion. In short, given a database X, a masking method ρ produces $X' = \rho(X)$ that is a sanitized version of X. This X' corresponds to a distorted version of X so that the sensitive information in X cannot be inferred, and at the same time the analysis we obtain from X' are similar to those we obtain from X.

Privacy models are computational definitions of privacy. Different privacy models exist taking into account the type of object being released, the type of disclosure under consideration, etc. Differential privacy [3], k-anonymity [10,11], privacy from reidentification [6,14] are some of these privacy models. When we are considering a database release (database publishing, database sanitization), k-anonymity and privacy from reidentification are two of the main models. They focus on identity disclosure. That is, we intend to avoid that intruders find a particular person in a database. Then, a database is safe against reidentification when it is not possible (or only possible to some extent) to identify a person in the published database. k-Anonymity has a similar purpose. That is, avoid

© The Author(s) 2022
K. Honda et al. (Eds.): IUKM 2022, LNAI 13199, pp. 226–238, 2022.
https://doi.org/10.1007/978-3-030-98018-4_19

reidentification and finding particular individual's data in a database. Nevertheless, the definition is different. k-Anonymity [10] requires that for each record in the database there are at least k-1 other records that are indistinguishable to it. In this way, there will be always confusion on which was the right link. Then, the goal of a masking method ρ is to produce $X' = \rho(X)$ that is compliant with one of these definitions.

Differential privacy is an alternative privacy model that focuses on the inferences from queries or functions when applied to a database. Then, we have $y = f(X)$ and we have that y satisfies differential privacy when an addition or a deletion of a record from X will not change much the result y. Local differential privacy is a variation of differential privacy that applies to individual records. Differential privacy can tackle some vulnerabilities from k-Anonymity when sensitive values in an equivalence class lack diversity or the intruder has background knowledge.

Since 2000, a significant amount of research has been done in the field of data privacy [5,13] about methods for databases. Some methods exist that provide a good trade-off between privacy and information loss. That is, research has been done to find methods that distort the database enough to avoid disclosure in some extent and at the same time keeping some of the interesting properties of the data for potential future usage. Interesting properties includes some statistics but also building models through machine and statistical learning.

Nowadays, there is increasing interest in database integration in order to build data-driven models. That is, for applying machine and statistical learning to large datasets in terms of both number of records and number of variables. Further virtual data integration (data federation) [17,18] have been explored, where data is accessed and virtually integrated in real-time across distributed data sources without copying or otherwise moving data from its system of records. The effects of masking into data integration is not well known. It is understood that masking will modify a database in a way that linkage between databases will not be possible. In contrast, masking has been proven not to be always a big obstacle for the correct application of machine learning algorithms. There are results that show that for some databases, masked data is still useful to build data-driven models.

In this paper we present a preliminary work on the analysis of the effects of masking with respect to database integration. We analyse the effects of two data masking strategies on databases. We show that while the number of correct linkages between two masked databases drop very quickly with respect to the amount of protection, the quality of data-driven models does not degrade so quickly.

The structure of this paper is as follows. In Sect. 2 we review some concepts that are needed in the rest of this work. In Sect. 3 we introduce our approach and in Sect. 4 we present our results. The paper finishes with some conclusions and directions for future research.

2   Preliminaries

In this paper we will consider two data protection mechanisms: microaggregation and rank swapping. They both permit to transform a database X into a database X' with some level of protection. Here, protection is against reidentification. We have selected these two masking methods because microaggregation and rank swapping have been proved to be two of the most effective masking methods against reidentification. See e.g. [1,2].

Microaggregation consists of building small clusters with the original datafile, compute the centroids of these clusters, and then replace the original data by the corresponding centroids. The clusters are all enforced to have at least k clusters. The number of records k is the privacy level. Small clusters represent a small perturbation, while large k imply large privacy guarantees.

When a database contains several attributes, and all these attributes are microaggregated together, the final file satisfies k-anonymity. Recall that a file satisfies k-anonymity when for each record there are $k-1$ other records with the same value. This will be the case of a microaggregated file when masking all the files at the same time.

There is a polynomial algorithm for microaggregation when we consider a single attribute [4]. Nevertheless, when more attributes are microaggregated together heuristic algorithms are used, as the problem is NP. See [9]. In this work we have used MDAV, one of these heuristic algorithms, and, more particularly, the implementation provided by the sdcMicro package in R. See [12] for details.

Rank swapping is a masking method that is applied attribute wise. For a given attribute, a value is swapped by another one also present in the file that is within a range. For example, consider that for an attribute V_1 we have in the file to be masked the following values $(1, 2, 4, 7, 9, 11, 22, 23, 34, 37)$. Then, if we consider a parameter $s = 2$ we can swap a value for another value situated either to two positions in the right or two on the left. For example, we can swap 9 with 4, 7, 11, or 22. Only one swap is allowed for each value.

Instead of giving an absolute number of positions (as $s = 2$ above) we may consider giving a percentage of positions in the file (say p).

The larger the p, the larger the distortion, and, thus, the larger the protection. In contrast, the smaller the p, the smaller the distortion, and, thus, the better the quality of the data, but also the larger the risk. Here risk is understood as a identity disclosure risk. In other words, we use the risk of reidentification.

In this paper we will not go into the details of computing disclosure risk for the files protected. See e.g. [7,8], for discussions on risk for microaggregation and rank swapping.

3   Evaluating Data-Driven Models with Data-Integration

Our approach to evaluate the effects of data protection for data-driven models when they are applied to data-integration consists of the following steps.

- Partition a database DB_0 horizontally in test and training. Let DB be the training part. Let DB_t be the testing part.
- Take the database DB and partition it vertically into two databases DB_1 and DB_2 sharing some attributes. Let nC be the number of attributes that both databases share.
- Mask independently using a masking method ρ the two databases DB_1 and DB_2 producing $DB_1' = \rho(DB_1)$ and $DB_2' = \rho(DB_2)$.
- Integrate DB_1' and DB_2' using the nC common attributes. Let DB' be the resulting database. That is, $DB' = integrate(DB_1', DB_2')$ where $integrate$ is an integration mechanism for databases.
- Compute a data-driven model for DB and the same data-driven model for DB'. Let us call them $m(DB)$ and $m(DB')$.
- Evaluate the models $m(DB)$ and $m(DB')$ using the test database DB_t.

In order to make this process concrete, some steps need further explanation. We will describe them below.

Database integration has been done applying distance-based record linkage. That is, for each record r_1 in DB_1' we compute the distance to each record r_2 in DB_2' and we select the most similar one. That is $r'(r_1) = \arg\min_{r_2 \in DB_2'} d(r_1, r_2)$. We use an Euclidean distance for d that compares the common attributes in both databases DB_1' and DB_2'.

As DB_1' and DB_2' both proceed from the same database DB through its partition and the process of masking the two parts, we can evaluate in what extent the database is correctly integrated. That is, we can count how many times $r'(r_1)$ is the correct link in DB_2' for r_1. As we will discuss in Sect. 4 the number of correct links drops very quickly with respect to the data protection level. That is, most of the links are incorrect even with a low protection.

The other steps that need to be described are the masking methods, and the computation of the model. In relation to masking, we apply the same method to both DB_1 and DB_2, and the methods are microaggregation (using MDAV) and rank swapping. We have explained these methods in Sect. 2. Then, in order to build the data-driven model we use a simple linear regression model. Comparison of the model is based on their prediction quality. More particularly, we compute the sum of squared errors of both $m(DB)$ and $m(DB')$ and compare them. The comparison is possible because we have saved some records of the original database for testing. DB_t has been kept unused from the original database DB_0.

4   Experiments and Results

In this section we first present the experiments performed, describing the datasets and giving details so that the experiments can be reproduced by those interested. Therefore, the description includes attributes used in each step, as well as the parameters of the masking methods. Then, we describe the results obtained.

4.1   Setting

We have applied our approach to two different datasets. They are

- CASC: This dataset consists of 1080 records and 13 numerical attributes. It has been extensively used to evaluate masking methods in data privacy. It was created in the EU project CASC, and it is provided by the sdcMicro package in R. We have used the version supplied by this sdcMicro package. See e.g. [5] for a description and for other uses of this dataset.
- Concrete Compressive Strength. This is a dataset consisting of 1030 records and 9 numerical attributes. It is provided by the UCI repository. We have selected this dataset because it is of small size, all data is numerical and it has been used in several works to study and compare several regression models, including linear regression. See e.g. [15,16].

The first step consists of partitioning the database into test and training sets. We have used 80% records for training and 20% for testing.

The second step is about the vertical partitioning of the databases. This is about selecting some attributes for building the first database DB_1 and some attributes for building the second one DB_2.

The attributes in the CASC file are AFNLWGT, AGI, EMCONTRB, FEDTAX, PTOTVAL, STATETAX, TAXINC, POTHVAL, INTVAL, PEARNVAL, FICA, WSALVAL, ERNVAL. We have splitted them in two databases considering $nC = 1, 2, 3, 4, 5, 6, 7, 8$ different sets of common variables. This corresponds to eight different pairs of databases. For the first pair, DB_1 includes attributes 1–7 in the list above, and DB_2 includes attributes 7–13 in the list above. The following pairs have been built as follows. The first databases include attributes 1—7, 1–8, 1–8, 1–9, 1–9, 1–10, 1–10, respectively. The second databases include, respectively, attributes 6–13, 6–13, 5–13, 5–13, 4–13, 4–13, 3–13.

The file related to the concrete problem has been used to generate 4 pairs of databases. The process of partitioning is similar to the case of the CASC file. In this case, we have $nC = 1, 2, 3, 4$ common attributes. The first databases consist of attributes 1–5, 1–6, 1–6, 1–7, respectively. The second databases consist of attributes 5–9, 5–9, 5–9, 5–9, respectively. For example, the first pair of databases DB_1 and DB_2 will be defined as follows: DB_1 contains the first 5 attributes, and DB_2 contains from attribute 5 to attribute 9.

Once data is partitioned, we have protected the two resulting databases using two different masking methods, each one with different parameterizations. We have used microaggregation (using MDAV as the microaggregation algorithm) and considering $k = 3, 4, 5, 6, 8, 10, 12, 15$, and rank swapping considering $p = 0.001, 0.01, 0.02, 0.03, 0.04, 0.05, 0.1$. In both cases, the larger the parameter, the larger the distortion. We have used the R package sdcMicro to protect the datasets.

As we have explained above, we have applied distance-based record linkage for database integration. We have used the Euclidean distance using the common attributes. We have used our own implementation of distance-based record linkage.

Finally, we have computed a model for both the original training dataset DB and the masked and integrated database DB'. We have used a linear model for one of the attributes. All other attributes were used as independent variables of the model. For the CASC dataset we have used the first variable as the dependent variable. For the Concrete dataset, we have used the last variable in the file as the dependent variable. In this second dataset, this variable is the one used as dependent in previous research. We have used the function lm in R for building the models.

In order to analyse the results of the models, we compute the sum of squared errors between the prediction of the linear model and the true values. To compute this error we use the test set that, as we have explained above, consists of 20% of the records of the original files.

In addition, in order to evaluate in what extent the database integration is good, once databases DB_1 and DB_2 have been masked into DB_1' and DB_2', we count the number of records that are correctly linked when we build DB' from them.

It is relevant to note that our approach contains some steps based on randomization. In particular, the partition of the original databases into training and testing is based on a uniform distribution. Then, rank swapping also uses a random element to determine how elements are swapped. Because of that, for each combination (partition, masking method, parameter) we have applied our approach 5 times and study the averages of these executions.

4.2   Results

On important element to take into account in our setting is the integration of the two databases DB_1' and DB_2' and whether the records in one database are correctly linked with the other database. Our experiments show that the number of correct links drops when the protection increases. More particularly, the number of correct links becomes very small very quickly except for the case of CASC dataset masked using rank swapping, where a larger distortion is needed for the same effect when the number of attributes is relatively large (when we are using more than 4 attributes in the linkage). This reduction on the number of correct links depends on the number of common attributes nC. The larger the number of common attributes, the larger the reidentifications. In Fig. 1, we display the number of correct links for the Concrete dataset (top figures) and CASC dataset (bottom figures), when the number of common attributes nC range from 1 to 6 and protection increases for both microaggregation and rank swapping (k or p, respectively, as above). The figures also include the number of correct links in the case of no protection at all. This is the first point on the left of each figure.

We can observe in the figures that the number of correct links when data is masked using rank swapping is always larger than the number of correct links when data is masked using microaggregation. This is so because in microaggregation we are masking all attributes at the same time. This produces a file that satisfies k-anonymity for a given k, and, thus, probability of correct linkage

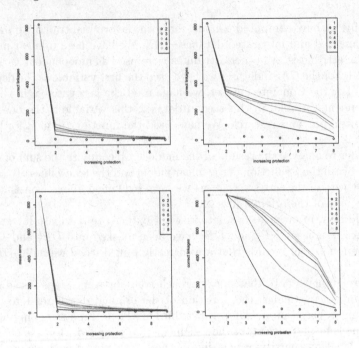

Fig. 1. Number of correct links for the Concrete dataset (top) and the CASC dataset (bottom) when data is protected using microaggregation (left) and rank swapping (right). Number of correct links decrease when protection is increased (i.e., k in microaggregation or p in rank swapping increase). Different curves correspond to different number of attributes in the reidentification (circles mean only one variable in reidentification). Number of links increase when the number of attributes increase (from 1 to 6 or 8).

becomes $1/k$. In contrast, in rank swapping each attribute is masked independently. When several attributes are considered in the linkage, noise of different attributes are independent, and then some records may have a larger probability of being correctly linked.

Figure 2 represents the mean squared error for the concrete dataset for the last attribute. Figures represent mean values after 10 runs. Top figures correspond to data protected using rank swapping and bottom figures correspond to data protected using microaggregation. From right to left (and then top to bottom) we have different number of attributes (from one to six). We can see that there is a trend of increasing error when we increase protection, and that the error is somehow smaller when the number of attributes used in the linkage increase.

Nevertheless if we compare these results about the error with the ones related to the number of correct links, we see that even the number of correct links can be very low, the error is not increasing so fast. We observe that there is still some quality in the models even when the files are not linked correctly.

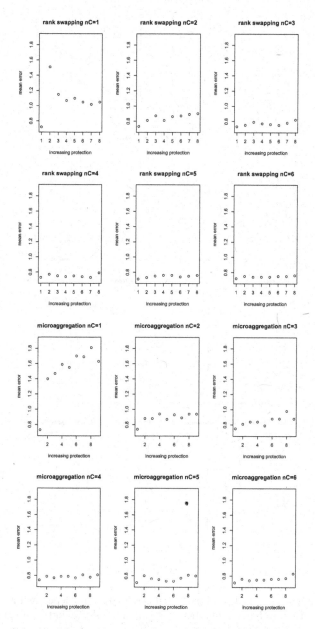

Fig. 2. Error of the models (mean squared error) for the Concrete dataset when data has been masked using rank swapping and microaggregation with different levels of protection (parameters p and k described in the text). From left to right and top to bottom number of common attributes in the integration process equal to $nC = 1, 2, 3, 4, 5,$ and 6.

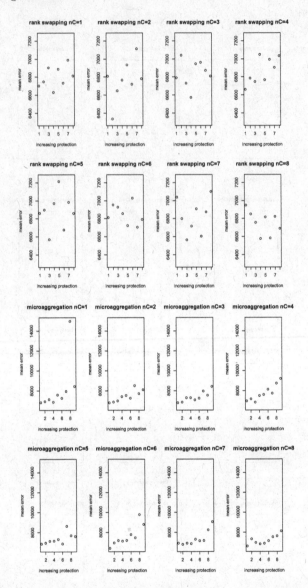

Fig. 3. Error of the models (mean squared error) for the CASC dataset when data has been masked using rank swapping and microaggregation with different levels of protection (parameters p and k). First attribute used as the dependent attribute. From left to right and top to bottom number of common attributes in the integration process equal to $nC = 1, 2, 3, 4, 5, 6, 7$, and 8.

Figure 3 provides the results for the CASC dataset. They are the results of mean squared error for the linear model of the first attribute. Figures on top refer to data masked using rank swapping and the figures in the bottom refer to

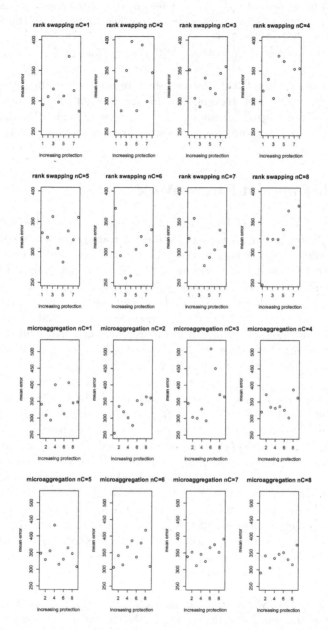

Fig. 4. Error of the models (mean squared error) for the CASC dataset when data has been masked using rank swapping and microaggregation with different levels of protection. Attribute 13th used as the dependent attribute. From left to right and top to bottom number of common attributes in the integration process equal to $nC = 1, 2, 3, 4, 5, 6, 7,$ and 8.

data masked using microaggregation. Then, figures correspond to different sets of common attributes. More particularly, from left to right and top to bottom we have $nC = 1, 2, 3, 4, 5, 6, 7, 8$.

Figure 4 also correspond to the CASC dataset for the last attribute (13th attribute).

These figures for the CASC dataset show in some cases this same trend of larger error for stronger protection, but this trend is not so clear in most of the figures. The clearer cases correspond to microaggregated files for models built with the first variable. Note that the results are the average of 5 runs for the CASC dataset.

5   Conclusions and Future Work

In this paper, we presented experimental results on how masking methods (microaggregation and rank swapping) protect privacy in respect of data integration. We applied both masking methods on two datasets and further evaluated how a data-driven model (a linear regression data model) behaves in respect of database integration with different privacy protection extents. Especially we have experimented with different number of common attributes between these two databases in terms of prediction performance on record linkage in data integration. We concluded, based on our preliminary results, that while the number of correct linkages between two masked databases drop very quickly with respect to the amount of protection, the quality of data-driven models does not degrade so quickly.

These results are in line of previous research in data privacy which shows that some data masking and, thus, data protection, can be achieved with low or even no cost for machine and statistical learning applications. See e.g. the discussion in [13]. The reason for this behavior need additional research.

The results in this paper suggested several additional interesting directions for future work. Firstly, we plan to extend the data integration scenario with privacy protection for more than two databases, which is critically important in big data era. Secondly, we intend to investigate semantic based data integration for privacy protection, which will have a hybrid consideration of string match and semantic match common attributes. Thirdly, we will further evaluate the privacy protection mechanisms in more machine learning prediction models (e.g., random forest, support vector machine, deep neural networks).

Acknowledgements. This work was partially supported by the Wallenberg AI, Autonomous Systems and Software Program (WASP) funded by the Knut and Alice Wallenberg Foundation. Partial support by Vetenskapsrådet project: "Disclosure risk and transparency in big data privacy" (VR 2016-03346, 2017-2020) is also acknowledged.

References

1. Domingo-Ferrer, J., Mateo-Sanz, J.M., Torra, V.: Comparing SDC methods for microdata on the basis of information loss and disclosure risk. In: Pre-proceedings of ETK-NTTS 2001, vol. 2, pp. 807–826. Eurostat (2001)
2. Domingo-Ferrer, J., Torra, V.: A quantitative comparison of disclosure control methods for microdata. In: Doyle, P., Lane, J.I., Theeuwes, J.J.M., Zayatz, L. (eds.) Confidentiality, Disclosure and Data Access: Theory and Practical Applications for Statistical Agencies, pp. 111–134. North-Holland (2001)
3. Dwork, C.: Differential privacy. In: Bugliesi, M., Preneel, B., Sassone, V., Wegener, I. (eds.) ICALP 2006. LNCS, vol. 4052, pp. 1–12. Springer, Heidelberg (2006). https://doi.org/10.1007/11787006_1
4. Hansen, S., Mukherjee, S.: A polynomial algorithm for optimal univariate microaggregation. IEEE Trans. Know. Data Eng. $15(4)$, 1043–1044 (2003)
5. Hundepool, A., et al.: Statistical Disclosure Control. Wiley, Hoboken (2012)
6. Jaro, M.A.: Advances in record-linkage methodology as applied to matching the 1985 Census of Tampa, Florida. J. Am. Stat. Assoc. $84(406)$, 414–420 (1989)
7. Nin, J., Herranz, J., Torra, V.: Rethinking rank swapping to decrease disclosure risk. Data Knowl. Eng. $64(1)$, 346–364 (2007)
8. Nin, J., Herranz, J., Torra, V.: On the disclosure risk of multivariate microaggregation. Data Knowl. Eng. 67, 399–412 (2008)
9. Oganian, A., Domingo-Ferrer, J.: On the complexity of optimal microaggregation for statistical disclosure control, statistical. J. U. N. Econ. Comm. Eur. $18(4)$, 345–354 (2000)
10. Samarati, P.: Protecting respondents' identities in microdata release. IEEE Trans. Knowl. Data Eng. $13(6)$, 1010–1027 (2001)
11. Samarati, P., Sweeney, L.: Protecting privacy when disclosing information: k-anonymity and its enforcement through generalization and suppression. SRI Intl. Technical report (1998)
12. Templ, M.: Statistical disclosure control for microdata using the r-package sdcmicro. Trans. Data Priv. $1(2)$, 67–85 (2008)
13. Torra, V.: Data Privacy: Foundations, New Developments and the Big Data Challenge. Springer, Heidelberg (2017)
14. Winkler, W.E.: Re-identification methods for masked microdata. In: Domingo-Ferrer, J., Torra, V. (eds.) PSD 2004. LNCS, vol. 3050, pp. 216–230. Springer, Heidelberg (2004). https://doi.org/10.1007/978-3-540-25955-8_17
15. Yeh, I.-C.: Modeling of strength of high performance concrete using artificial neural networks. Cem. Concr. Res. $28(12)$, 1797–1808 (1998)
16. Yeh, I.-C.: Analysis of strength of concrete using design of experiments and neural networks. J. Mater. Civ. Eng. ASCE $18(4)$, 597–604 (2006)
17. Leymann, F.: A practitioners approach to data federation. In 4. In: Workshop Föderierte Datenbanken, Berlin, Germany. CEUR-WS/Vol.25 (1999)
18. Vu, X.S., Mlouk, A., Elmroth, E., Jiang, L.: Graph-based interactive data federation system for heterogeneous data retrieval and analytics. WWW $19(5)$, 3595–3599 (2019)

Machine Learning

Toward Latent Cognizance on Open-Set Recognition

Pisit Nakjai[1]([✉])[iD] and Tatpong Katanyukul[2][iD]

[1] Uttaradit Rajabhat University, Uttradit, Thailand
pisit.nak@uru.ac.th
[2] Khon Kaen University, Khon Kaen, Thailand
tatpong@kku.ac.th

Abstract. Open-Set Recognition (OSR) has been actively studied recently. It attempts to address a closed-set paradigm of conventional object recognition. Most OSR approaches are quite analytic and retrospective, associable to human's system-2 decision. A novel bayesian-based approach Latent Cognizance (LC), derived from a new probabilistic interpretation of softmax output, is more similar to natural impulse response and more associable to system-1 decision. As both decision systems are crucial for human survival, both OSR approaches may play their roles in development of machine intelligence.

Although the new softmax interpretation is theoretically sound and has been experimentally verified, many progressive assumptions underlying LC have not been directly examined. Our study clarifies those assumptions and directly examines them. The assumptions are laid out and tested in a refining manner. The investigation employs AlexNet and VGG as well as ImageNet and Cifar-100 datasets.

Our findings support the existence of the common cognizance function, but the evidence is against generality of a common cognizance function across base models or application domains.

Keywords: Latence cognizane · Penultimate information · Open-set recognition · Pattern recognition · Neural network

1 Introduction

Deep learning has excellent capabilities for learning high-dimensional features of complex data and assigning decision-making boundaries between the classes. Deep Neural Networks (DNNs) can learn high-level features that allow them to achieve outstanding performance on identifying the number of categories and overcoming many of the challenges associated with recognition/classification tasks [1,2]. The high-dimensional features extracted from these networks are widely used in various challenges, including computer vision recognition [3], autonomous driving [4], language transcription [5], sign recognition [6], and biomedical applications [7].

© Springer Nature Switzerland AG 2022
K. Honda et al. (Eds.): IUKM 2022, LNAI 13199, pp. 241–255, 2022.
https://doi.org/10.1007/978-3-030-98018-4_20

Although acceptable efficiency in DNNs requires robustness from turbulence input, even small perturbations could alter the classification results. The "adversarial image" problem, where it is possible to disguise an image with a fake label, has recently been demonstrated [8,9]. Moreover, the phenomenon of "fooling image", where DNNs miss-classify images unrecognizable by humans with a high level of confidence, has recently been reported [10]. These aforementioned problems have come about because traditional DNNs usually use the Softmax function that forces them to choose one of the seen classes for all input. Thus, the DNNs consider only a finite set (seen classes). It will make a strongly wrong prediction when the input samples come from a set of out-interesting (unseen classes).

Moreover, most studies on recognition have been based on Closed-Set Recognition in which the training set is assumed to be included in the environment [3,6,11]. However, in the real world, an unseen class could appear in the testing phase, which leads to the Open-Set Recognition (OSR) situation. In this scenario, the classifier model not only effectively classifies the seen classes but also detects an unseen ones when they appear. For example, a company should be able to recognize both employees and non-employees from its security camera footage.

Researchers have recently proposed the Latent Cognizance (LC) mechanism that provides a new probability interpreter of Softmax inference based on Bayes' theorem [12]. LC interpretation has been employed for a sign language recognition task and is believed to have a potential for the OSR applications.

Our goal in this study is to investigate the Latent Cognizance mechanism that can classify seen and unseen classes. We made the three following assumptions:

1. For a well-trained classification using a neural network—multilayer perceptron—, there exist differentiable functions h_k such that $\sum_{k=1}^{K} h_k(a)$ is proportional to the probability of seen classes $p(s|x)$, when a is a vector output from a neural network. Note that notion $h_k(\cdot)$ is a cognizance function for $k = 1, \ldots, K$.
2. For rather well-trained classification, there exists a common function $h(\cdot)$ that indicates the probability of seen classes $p(s|x)$.
3. Cross identification—testing $\sum_{k=1}^{K} h(a_k)$ on one dataset where $h(\cdot)$ is obtained by fitting to another dataset—attains an area under curve (AUC) score over 0.5. The $AUC > 0.5$ mean that the classifier can distinguish between seen and unseen class. This implies that a common function $h(\cdot)$ is common across all datasets and models.

These assumptions are derived from a level of progressiveness in LC development. Assumption 1 (there exists cognizance functions) is the starting point. It is the easiest to satisfy, but perhaps the most difficult to use in practice, since it means that for K categories in the worst case we may need to determine K different function one for each category. This assumption is the core to LC and disproving it would virtually disproving LC itself. Assumption 2 (there exists a common cognizance function among all categories) is a tighter assumption and more progressive. Disproving it may not jeopardize the entire idea of LC, but it will render LC impractical or at least very difficult to apply. If assumption 2 is proved, it will simplify a task so that we can find only one function that is appropriate for a cognizance function for the given task. Assumption 3 (The common cognizance function is general across different domains) is the tightest assumption we examine. If it is proved true, the application of LC will be generally simple. We can just find one good cognizance function and then we can apply it to any domain. This will promote the utilization of LC with great convenience.

2   Related Work

Whereas closed-set classification has been accepted for quite some time [6,13]. OSR is not regarded as a classification task. Nevertheless, the OSR challenge has been explored by many researchers.

The centroid of the seen classes is considered by applying the Distance Center-Based Space model [14,15], in which the distance-based assumption is that the seen classes cluster closer to the positive data—training data—than the unseen classes.

Thus, data that is distant from the positive data might belong to an unseen class. Mendes-Junior et al. [16] proposed the Nearest Neighbor Distance Ratio (NNDR) multiclass open-set classifier to overcome the open-set problem. This is based on the Nearest Neighbor (NN) classifier modified to cope with unseen classes by considering the distance ratio; a class is classified as unseen when the distance ratio is over 1.

Another approach, the Nearest Non-Outlier (NNO) classifier, was proposed by [17]. It was developed based on a classic pattern recognition algorithm, the Nearest Class Mean (NCM) classifier [9,18]. In NCM classification, the Mahalanobis distance of each sample is calculated and transformed to represent its class with a mean feature vector [19]. However, NNDR and NNO consider the thresholding value without using the distribution information and in general, distance center-based methods consider little or no distribution information of the data and lack a strong theoretical background.

To deal with this, Rudd et al. [20] proposed the Extreme Value Machine (EVM), which was adapted from the statistical Extreme Value Theory (EVT) [21] with the concept of margin distributions. The Ψ-model of EVM defines a radial inclusion function that is an EVT rejection model where the probability of inclusion corresponds to the probability that the sample does not align in the margin. The EVM was tested on the OLETTER dataset [22] and compared with the state-of-the-art Weibull-calibrated SVM (W-SVM) [22]. Although both EVM and W-SVM had favorable performances, EVM had a lower training cost. The ImageNet dataset was used as a source of computer vision data to test EVM and compare its performance with the NNO algorithm; EVM outperformed NNO in terms of the F1 Measure for accuracy. In addition, the OSR problem can be handled by applying the DNN approach.

There have been several studies on classifier rejection of unseen classes, one of the most well-known approaches being OpenMax [23]. An extreme value distribution is used in this algorithm to re-calculate the probability scores of multiple classes, including unseen ones, the assumption being that images from unseen classes will have a lower probability value in the classification results.

The Open-max presents a new layer called "Open-max" that can efficiently handle unseen classes by using the score from the last fully connected layer before the Softmax activation function penultimate layer to predict the input. If the input is far from the training data, it will be rejected and marked as belonging to an unseen class. The Open-max relaxes a restriction of the classification task, by setting the probabilities of the seen classes to 1 and recalculating new probabilities for both seen and unseen classes using the intra-distance aggregation calibrated score for the known classes and a Weibull distribution to estimate new probabilities for the unseen classes. The Open-max also uses the penultimate layer score to estimate a parameter of the Weibull distribution. The rationale is that Open-max normalizes or rescales the output values to follow a logistic distribution to retrieve the information lost during the Softmax processing.

A new interpretation of the Softmax function called Latent Cognizance (LC) was investigated by Nakjai et al. [24]. Their work employs the LC technique to distinguish non-sign postures in hand-sign recognition tasks for sign language interpretation. Their LC approach was shown to be effective at identifying sign postures for which categories have not been assigned during the training process. Moreover, its potential and implications could reach far beyond this specific application and move object recognition toward scalable OSR.

In the interests of exploring the potential of LC, we investigated whether (1) the LC function exists, (2) it can be a common function, and (3) it can be applied in the model training phase for different environments. Its performance was evaluated by applying it to the ImageNet and CIFAR-100 datasets. Section 3 provides background on LC, while the study assumptions are provided in Sect. 4. The final section offers conclusions on this study.

3   LC Background

Artificial neural networks a deep network often employ a softmax function in the final layer for multi-class classification task. Softmax takes an output vector of the previous layer and converts each of them into K classes probabilities that sum to one. The softmax formula is as follows,

$$y = softmax(a) = \frac{exp(a_k)}{\sum_{i=1}^{K} exp(a_i)}, \tag{1}$$

where a is a previous layer output vector that is called penultimate output vector. The penultimate vector $a = [a_1, \ldots, a_K]^T = f'(x, w)$, where f' is a computational output of neural network before the softmax, with input x and weight parameter w of the neural network.

Softmax is such a powerful mechanism and allows numerous impressive classification performance. However, when it comes to an unseen input—an input of a class out of a set of classes in the training process—, softmax output is known to be virtually meaningless. This observation contradicts the conventional interpretation: $y_k \equiv p(y = k|\mathbf{x})$. Nakjai et al. [24] have reinterpreted the softmax output as a conditional probability that the given seen input $\mathbf{x}$ belongs to class k,

$$y_k \equiv p(y = k|\mathbf{x}, s), \tag{2}$$

where s indicates that $\mathbf{x}$ belongs to one of the seen class or $1, \ldots, K \in s$. This interpretation emphasizes the condition s, which has been faint and inattention.

Based on that, given Bayes' theorem, the softmax output

$$y_k = p(y = k|\mathbf{x}, s) = \frac{p(y = k, s|\mathbf{x})}{\sum_{i=1}^{K} p(y = i, s|\mathbf{x})}. \tag{3}$$

The term $p(y = k, s|\mathbf{x})$ can be of great benefit. Supposed that $p(y = k, s|\mathbf{x})$ is known, seen and unseen patterns can be differentiated through marginal probability

$$p(s|\mathbf{x}) = \sum_{k=1}^{K} p(y = k, s|\mathbf{x}). \tag{4}$$

A low value of $p(s|\mathbf{x})$ indicates a high chance that $\mathbf{x}$ is of an unseen class and vice versa. Therefore, obtaining a value of $p(y = k, s|\mathbf{x})$ can solve the open-set problem. Determining a value of $p(y = k, s|\mathbf{x})$ can be very difficult in practice, but under an objective to identify the unseen a good estimation should be sufficient.

Conferring Eq. 1 to Eq. 3, the following relation is found:

$$\frac{\exp(a_k)}{\sum_i \exp(a_i)} = \frac{p(y = k, s|\mathbf{x})}{\sum_i p(y = i, s|\mathbf{x})}. \tag{5}$$

Consider similar patterns on both sides of Eq. 5. The similar patterns indicate some kind of relation between penultimate values on the left side and the probabilities on the right side. In order to investigate approaches for good estimation of $p(y = k, s|\mathbf{x})$, the following is deduced. Given a well-trained softmax inference $f : \mathbf{x} \mapsto \mathbf{y}$ that f is internally composed of $f' : \mathbf{x} \mapsto \mathbf{a}$ and softmax $: \mathbf{a} \mapsto \mathbf{y}$, the penultimate vector $\mathbf{a}$ relates to posterior probability $p(y = k, s|\mathbf{x})$ through function $\tilde{h}_k(\mathbf{a}) = p(y = k, s|\mathbf{x})$. As it is sufficient under a task of identifying an unseen and as to lessen a burden on enforcing probabilistic properties, it is more convenient to work with a function whose value just correlates to the probability. [12] denote cognizance function $g(\cdot)$ such that $g(a_k(x))) \propto p(y = k, s|x)$, where $a_k(x)$ represents the kth penultimate value corresponding to input x. Thus, an unseen input x can be identified by a low value of

$$\sum_{k=1}^{K} g(a_k(x)) \propto p(s|x). \tag{6}$$

[12] called this Latent Cognizance (LC) approach. LC approach follows a new softmax interpretation based on [12], i.e., $y_k = p(y = k|\mathbf{x}, s)$, where s represents a state of being a seen class. Investigation of all major assumptions underlying LC development is discussed in Sect. 4.

4   Latent Cognizance Assumptions

Since LC is recently discovered, its full potential is yet to be realized. We contemplate that a progressive examination where its key assumptions are examined from the most rudimentary to more refiner ones will give a better insight. In this progressive manner, hopefully LC nature will be revealed in depth for what it is and what it is not so that we do not rush to a conclusion and perhaps these findings can later be better utilized.

4.1   Assumption 1: There Exists Cognizance Functions

Definition. Assumption 1 assume that there exists differentiable functions h_k's for $k = 1, \ldots, K$ such that, given $\mathbf{a} = f'(\mathbf{x})$ and f' is well-tuned, each $h_k(\mathbf{a})$ correlates to $p(y = k, s|\mathbf{x})$. Therefore, $p(s|\mathbf{x})$ can be deduced from $\sum_{k=1}^{K} h_k(\mathbf{a})$, noted as $p(s|\mathbf{x})$ correlates to $\sum_{k=1}^{K} h_k(\mathbf{a})$. Another word, a value of $\sum_{k=1}^{K} h_k(\mathbf{a})$ indicates a degree of input being of a seen class. This assumption poses the most rudimentary premise that there exists the cognizance function $h_k(\mathbf{a})$ correlates to $p(y = k, s|\mathbf{x})$. Noted that this notion of cognizance function $h_k : \mathbb{R}^K \mapsto \mathbb{R}$ is broader than [24]'s $g : \mathbb{R} \mapsto \mathbb{R}$. [24]'s cognizance is examined here under Assumption 2. Figure 1a illustrates the logical view based on Assumption 1.

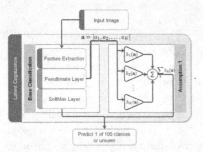
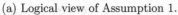
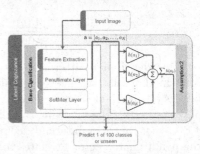

(a) Logical view of Assumption 1.   (b) Logical view of Assumption 2.

Fig. 1. An our logical view of Assumtion1 and Assumtion2.

Assumption 1 Experiment. To prove the existence of $h_k(\mathbf{a})$, a 2-hidden-layer Multi-layer percepton (MLP) was learned with a dedicated set of data. This learning process was to tune a 2-hidden-layer MLP to approximate $h_k(\mathbf{a})$, and this process is called "learning" to distinguish it from "training", which is used to refer to a fine-tuning process of the base classifier, e.g., Alexnet. To assure the LC premise, i.e., the model is well trained, the base classifier Alexnet was trained on only 100 classes, previously found to be the top 100 performing classes(with each class having over 70% accuracy). Specifically, AlexNet with pretrained weights was trained with samples from the chosen 100 classes from ILSVRC 2012 training dataset. AlexNet was trained to achieved the accuracy of 0.889 (on validation data, randomly chosen 10% from the training data). To learn $h_k(\mathbf{a})$, both "seen" and "unseen" samples were used. The "seen" samples were samples from the chosen 100 classes, which also used in training. The seen learning set made up of 129,264 samples. The "unseen" sample were samples from the other 900 classes. The total number of unseen samples was 1,150,802 samples. To balance the seen and unseen data sizes, only 143 samples per class were chosen for the unseen learning set, which totally made up to 128,700 samples. Noted that, 10% of these datasets was used for validation. Since ILSVRC 2012 test set did not have the class labels, ILSVRC 2010 training set was used for testing the seen/unseen identification capability. The ILSVRC 2010 training dataset contained 74,488 samples of the seen classes and 1,186,873 samples of the unseen classes. The classes in both ILSVRC 2010 and ILSVRC 2012 were inspected for class integrity.

The MLP used in this experiment had 2 hidden layers. Each layer had 2048 nodes and used hyperbolic tengent as its activation function. The output layer had sigmoid function as its activation function. Specifically, given input $\mathbf{x} \in \mathbb{R}^{100}$, the output $\mathbf{y} \in \mathbb{R}^{100}$ has computed from: $\mathbf{z}^{(1)} = \mathbf{h}^{(1)} \cdot (\mathbf{w}^{(1)} \cdot \mathbf{x} + \mathbf{b}^{(1)})$, $\mathbf{z}^{(2)} = \mathbf{h}^{(2)} \cdot (\mathbf{w}^{(2)} \cdot \mathbf{z}^{(1)} + \mathbf{b}^{(2)})$, $\mathbf{y} = \mathbf{h}^{(3)} \cdot (\mathbf{w}^{(3)} \cdot \mathbf{z}^{(2)} + \mathbf{b}^{(3)})$, where $\mathbf{w}^{(1)} \in \mathbb{R}^{2048 \times 100}$, $\mathbf{w}^{(2)} \in \mathbb{R}^{2048 \times 2048}$ and $\mathbf{w}^{(3)} \in \mathbb{R}^{100 \times 2048}$ are model weights; $\mathbf{b}^{(1)}, \mathbf{b}^{(2)} \in \mathbb{R}^{2048}$ and $\mathbf{b}^{(3)} \in \mathbb{R}^{100}$ are model biases; $\mathbf{h}^{(1)}, \mathbf{h}^{(2)}$, and $\mathbf{h}^{(3)}$ are element-wise activation functions of 2 hidden layers and the output layer, respectively. The weights and biases are learned to minimize the separation loss—inspired by Tripet Loss [25]—

$$\mathcal{L} = \left[\frac{1}{M} \sum_{m \in U} z_m^{(u)} - \frac{1}{N} \sum_{n \in S} \log z_n^{(s)} + \alpha \right]^+, \tag{7}$$

where S is a set of seen examples, U is a set of unseen examples, $z_m^{(u)}$ is the mth unseen example, $z_n^{(s)}$ is the nth seen example, and $M, N > 0$. User specific value α is a constant to set the separation margin, which is set to 1 in our experiment. The learning process employed ADAM optimization with learning rate of 10^{-4}, went through 1,000 epochs and took the best performing set of weights and biases (with loss $<10^{-4}$).

Results. Table 1 shows the evaluation results (on the 1st row of Table 1). The table emphasizes Assumption 1 approximation function $\sum_k h_k(\mathbf{a})$ whose penultimate vector $\mathbf{a}$ obtained from the base classifier—AlexNet as indicated. The approximation $h_k(\cdot)$ is learned using ImageNet data (ILSVRC 2012). The Area Under Curve (AUC) indicates how well the LC with the approximation $h_k(\cdot)$ distinguishes between seen and unseen samples, tested on ImageNet data (ILSVRC

2010). Significance test—Kruskal-Wallis Test—also confirms the separation with p-value $<2.2 \times 10^{-16}$. This result supports Assumption 1.

Figure 2 shows boxplots and Precision-Recall curve (P-R curve) using LC with $\sum_k h_k(\mathbf{a})$. Figure 2a has the y-axis representing LC values. The boxplot, P-R curve, and the AUC (shown in Table 1) all strongly support Assumption 1.

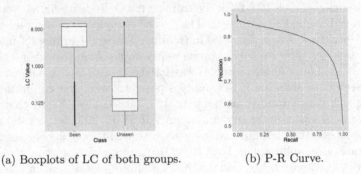

(a) Boxplots of LC of both groups.   (b) P-R Curve.

Fig. 2. Assumption 1: LC with $\sum_k h_k(\mathbf{a}), \mathbf{a} \in \mathbb{R}^K$.

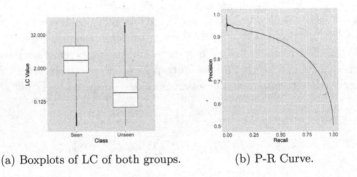

(a) Boxplots of LC of both groups.   (b) P-R Curve.

Fig. 3. Assumption 2: LC with $\sum_k h(a_k), a_k \in \mathbb{R}$. The cognizance h learns from AlexNet model and ImageNet dataset.

Given that each h_k maps penultimate vector to the corresponding seen class probability, to have K seen class probabilities is to have K functions. This requires quite resources and it would be much more pragmatic if there is only one common cognizance function, like what originally proposed by [24].

4.2   Assumption 2: There Exists a Common Cognizance Function

Definition. Assume that given $[a_1, \ldots, a_K] = \mathbf{a} = f'(\mathbf{x})$ and f' is well-tuned, there exists a differentiable function g such that, $g(a_k)$ correlates to $p(y = k, s|\mathbf{x})$. Thus, $p(s|\mathbf{x})$ correlates to $\sum_{k=1}^{K} g(a_k)$.

Table 1. Evaluation of Assumptions 1 and 2.

Assumption	Base Classifier	Learning Seen/Unseen	AUC on ImageNet	Kruskal-Wallis Test (p-value)
Assumption 1: $\sum_k h_k(\mathbf{a})$	Alexnet	ImageNet	0.890	$<2.2 \times 10^{-16}$
Assumption 2: $\sum_k h(a_k)$	Alexnet	ImageNet	0.849	$<2.2 \times 10^{-16}$

This experiment followed Assumption 1, but replacing functions $h_k(\mathbf{a})$'s with a common function $h(a_k)$ for $k = 1, \ldots, K$. Figure 1b illustrates a logical view of Assumption 2 conferring to Fig. 1a. The function $h(a_k)$ was approximated with an MLP. The MLP had 2,000 hidden nodes. Specifically, given input $\mathbf{x} = [x_1, \ldots, x_{100}] \in \mathbb{R}^{100}$, output $\mathbf{y} = [y_1, \ldots, y_{100}] \in \mathbb{R}^{100}$ was computed from $z_i^{(1)} = h^{(1)}(w^{(1)} \cdot x_i + b^{(1)})$ and $y_i = h^{(2)}(w^{(2)} \cdot z_i^{(1)} + b^{(2)})$, where $w^{(1)} \in \mathbb{R}^{2000 \times 1}$, $w^{(2)} \in \mathbb{R}^{1 \times 2000}$, $b^{(1)} \in \mathbb{R}^{2000}$, $b^{(2)} \in \mathbb{R}$, $h^{(1)}(\cdot) = \tanh(\cdot)$ and $h^{(2)}(\cdot) = \text{sigmoid}(\cdot)$. The model was learned in a similar manner described in Assumption 1 experiment. The best performing set of weights and biases were found at loss <0.5.

Results. The second row of Table 1 shows the evaluation results of this assumption. Boxplots and P-R curve are shown in Fig. 3. The results show strong support for Assumption 2.

Assumption 1 and 2 Comparison. Boxplot of the seen group is very well separated from the unseen group. The separation margin of Fig. 2 is wider than one of Fig. 3, but both show strong supports for their respective assumptions. It should be noted that assumption 1 is somewhat more general than assumption 2 as in assumption 1: cognizance function h_k of different category k can be different, while assumption 2: the cognizance function h has to be the same for all categories. While assumption is more general, the application of assumption 2 is more manageable since the task of finding an appropriate cognizance function is to find only one cognizance function for all categories (c.f., finding K cognizance functions each for each category).

To simplify the experiment, instead of finding the common cognizance function g, we find an approximate cognizance based on the model and domain and check if the found cognizance works with relatively consistent performance across environments (models and domains).

4.3 Assumption 3: The Common Cognizance Function is General Across Different Domains

Definition. Given the models are well-trained, there exists a common cognizance function g, i.e., it is general across models and domains. Specifically, given there exists $g(a_k^{(1)})$ correlates to $p^{(1')}(y = k, s|\mathbf{x})$ when g is a well-established cognizance function for a well-tuned model (1) under domain (1'), $a_k^{(1)}$ is a penultimate value obtained from a well-tuned model (1), and $p^{(1')}$ is a probability regarding domain (1'), then $g(a_k^{(2)})$ correlates to $p^{(2')}(y = k, s|\mathbf{x})$

when $a_k^{(2)}$ is a penultimate value obtained from a well-tuned model (2) under domain (2′), and $p^{(2′)}$ is a probability regarding domain (2′). The consequence of this assumption is that if holds, cognizance g once found can be used on any occasion and its implication and application can be profound and wide. However, if this assumption is disproved, it means that for every scenario (domain, task, or model), there has to be a learning process to determine this cognizance function. Another word, if disproved, the applicability of cognizance is quite limited.

To simplify the experiment, instead of finding the common cognizance function g, we find an approximate cognizance based on the model and domain and check if the found cognizance works with relatively consistent performance across environments (models and domains).

Assumption 3 Experiment. To verify Assumption 3, two cognizance functions $h(\cdot)$ and $\hat{h}(\cdot)$ are obtained using two different environments (using different models and learning from different data). Then, each cognizance function is put to the test both on its native environment and on its foreign context.

Our experiment employs AlexNet model and ImageNet dataset as the first environment (Model 1: AlexNet, Domain 1′: ImageNet). The first cognizance $h(\cdot)$ then is obtained using the first environment as described in Assumption 2 Experiment.

A modified version of Visual Geometry Group version 16 layers (VGG16, [26]) is chosen as our model of the second environment (Model 2: Our VGG, Domain 2′: Cifar-100). Our VGG model is shown in Fig. 4. Cifar-100 dataset [27] is chosen as our domain of the second environment. Ten classes out of total 100 classes are randomly chosen to be seen classes. To train our VGG model, 5000 images of seen classes (500 images each class) from Cifar-100 training set are used. To learn the second cognizance $\hat{h}(\cdot)$, 5000 images of seen classes along with 5000 (unseen) images randomly chosen from the remaining of the Cifar-100 training set are used. The second cognizance $\hat{h}(\cdot)$ is obtained using the second environment.

To test each cognizance function, the first environment uses AlexNet model to produce penultimate vectors corresponding to data from ImageNet ILVSRC 2010 as described in Assumption 1 Experiment. The second environment uses VGG16 model to produce penultimate vectors corresponding to data from Cifar-100: 1000 images of the seen classes (100 images each class) and 1000 (unseen) images randomly chosen from the remaining of Cifar-100 test set are used.

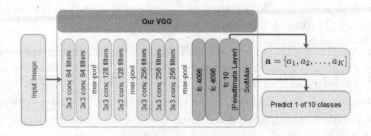

Fig. 4. Our modified VGG Architecture

Results. Table 2 shows evaluation results. The first row represents the first cognizance function (learned from the first environment: AlexNet model and ImageNet dataset). Its AUC and p-value of Kruskal-Wallis Test on the first environment reveal its performance on its native ground, while ones on the second environment reveal its performance on a foreign context. The second row represents the second cognizance function (learned from the second environment: our VGG model and Cifar-100 dataset). Its AUC and p-value of Kruskal-Wallis Test on the first environment reveal its performance on its foreign environment, while ones on the second environment reveal its performance on its native ground.

Table 2. Evaluation of Assumption 3. Function 1 learns from Environment 1. Function 2 learns from Environment 2. Environment 1 uses AlexNet model and ImageNet dataset. Environment 2 uses our VGG16 model and Cifar-100 dataset.

Cognizance	Tested on			
	Environment 1		Environment 2	
	AUC	Kruskal-Wallis Test (p-value)	AUC	Kruskal-Wallis Test (p-value)
Function 1: h	0.849	$<2.2 \times 10^{-16}$	0.647	$<2.2 \times 10^{-16}$
Function 2: $\hat{h}$	0.385	$<2.2 \times 10^{-16}$	0.661	$<2.2 \times 10^{-16}$

Figure 3 and 5 show boxplots and P-R curves of both cognizance functions on their native environments. Figure 6 and 7 show boxplots and P-R curves of both cognizance functions on their foreign environments.

Given results from Table 2 along with Fig. 7, evidence found here is against the generality of both cognizance functions.

To interpret results shown in Table 2, recall that h (function 1) is derived from environment 1 and $\hat{h}$ (function 2) is derived from environment 2, therefore, native performances (h on environment 1 and $\hat{h}$ on environment 2) are seen along the diagonal entries (AUCs of 0.849 and 0.661). The foreign performance shows how robust the cognizance function is when it is applied to a different setting and this is shown as AUCs of 0.647 and 0.385 (h on environment 2 and $\hat{h}$ on environment 1, respectively).

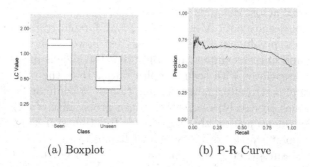

(a) Boxplot   (b) P-R Curve

Fig. 5. The second cognizance function $\hat{h}$ (learned from Environment 2: our VGG model and Cifar-100 dataset) tested on its native environment (Environment 2).

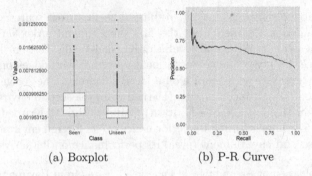

(a) Boxplot   (b) P-R Curve

Fig. 6. The first cognizance function h (learned from Environment 1) tested on its foreign ground (Environment 2).

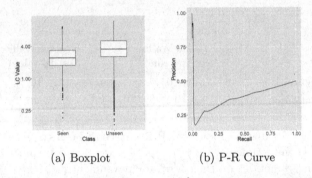

(a) Boxplot   (b) P-R Curve

Fig. 7. The second cognizance function $\hat{h}$ (learned from Environment 2) tested on its foreign ground (Environment 1).

Both cognizance functions show significantly drop in their performances when used in the foreign domains. For example, the cognizance function $\hat{h}$ is obtained with our VGG16 model and trained on the Cifar-100 dataset. The cognizance function $\hat{h}$ achieved AUC 0.661 on the Cifar testing dataset, but it cannot keep their performance and has AUC drop to 0.385 when testing on another environment. Therefore, cognizance functions obtained as in our experiments are shown not to be general across domains.

Surprisingly, despite poor AUC (0.385) of $\hat{h}$ on foreign ground, the boxplot in Fig. 7a and a low p-value on Kruskal-Wallis Test however show well separation between seen and unseen samples. It is only that the logic is reverse: a low value of cognizance associates with seen samples rather than the other way around. This observation is strongly against Assumption 3—a common cognizance function whose value (positively) correlates to seen probability $p(s|\mathbf{x})$—, but regard to seen/unseen identification this cognizance function still shows some potential. Another point worth noting is that although our evidence here is apparently against Assumption 3, the evidence is acquired based on the cognizance functions obtained through a process as described in our experiments. A proper investigation on this issue may deserve another dedicated study.

The ramification of disapproval of Assumption 3 is that the cognizance function is not general across environments. Changing model or changing domain requires an extra process to figure out a proper cognizance function or at least to test if the commonly used function is still suitable. This limits the convenience of applying LC in a significant way, otherwise one good general cognizance function could be examined and once found it can be readily available. However, as our result has shown, some cognizance function, e.g., h (Table 2), seems to be robust to some degree. A criteria for good cognizance function and the criticality when it is used beyond its native environment may deserve further study.

5   Conclusion

Latent Cognizance (LC) has been shown a great potential for an open-set capacity. Its development is based on a new interpretation of a softmax inference along with progressive assumptions. The new interpretation of a softmax inference has been verified in [24]. Our work thoroughly examines those three progressive assumptions underlying LC. Assumptions 1 and 2, i.e., there exists the cognizance function and a cognizance function is common across class labels, are supported by the experimental results, respectively. Assumptions 1 and 2 are supported by that both $h_k(\cdot)$ and $h(\cdot)$ can differentiate the samples of seen and unseen classes with p-value $<2.2 \times 10^{-6}$ on Kruskal-Wallis test. However, our experimental evidence shows contradiction to Assumption 3 and implies that (1) a cognizance function $h(\cdot)$ learned from one model cannot be generalized to another model and (2) a cognizance function $h(\cdot)$ learned for one task cannot be generalized to another task.

LC has been shown effective. Its cognizance function is shown to exist and be common across class label, but its generality across models and domains may come naturally.

Acknowledgement. The authors are grateful to the reviewers for their valuable comments and suggestions which help to improve this manuscript. This work was financially supported by Thailand Science Research and Innovation (TSRI) and Uttaradit Rajabhat University.

References

1. Deng, J., Dong, W., Socher, R., Li, L.J., Li, K., Fei-Fei, L.: Imagenet: a large-scale hierarchical image database. In: CVPR (2009)
2. Russakovsky, O., et al.: ImageNet large scale visual recognition challenge. Int. J. Comput. Vis. (IJCV) **115**(3), 211–252 (2015)
3. GLin, G., Shen, C., Van Den Hengel, A., Reid, I.: Efficient piecewise training of deep structured models for semantic segmentation. In: 2016 IEEE Conference on Computer Vision and Pattern Recognition (CVPR), pp. 3194–3203. IEEE Computer Society (2016)

4. Okuyama, T., Gonsalves, T., Upadhay, J.: Autonomous driving system based on deep Q learnig. In: International Conference on Intelligent Autonomous Systems (ICoIAS), 2018, pp. 201–205 (2018)
5. Sutskever, I., Vinyals, O., Le, Q.V.: Sequence to sequence learning with neural networks. Adv. Neural Inf. Process. Syst. **4**(January), 3104–3112 (2014). arXiv:1409.3215
6. Nakjai, P., Katanyukul, T.: Hand Sign recognition for thai finger spelling: an application of convolution neural network. J. Sign. Process. Syst. **91**, 131–146 (2019)
7. Esteva, A., et al.: A guide to deep learning in healthcare. Nat. Med. **25**(1), 24–29 (2019)
8. Goodfellow, I.J., Shlens, J. and Szegedy, C.: Explaining and Harnessing Adversarial Examples, arXiv:1412.6572 [cs, stat] (December 2014)
9. Keinosuke, F.: Introduction to Statistical Pattern Recognition - 2nd edn. Academic Press, Cambridge (1990)
10. Nguyen, A., Yosinski, J., Clune, J.: Deep Neural Networks are Easily Fooled: High Confidence Predictions for Unrecognizable Images, arXiv:1412.1897 [cs] (December 2014)
11. Ahmed, T., Sabab, N.H.N.: Classification and understanding of cloud structures via satellite images with EfficientUNet, arXiv:2009.12931 [cs, eess] (May 2021)
12. Nakjai, P., Ponsawat, J., Katanyukul, T.: Latent cognizance: what machine really learns. In: ACM International Conference Proceeding Series, pp. 164–170. ACM, New York, USA (2019)
13. Krizhevsky, A., Sutskever, I., Hinton, G.E.: ImageNet classification with deep convolutional neural networks. Commun. ACM **60**(6), 84–90 (2017)
14. Miller, D., Sünderhauf, N., Milford, M., Dayoub, F.: Class Anchor Clustering: A Distance-based Loss for Training Open Set Classifiers, arXiv:2004.02434 [cs] (July 2020)
15. Fei, G., Liu, B.: Breaking the closed world assumption in text classification. In: Proceedings of the 2016 Conference of the North American Chapter of the Association for Computational Linguistics: Human Language Technologies, pp. 506–514. Association for Computational Linguistics, San Diego, California (2016)
16. Mendes Júnior, P.R., et al.: Nearest neighbors distance ratio open-set classifier. Mach. Learn. **106**(3), 359–386 (2016)
17. Bendale, A., Boult, T.: Towards open world recognition. In: 2015 IEEE Conference on Computer Vision and Pattern Recognition (CVPR), pp. 1893–1902 (2015). arXiv:1412.5687
18. Webb, A.R., Copsey, K.D.: Statistical Pattern Recognition, 3rd edn. Wiley, West Sussex. England, New Jersey (2011)
19. Mclachlan, G.: Mahalanobis distance. Resonance **4**, 20–26 (1999)
20. Rudd, E.M., Jain, L.P., Scheirer, W.J., Boult, T.E.: The extreme value machine. IEEE Trans. Pattern Anal. Mach. Intell. **40**(3), 762–768 (2018)
21. De Haan, L., Ferreira, A., Ferreira, A.: Extreme Value Theory: An Introduction, Springer Series in Operations Research and Financial Engineering. Springer-Verlag, New York (2006)
22. Scheirer, W.J., Jain, L.P., Boult, T.E.: Probability models for open set recognition. IEEE Trans. Pattern Anal. Mach. Intell. **36**(11), 2317–2324 (2014)
23. Bendale, A., Boult, T.: Towards Open Set Deep Networks, arXiv:1511.06233 [cs] (November 2015)
24. Nakjai, P., Katanyukul, T.: Automatic hand sign recognition: identify unusuality through latent cognizance. In: Pancioni, L., Schwenker, F., Trentin, E. (eds.)

ANNPR 2018. LNCS (LNAI), vol. 11081, pp. 255–267. Springer, Cham (2018). https://doi.org/10.1007/978-3-319-99978-4_20

25. Schroff, F., Kalenichenko, D., Philbin, J.: FaceNet: a unified embedding for face recognition and clustering. In: IEEE Conference on Computer Vision and Pattern Recognition (CVPR), 2015, pp. 815–823 (2015)

26. Simonyan, K., Zisserman, A.: Very deep convolutional networks for large-scale image recognition. In: 3rd International Conference on Learning Representations, ICLR 2015 - Conference Track Proceedings, pp. 730–734. IEEE, Kuala Lumpur, Malaysia (2015). arXiv:1409.1556

27. Krizhevsky, A.: Learning multiple layers of features from tiny images, Master's thesis, Department of Computer Science, University of Toronto (2009)

Noise Fuzzy Clustering-Based Robust Non-negative Matrix Factorization with I-divergence Criterion

Akira Okabe, Katsuhiro Honda$^{(\boxtimes)}$ ⓘ, and Seiki Ubukata ⓘ

Osaka Prefecture University, Sakai, Osaka 599-8531, Japan
{honda,subukata}@cs.osakafu-u.ac.jp

Abstract. Non-negative Matrix Factorization (NMF) is a technique for factorizing a non-negative matrix into the products of non-negative component matrices and has been used in such applications as air pollution analysis. In order to make NMF robust against noise, noise clustering-based approach was proposed with least square criterion, where NMF model estimation was performed in conjunction with noise rejection under the iterative optimization principle. In this paper, another robust NMF model was proposed supported by I-divergence criterion, which considers asymmetric distance measures rather than symmetric ones in the least square model. The updating formula of fuzzy memberships for non-noise degrees of objects are also constructed based on I-divergence criterion. The characteristic features of the proposed method are compared with the conventional one through numerical experiments using an artificial dataset.

Keywords: Non-negative matrix factorization · Noise fuzzy clustering · I-divergence

1 Introduction

Non-negative Matrix Factorization (NMF) [1] is a technique for factorizing a non-negative matrix into the products of non-negative component matrices and has been used in such applications as air pollution analysis and audio source separation. In order to evaluate the deviation between the original observations and their lower-rank reconstructions, NMF models adopt not only the symmetric least square criterion but also other asymmetric measures such as I-divergence [2] and Itakura-Saito divergence [3].

When datasets include noise, we should estimate NMF models by rejecting their influences. Besides element-wise noise rejection [4,5], object-wise noise rejection was proposed, where the non-noise degree of each object was estimated supported by noise fuzzy clustering concept [6]. Fuzzy c-means (FCM) [7,8] is a basic unsupervised classification method, which partitions objects into several fuzzy clusters with prototypical centroids. Noise fuzzy clustering [9,10] tried

© Springer Nature Switzerland AG 2022
K. Honda et al. (Eds.): IUKM 2022, LNAI 13199, pp. 256–266, 2022.
https://doi.org/10.1007/978-3-030-98018-4_21

to make FCM robust against noise by introducing an additional noise cluster, where fuzzy membership degrees of each object are simultaneously estimated not only for normal clusters but also for the noise cluster. From the robust model estimation viewpoint, noise fuzzy clustering can be identified with iteratively reweighted least square method [11], and then, the noise rejection approach has been utilized in robust data analysis such as robust PCA-based k-means [12].

Considering the algorithmic similarity among NMF and fuzzy c-varieties (FCV) with least square criteria [13], a noise fuzzy clustering-based robust NMF [6] was proposed, which iteratively performs NMF modeling and non-noise degree estimation until convergence. Because the least square-type criterion of NMF has a similar feature with the FCV criterion, it was easily reused in fuzzy membership estimation with noise fuzzy membership updating formula.

In this paper, another robust NMF model was proposed supported by I-divergence criterion, which considers asymmetric distance measures rather than symmetric ones in the least square model. The updating formula of fuzzy memberships for non-noise degrees of objects are also constructed based on I-divergence criterion. The characteristic features of the proposed method are compared with the conventional one through numerical experiments using an artificial dataset. The remaining parts of this paper are organized as follows: Sect. 2 briefly reviews NMF and noise fuzzy clustering-induced robust modeling and Sect. 3 proposes a novel robust NMF model with I-divergence criterion. The experimental results are presented in Sect. 4 and the summary conclusion is given in Sect. 5.

2 Brief Review on NMF and Noise Fuzzy Clustering

2.1 NMF

NMF [1] is a technique for decomposing an $n \times m$ matrix $X = \{x_{ij}\}$ composed of only non-negative elements into the product of two lower-order non-negative matrices of $n \times p$ matrix $W = \{w_{ik}\}$ and $p \times m$ matrix $H = \{h_{kj}\}$ as follows:

$$X \approx WH, \tag{1}$$

where intrinsic dimension p is constrained to $n > p$ and $m > p$, respectively.

In order to achieve minimum error reconstruction, NMF adopts some kinds of error measures in building the objective function to be minimized. When we adopt the least square criterion, the NMF objective function is given as:

$$J_{nmf1} = \sum_{i=1}^{n} \sum_{j=1}^{m} \left(x_{ij} - \sum_{k=1}^{p} w_{ik} h_{kj} \right)^2. \tag{2}$$

The lower-dimensional factors are available for making it easier for humans to understand the intrinsic features of multi-dimensional data.

The algorithm is based on the alternating optimization of the elements of matrices W and H, where their updating formulas are derives as [2]:

$$h_{kj} = h_{kj} \frac{(W^\top X)_{kj}}{(W^\top W H)_{kj}}, \tag{3}$$

$$w_{ik} = w_{ik} \frac{(XH^\top)_{ik}}{(WHH^\top)_{ik}}. \tag{4}$$

Besides the above symmetric least square criterion, other asymmetric measures have been also proved to be useful in many application fields. For example, when we adopt I-divergence measure, i.e., Kullback-Leibler (KL) divergence, the NMF objective function is modified as:

$$J_{nmf2} = \sum_{i=1}^{n} \sum_{j=1}^{m} \left(x_{ij} \log \frac{x_{ij}}{\sum_{k=1}^{p} w_{ik} h_{kj}} - x_{ij} + \sum_{k=1}^{p} w_{ik} h_{kj} \right). \tag{5}$$

Then, the updating formulas are revised as [2]:

$$h_{kj} = h_{kj} \frac{\sum_{i=1}^{n} \frac{x_{ij} w_{ik}}{\sum_{k=1}^{p} w_{ik} h_{kj}}}{\sum_{i=1}^{n} w_{ik}}, \tag{6}$$

$$w_{ik} = w_{ik} \frac{\sum_{j=1}^{m} \frac{x_{ij} h_{kj}}{\sum_{k=1}^{p} w_{ik} h_{kj}}}{\sum_{j=1}^{m} h_{kj}}. \tag{7}$$

Here, these NMF models have non-uniqueness features by transforming the two factorization matrices by a monomial matrix T as:

$$WH = WTT^{-1}H = (WT)(T^{-1}H) = W'H', \tag{8}$$

where a simple example can be constructed with a scaling and/or a permutation.

2.2   Noise Fuzzy Clustering

FCM [7,8] is a fuzzy extension of k-means clustering [14], whose goal is to partition n objects with m-dimensional observation x_i, $i = 1, \ldots, n$ into C fuzzy clusters represented by their prototypical centroids b_c, $c = 1, \ldots, C$. The FCM objective function to be minimized is given as the following weighted within-cluster errors:

$$J_{fcm} = \sum_{c=1}^{C} \sum_{i=1}^{n} u_{ci}^{\theta} ||x_i - b_c||^2, \tag{9}$$

where u_{ci} ($u_{ci} \in [0,1]$) is the fuzzy membership of object i to cluster c and represents the degree of belongingness under the probabilistic context with $\sum_{c=1}^{C} u_{ci} = 1$. θ ($\theta > 1$) is the fuzzification penalty such that a large θ brings fuzzier cluster boundaries while $\theta \to 1$ reduces to the crisp k-means. Starting from a random initial partition, cluster prototypes b_c and fuzzy memberships u_{ci} are iteratively updated under the alternating optimization scheme.

In order to improve the noise sensitive feature of FCM-type least square criterion, Davé proposed noise fuzzy clustering [9], which introduces an additional noise cluster to dump all noise objects into it. Considering C normal clusters and $C + 1$-th noise cluster, the objective function of noise FCM was defined as:

$$J_{nfcm} = \sum_{c=1}^{C} \sum_{i=1}^{n} u_{ci}^{\theta} ||x_i - b_c||^2 + \gamma \sum_{i=1}^{n} u_{C+1,i}^{\theta}, \tag{10}$$

where $u_{C+1,i}$ is the fuzzy membership to the noise cluster and the probabilistic constraint is modified as $\sum_{c=1}^{C+1} u_{ci} = 1$. γ is the distance between each object and the noise cluster, and is constant for all objects. If an object is distant from all C normal clusters more than γ, it is dumped into the noise cluster and the remaining C memberships are $\sum_{c=1}^{C} u_{ci} < 1$.

The updating formulas are derived as:

$$u_{ci} = \frac{||x_i - b_c||^{\frac{2}{1-\theta}}}{\gamma^{\frac{1}{1-\theta}} + \sum_{\ell=1}^{C} ||x_i - b_\ell||^{\frac{2}{1-\theta}}} \quad (\text{if } c \leq C), \tag{11}$$

$$u_{C+1,i} = 1 - \sum_{\ell=1}^{C} u_{\ell i}, \tag{12}$$

$$b_c = \frac{\sum_{i=1}^{n} u_{ci}^{\theta} x_i}{\sum_{i=1}^{n} u_{ci}^{\theta}}. \tag{13}$$

Here, when $C = 1$, the above model is reduced to a robust average estimator, where the average value is calculated in the boundary of γ radius [10]. This noise rejection scheme was also utilized in robust least square-type data analysis, where the FCM clustering criterion is replaced with other least square criteria. For example, if it is introduced into principal component analysis (PCA)-induced k-means, robust k-means clustering is achieved through robust PCA considering non-noise membership degrees [12]. In the followings, the noise clustering concept is introduced into the robust NMF context.

3   Noise Fuzzy Clustering-Induced Robust NMF

In this paper, two approaches for robustifying NMF are considered induced by noise fuzzy clustering concept. After a brief review of least square-type model, a novel model of I-divergence-based robust NMF is proposed.

3.1   Robust NMF with Least Square Criterion

In a previous research [6], a robust extension of least square criterion-based NMF was proposed by introducing non-noise fuzzy memberships with a noise cluster, whose objective function was defined by replacing the FCM criterion with the reconstruction error as:

$$J_{rnmf1} = \sum_{i=1}^{n} u_i^{\theta} \sum_{j=1}^{m} \left(x_{ij} - \sum_{k=1}^{p} w_{ik}h_{kj} \right)^2 + \gamma \sum_{i=1}^{n} (1 - u_i)^{\theta}, \qquad (14)$$

where u_i $(0 \le u_i \le 1)$ is the fuzzy membership representing the non-noisiness degree of individual i and θ $(\theta > 1)$ is the fuzzification weight. γ is a constant representing the noise sensitivity, which can be identified with the distance between the individual and the noise cluster. An individual is rejected as noise if the sum of squared errors in lower-rank approximation is larger than γ.

Under the alternating optimization principle, the updating rules for u_i and h_{kj} are derived as:

$$u_i = \left(1 + \left(\frac{d_i}{\gamma} \right)^{\frac{1}{\theta-1}} \right)^{-1}, \qquad (15)$$

$$d_i = \sum_{j=1}^{m} \left(x_{ij} - \sum_{k=1}^{p} w_{ik}h_{kj} \right)^2, \qquad (16)$$

and

$$h_{kj} = h_{kj} \frac{(W^{\top}UX)_{kj}}{(W^{\top}UWH)_{kj}}, \qquad (17)$$

where U is the diagonal matrix, whose ith diagonal element is u_i^{θ}. On the other hand, w_{ik} is still updated by Eq. (4).

3.2 Robust NMF with I-divergence Criterion

In this section, a novel robust NMF model is proposed by introducing the noise clustering concept into NMF with I-divergence criterion. Besides the symmetric measure of Eq. (16), the I-divergence NMF criterion

$$\hat{d}_i = \sum_{j=1}^{m} \left(x_{ij} \log \frac{x_{ij}}{\sum_{k=1}^{p} w_{ik}h_{kj}} - x_{ij} + \sum_{k=1}^{p} w_{ik}h_{kj} \right), \qquad (18)$$

is utilized with an asymmetric measure in some application areas, whose feature is compared with a symmetric one as in Fig. 1.

Then, the proposed objective function for robust NMF with I-divergence is defined as:

$$J_{rnmf2} = \sum_{i=1}^{n} u_i^{\theta} \sum_{j=1}^{m} \left(x_{ij} \log \frac{x_{ij}}{\sum_{k=1}^{p} w_{ik}h_{kj}} - x_{ij} + \sum_{k=1}^{p} w_{ik}h_{kj} \right) + \gamma \sum_{i=1}^{n} (1 - u_i)^{\theta},$$

$$\qquad (19)$$

where u_i, θ and γ play the same roles with those in Eq. (14).

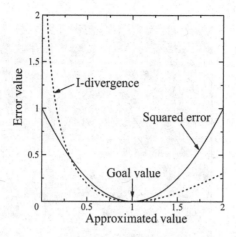

Fig. 1. Comparison of symmetric/asymmetric error measures with goal value = 1.

Considering fuzzy memberships u_i, the updating formula for h_{kj} is modified as:

$$h_{kj} = h_{kj} \frac{\sum_{i=1}^{n} \frac{u_i^\theta x_{ij} w_{ik}}{\sum_{k=1}^{p} w_{ik} h_{kj}}}{\sum_{i=1}^{n} u_i^\theta w_{ik}}, \qquad (20)$$

while that for w_{ik} is still Eq. (7).

Next, the updating formula for u_i is given as:

$$u_i = \left(1 + \left(\frac{\hat{d}_i}{\gamma}\right)^{\frac{1}{\theta-1}}\right)^{-1}. \qquad (21)$$

A sample procedure of the proposed algorithm is written as follows:

Algorithm: Noise Fuzzy Clustering-based Robust Non-negative Matrix Factorization with I-divergence Criterion

Step 1. Initialize elements of matrices W and H with random non-negative values.

Step 2. Initialize memberships u_i by Eq. (21).

Step 3. Update elements of matrix H by Eq. (20).

Step 4. Update elements of matrix W by Eq. (7).

Step 5. Update memberships u_i by Eq. (21).

Step 6. If memberships u_i converge, stop. Otherwise, return to Step 3.

4   Numerical Experiments

In order to compare the characteristic of two robust NMF models, this section presents some experimental results.

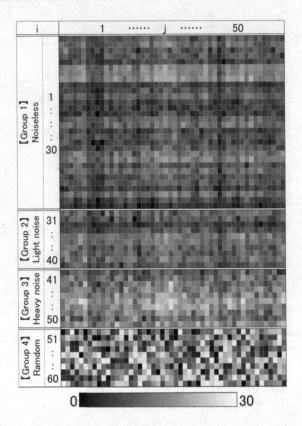

Fig. 2. 60×50 artificial observation matrix X depicted in grayscale.

4.1 Data Set

The artificially generated 60×50 observation matrix X shown in Fig. 2 was used, which is the mixture of four sub-groups. Group 1 composed of 30 individuals completely follows a generative model $\tilde{W}$ and $\tilde{H}$ such that $X = \tilde{W}\tilde{H}$. Group 2 and Group 3 composed of 10 individuals each follow the same generative model with light or heavy noise as $X = \tilde{W}\tilde{H} + E$, where $E = \{\varepsilon_{ij}\}$ is a random noise matrix like $\varepsilon_{ij} \in [0, 5)$ for Group 2 and $\varepsilon_{ij} \in [0, 10)$ for Group 3, respectively. Group 4 composed of 10 individuals has just random observation as $X = E$ with $\varepsilon_{ij} \in [0, 30)$. The detailed generation process can be found in [15].

4.2 Comparison of Noise Rejection Features

In order to compare the matrix decomposition ability of the two robust NMF models, they were applied with various noise sensitivity weights γ ($\gamma \in \{0.1, 0.2, 0.5, 1, 2, 5, 10, 20, 50, 100, 200, 500, 1000, 2000, 5000, 10000 \}$) with a fixed fuzziness penalty $\theta = 2$, which is a standard setting in FCM [7].

Figure 3 compares the average non-noise fuzzy memberships u_i derived with various γ in 100 different random initialization. The right end with $\gamma = 10000$

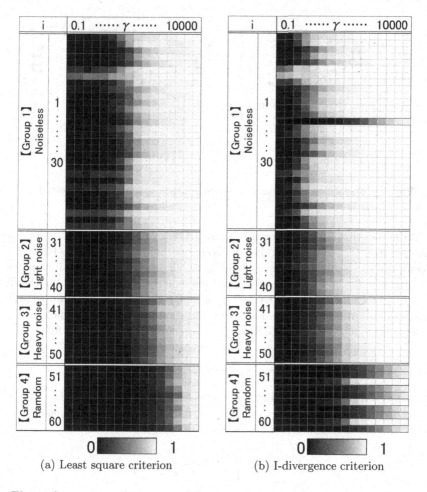

Fig. 3. Comparison of fuzzy memberships u_i with various noise sensitivity γ.

is almost equivalent to the conventional NMF without noise rejection, where almost all memberships are $u_i \approx 1$, while the models become sensitive to noise by rejecting many noise individuals as γ becomes smaller.

Looking from the right end to left, both models first rejected Group 4 of complete random observation, and then, Group 3 of heavy noisy data and Group 2 of light noisy data are secondly and thirdly rejected. That is, both models could successfully achieve gradual noise rejection by tuning the sensitivity weight γ.

Here, the proposed robust NMF with I-divergence had a slightly different feature for some individuals as emphasized in red rectangles such that they were rejected even when γ was very large, e.g., $\gamma = 1000$. Especially, individual 14 was first rejected although it belongs to Group 1, i.e., completely non-noise. By carefully checking their observation, they were found to have one or more zero elements $x_{ij} = 0$.

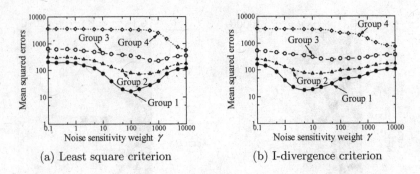

(a) Least square criterion   (b) I-divergence criterion

Fig. 4. Comparison of approximation errors with various noise sensitivity γ.

Be noted that I-divergence is an asymmetric measure that is much more sensitive to negative errors against positive one as shown in Fig. 1. Then, NMF decomposition tends to achieve slightly higher approximation values while a very small element like $x_{ij} = 0$ cannot be exactly reconstructed for avoiding negative errors for other positive attributes. By the way, in some applications, zero observation may be caused by error and should be rejected as noise. In this sense, the proposed robust NMF with I-divergence seems to be useful.

4.3   Comparison of Approximation Errors

Next, the approximation ability of the two robust NMF models is compared. Figure 4 compares the trajectories of the 100 trials averages of mean square errors in NMF approximation for each group such that

$$\frac{1}{n_g} \sum_{i \in G_g} \sum_{j=1}^{m} \left(x_{ij} - \sum_{k=1}^{p} w_{ik} h_{kj} \right)^2, \tag{22}$$

where n_g is the number of individuals of each group G_g ($g = 1, \ldots, 4$).

Looking from the right end to left, in both models, Group 4 was first rejected at around $\gamma = 1000$ and other groups' errors were decreased by ignoring Group 4, i.e., the approximation ability was improved by rejecting complete random noise. Then, Group 3 and Group 2 were secondly and thirdly ignored and the models fairly fitted to the non-noise Group 1 around $\gamma = 100$ in Fig. 4(a) and $\gamma = 5$ in Fig. 4(b), respectively. Although the two models had slightly different membership features as shown in Fig. 3, both models successfully improved their approximation ability by focusing on the non-noise groups. Finally, errors for Group 1 also became larger when γ is very small like $\gamma = 1$ or smaller because the models focused only on a few individuals by rejecting almost all other ones.

On the other hand, Fig. 5 compares the trajectories of the 100 trials averages of I-divergence distances in the I-divergence-based NMF approximation for each group such that

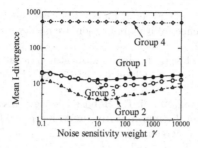

Fig. 5. Trajectory of I-divergence with various noise sensitivity γ.

$$\frac{1}{n_g} \sum_{i \in G_g} \sum_{j=1}^{m} \left(x_{ij} \log \frac{x_{ij}}{\sum_{k=1}^{p} w_{ik} h_{kj}} - x_{ij} + \sum_{k=1}^{p} w_{ik} h_{kj} \right). \tag{23}$$

In contrast to Fig. 4(b) with squared errors, Fig. 5 emphasizes the errors of Group 4 composed of random observations only and Group 1 composed of non-noise individuals only because the two groups include one or several individuals having zero values. So, the second measure of I-divergence can contribute to distinguishing such small noise observations.

From the above results, we can see that the two robust NMF models are useful in matrix decomposition with noise rejection and the adjustable weight γ works for effectively tuning their noise sensitivity while I-divergence-induced model is additional feature of rejecting very small elements.

5   Conclusions

In this paper, a novel robust NMF model was proposed by adopting I-divergence criterion, which is an asymmetric distance measure, and its characteristics were demonstrated through a comparison with the conventional least square criterion-induced model. Besides a similar approximation ability, the proposed model has additional feature of rejecting very small elements, which may be caused by error in some applications.

Possible future works include development of automatic selection mechanism for noise sensitivity weight γ.

Acknowledgment. This work was supported in part by JSPS KAKENHI Grant Number JP18K11474.

References

1. Lee, D.D., Seung, H.S.: Learning the parts of objects by non-negative matrix factorization. Nature **401**(6755), 788–791 (1999)
2. Lee, D.D., Seung, H.S.: Algorithms for nonnegative matrix factorization. Adv. Neural Inf. Process. Syst. **13**, 556–562 (2000)

3. Févotte, C., Bertin, N., Durrieu, J.-L.: Nonnegative matrix factorization with the Itakura-Saito divergence. With application to music analysis. Neural Comput. **21**(3), 793–830 (2009)
4. Zhang, L., Chen, Z., Zheng, M., He, X.: Robust non-negative matrix factorization. Front. Electr. Electron. Eng. China **6**, 192–200 (2011). https://doi.org/10.1007/s11460-011-0128-0
5. Shen, B., Liu, B., Wang, Q., Ji, R.: Robust nonnegative matrix factorization via L_1 norm regularization by multiplicative updating rules. In: Proceedings of 2014 IEEE International Conference on Image Processing, pp. 5282–5286 (2014)
6. Ueno, M., Honda, K., Ubukata, S., Notsu, A.: Robust non-negative matrix factorization based on noise fuzzy clustering mechanism. In: Proceedings of 2019 2nd Artificial Intelligence and Cloud Computing Conference and 2019 Asia Digital Image Processing Conference, pp. 1–5 (2019)
7. Bezdek, J.C.: Pattern Recognition with Fuzzy Objective Function Algorithms. Plenum Press, New York (1981)
8. Miyamoto, S., Ichihashi, H., Honda, K.: Algorithms for Fuzzy Clustering. Springer, Heidelberg (2008). https://doi.org/10.1007/978-3-540-78737-2
9. Davé, R.N.: Characterization and detection of noise in clustering. Pattern Recognit. Lett. **12**(11), 657–664 (1991)
10. Davé, R.N., Krishnapuram, R.: Robust clustering methods: a unified view. IEEE Trans. Fuzzy Syst. **5**, 270–293 (1997)
11. Holland, P.W., Welsch, R.E.: Robust regression using iteratively reweighted least-squares. Commun. Stat. **A6**(9), 813–827 (1977)
12. Honda, K., Notsu, A., Ichihashi, H.: Fuzzy PCA-guided robust k-means clustering. IEEE Trans. Fuzzy Syst. **18**(1), 67–79 (2010)
13. Honda, K., Ichihashi, H.: Linear fuzzy clustering techniques with missing values and their application to local principal component analysis. IEEE Trans. Fuzzy Syst. **12**(2), 183–193 (2004)
14. MacQueen, J. B.: Some methods of classification and analysis of multivariate observations. In: Proceedings of the 5th Berkeley Symposium on Mathematical Statistics and Probability, pp. 281–297 (1967)
15. Honda, K., Ueno, M., Ubukata, S., Notsu, A.: Robust non-negative matrix factorization based on noise fuzzy clustering mechanism and application to environmental observation data analysis. J. Japan Soc. Fuzzy Theory Intell. Inform. **33**(2), 593–599 (2021). (in Japanese)

Topic Modeling of Political Dynamics with Shifted Cosine Similarity

Yifan Luo[1], Tao Wan[2], and Zengchang Qin[1(✉)]

[1] Intelligent Computing and Machine Learning Lab, School of Automation Science and Electric Engineering, Beihang University, Beijing 100191, China
{luoyifan,zcqin}@buaa.edu.cn
[2] School of Biological Science and Medical Engineering, Beihang University, Beijing 100191, China
taowan@buaa.edu.cn

Abstract. Topic modeling with community detection can be used to explore the latent semantic structure of documents, we can utilize a network, i.e., a graph to depict the semantic relation between words. In some network based topic models, in order to obtain a network with obvious community structure, the similarity between words (vertices) is essential. Word embeddings trained from a large corpus empirically perform as well as in rich semantic representation, thus this research is intended to construct a novel similarity in a network based topic model (NAM). In this paper, we first intuitively propose a similarity measure based on shifted cosine similarity between word embeddings. This similarity is exploited to replace the similarity based on typical point-wise mutual information (PMI). Secondly, based on different similarity measures, topics of corpus in a global period are induced by NAM. Finally, we use NAM to capture the dynamic changes of political topics in China and interpret the dynamic processes using historical background. Although our similarity measure introduces semantic differences caused by the difference between data sets and has one more parameter, the experimental results show the effectiveness of our new proposed measure.

Keywords: Topic model · Network analysis · Word embeddings

1 Introduction

Topic models can conveniently discover latent semantic structure from textual data, for which they play an important role in many fields that involve Natural Language Processing. Conventional topic models, such as Probabilistic Latent Semantic Analysis (PLSA) [7] and Latent Dirichlet Allocation (LDA) [2], have been widely applied. Based on the bag-of-words model, LDA [2] represents each given document as a multinomial distribution over topics and represents each topic as a multinomial distribution over words. However, in LDA [2], the number of topics is set manually.

K. Honda et al. (Eds.): IUKM 2022, LNAI 13199, pp. 267–278, 2022.
https://doi.org/10.1007/978-3-030-98018-4_22

Network analysis based topic models are also developed. Li et al. [9] introduce community and dynamic into topic models. Rule et al. [14] apply a network analysis based method in exploring the topics and the changes of them in State of the Union discourse. Cointet et al. [5] utilize the same method to analyze the research topics in gene expression profiling. The method and technique used by Rule et al. and Cointet et al. mentioned above can actually be provided by CorTexT platform[1], which is abbreviated as NAM (network analysis based model) in this paper. More specifically, they first project the semantic relation between frequent words into a graph (network) and filter the network until it presents a community structure [12]. Communities detected by a community detection algorithm are regarded as topics, thus the number of topics doesn't need to be set manually. Moreover, based on a measure to quantify the proximity between topics in different period, the dynamic topic model (DTM) can also be induced.

Word embeddings have proven to be powerful semantic representations in practice. Effective neural network models like Continuous Bag-of-Words (CBOW) and Skip-gram models [10], learn word embeddings from contextual information and project words into a continuous linear space. Therefore, it should be useful to combine word embeddings with NAM. When other conditions remain unchanged, the final network of NAM is only determined by similarity between words. We note that since Rule et al. [14] only use the word pair with the Pointwise Mutual Information (PMI) value above zero to calculate the distributional similarity, it's possible to use cosine values between word embeddings after subtracting a threshold to execute the same calculation, which can introduce word embeddings into NAM and make full advantages of their ability to capture semantics.

We aim to apply NAM in analyzing the political topics and their changes in the political reports from government and party in China (RGPC). Meanwhile, to make use of word embeddings, we prepare reports from The People's Daily (RPD), which may be semantically similar to RGPC, because RPD usually reflect the official political tendency.

In this paper, we first overview NAM by some more rigorous definitions. Second, we propose a distributional similarity based on shifted cosine values (cosine values subtracting a constant) between word embeddings. Thirdly, measured by coherence values [4,11,13] on RGPC in a global period, we observe that shifted cosine values can achieve comparable performance with PMI, which prove the effectiveness of new similarity in terms of the disadvantage that word embeddings are learned from RPD rather than RGPC. Finally, based on coherence values, the dynamic political topics in China determined by two kinds of similarity are visualized and interpreted in terms of social and history background. The experimental results show that although our similarity introduces semantic bias caused by data set differences and adds a parameter to be adjusted, it can still capture the changes of political themes of historical significance.

---

[1] https://www.cortext.net/.

2 Related Works

Over the past decade, various network analysis based approaches have been proposed to generate topics from corpus. Li et al. [9] elaborately combine community detection with topic analysis to help to understand social networks. Cointet et al. [5] showed a semantic network to interpret the research topics of gene expression profiling. This network shows an obvious community structure [12], and is intuitive and easy to interpret. This network analysis based topic model is also utilized by rule et al.[14] to interpret the political theme shifts of State of the Union discourses. The method used by rule et al. and cointet et al. [5,14] is supported by CorTexT platform. However, the platformcan not directly meet our experimental needs, such as calculating new similarity and preprocessing Chinese corpus. Therefore, instead of using this platform, we wrote our own scripts to execute our experiments.

Word embeddings improved by mikolov et al. [10] have proven to be powerful to capture semantics from corpus. The advantage of word embeddings over semantics is also made use of to improve the performance of topic models. Das et al. [6] combine multivariate Gaussian distributions with word embeddings. Li et al. [8] apply auxiliary word embeddings in a topic model for short texts.

We introduce word embeddings into NAM, exploiting not only the rich semantic from word embeddings, but also the intuitive and clear network structures to interpret the produced topics.

3 Methods

3.1 Preliminaries

Two definitions are given as follows:

Definition 1: Given a directed (or undirected) graph G, if a simple graph g satisfies the following requirements: 1, g has the same vertices as G;

2, For any given vertices i and $j (i \neq j)$ of G, if there is at least one edge between these two vertices, then there is only one undirected edge between vertices i and j of graph g;

3, For any given vertices i and $j (i \neq j)$ of G, if there is no edge between these two vertices, then there is also no edge between vertices i and j of graph g;

Then we call the simple graph g as a simplified graph of graph G. Notably, if G is an undirected graph, then G is a simplified graph of itself.

Definition 2: If for any given pair i and j, similarity $s(i,j)$ is equal to $s(j,i)$, we call that $s(i,j)$ is symmetrical, else we call $s(i,j)$ asymmetrical.

Notations: In this paper, the simplified graph of a graph G is typically denoted as lower case g, e.g., g' and G', g_p^T and G_p^T, etc. For simplicity, we use i or j to refer to terms of W_p, although they are the indexes of terms.

3.2   Network Analysis Model

In this subsection, we describe the network analysis based model (NAM) [14] with graph theory knowledge and supplementary definitions in Subsect. 3.1. Our settings may not be exactly the same as those actually used by Rule et al.

Linguistic Processing and Similarity Calculation. N most frequent noun terms are extracted from corpus produced during the period p, and the set consists of these terms is denoted as W_p. Similarity $s(i, j)$ between any term pair i and j can be calculated by various methods (see Subsect. 3.3). Here, terms i and j are from W_p. In this paper, the similarity will be projected into range $[0, 1]$ to be compatible with the filtering process.

Construction and Filtering of Network. If $s(i, j)$ is asymmetrical, then a directed graph G_p is built according to the following procedure: First, N terms in the set W_p are used as vertices. Second, for any vertex pair i and j ($i \neq j$), a edge from i to j is built, with its weight being $s(i, j)$. Notably, loops will not be introduced into G_p, so the number of edges of directed graph G_p is actually $N(N - 1)$. Else if $s(i, j)$ is symmetrical, then an undirected graph G_p is built using similar process. But in this situation, there are only $\frac{N(N-1)}{2}$ edges of G_p.

After G_p is constructed, its filtering process begins. At the beginning, a threshold v is initialized as 0, and we remove all edges below v from G_p, thus we obtain a temporary graph G_p'. We then explore the connectedness of graph g_p^v (the simplified graph of graph G_p'). All vertices of components except those of the main connected component of graph g_p^v are removed from G_p', along with the edges connecting these vertices, from which we obtain graph G_p^v. Then v is added by a incremental Δv. Similar to the process above, we can obtain G_p^v from $G_p^{v-\Delta v}$ and update v until the termination condition.

We approximatively follow the termination condition of filtering algorithm as in reference [14], that is, when a component larger than 2 vertices is separated from the principle component of simplified graph g_p^v, the filtering process is stopped. The termination condition is ignored only when the threshold v is 0. The final threshold v is denoted as T. Basing on the filtering process above, the final semantic network G_p^T can be built from corpus produced in any given period p. The set of the vertices of G_p^T is denoted as W_p^T.

Note that the edge between any given vertex pair i and j may be removed (even if i and j belong to the same community), so we define $e(i, j)$ to replace $s(i, j)$. The definition of $e(i, j)$ is as followings: if there exist an edge from vertex i to vertex j in G_p^T, then $e(i, j) = s(i, j)$; otherwise $e(i, j) = 0$.

Community Detection and Topic Representation. The vertices connected with each other closely in the filtered networks G_p^T are thought to belong to the same community, i.e., cluster. Louvain algorithm [3] is applied to identify communities of G_p^T. The communities are denoted as ϕ_p^l, where $l \in \{1, 2, 3...L\}$, and L is the number of communities of G_p^T. The communities detected above are actually not intersect, and their union is W_p^T. Every community is made of vertices, i.e., noun terms. Due to the asymmetry of $e(i,j)$, we practically use $v(w, \phi_p^l)$ given as follows to measure the importance of vertex w to ϕ_p^l:

$$v(w, \phi_p^l) = \begin{cases} \dfrac{1}{2} \dfrac{\sum\limits_{c \in \phi_p^l} (e(w,c)+e(c,w))}{\sum\limits_{(c1,c2) \in \phi_p^l{}^2} e(c1,c2)}, & w \in \phi_p^l \\ 0, & w \notin \phi_p^l \end{cases} \tag{1}$$

In fact, a community is regarded as a topic. Intuitively, for a given community ϕ_p^l, a large $v(w, \phi_p^l)$ means the great importance of word w to ϕ_p^l.

3.3 Distributional Similarity

We show a distributional similarity in a more general form as follows:

$$s(i,j) = \dfrac{\sum\limits_{\substack{c \in W_p/\{i,j\} \\ w(i,c) > 0 \\ w(j,c) > 0}} min\{w(i,c),\ w(j,c)\}}{\sum\limits_{\substack{c \in W_p/\{i,j\} \\ w(i,c) > 0}} w(i,c)} \tag{2}$$

The formula above was actually proposed by weeds et al. [16] earlier. Here, $w(i,j)$ is just a weighting function, and different functions can be used to act as $w(i,j)$, such as PMI. If $w(i,j) > 0$ means that j is a feature word of i, then formula 2 measure the similarity between different words by comparing their feature words and $w(i,j)$ serves as weight. $s(i,j)$ is asymmetric and lines in range $[0,1]$. When there is no term j such that $w(i,j) > 0$, the denominator of formula 2 is set as a small constant, i.e. 10^{-6} in practice to avoid denominator becoming 0. In the next three sub sections, three kinds of similarity are introduced.

PMI Based Distributional Similarity. Point-wise mutual information $I(i,j)$ between term i and j is defined as follows:

$$I(i,j) = log_2 \dfrac{p(i,j)}{p(i) \cdot p(j)} \tag{3}$$

Probability $p(i,j)$ and $p(i)$ can be derived from a co-occurrence matrix[2]. When $w(i,j)$ in formula 2 is pointwise mutual information (PMI) $I(i,j)$, $s(i,j)$ is actually the similarity utilized by rule et al. [14]. PMI based distributional similarity in the form of formula 2 is denoted as s_{PMI} in this paper.

Shifted Cosine Based Distributional Similarity. Note that in s_{PMI}, the weight can actually be thought as the difference between PMI and a special threshold, i.e., zero. The zero PMI threshold corresponds to a situation that two

---

[2] https://docs.cortext.net/metrics-definitions/.

terms occur independently in the sense of probability. Similarly, for introducing word embeddings into distributional similarity and making use of their effective semantic representation ability, we propose to use the difference between cosine similarity and a threshold to be adjusted as weight to calculate the distributional similarity. We call the weight function proposed above as shifted cosine (SCS), exactly the formula is given by follows:

$$SCS(i,j) = cos(i,j) - \theta \tag{4}$$

where $cos(i,j)$ is the cosine value between word embeddings of term i and j and θ is a constant. However, unlike the unique zero PMI value based on probabilistic theory, θ will be decided by experiments. SCS based distributional similarity in the form of formula 2 is denoted as s_{SCS} in this paper.

Translated and Rescaled Cosine Similarity. Translated and rescaled cosine similarity is also utilized as a distributional similarity. For compatibility with filtering process, cosine similarity are projected into range $[0,1]$ by an affine transformation as follows:

$$TRc(i,j) = (cos(i,j) + 1)/2 \tag{5}$$

where TRc is the translated and rescaled cosine similarity.

3.4 Dynamic Topic Over Time

Basing on the contribution score $v(w, \phi_p^l)$, proximity measure of topics can also be defined. Rule et al. [14] only consider the proximity between any pair of topics in two successive period. Similar with the Bhattacharyya distance, proximity measure of topics is defined as follows [14]:

$$\rho(\phi_p^i, \phi_{p-1}^j) = 1 - \frac{1}{\sqrt{2}} \sqrt{\sum_{w \in \left(\phi_p^i \cup \phi_{p-1}^j\right)} \left(\sqrt{v(w, \phi_p^i)} - \sqrt{v(w, \phi_{p-1}^j)}\right)^2} \tag{6}$$

4   Experimental Setup

4.1   Data-sets and Pre-processing

We collect data-sets RGPC and RPD produced in period 1950-2003, which is called as global period in this paper. Relatively, any sub period of global period are called local period. Three local periods are set as follows: [1950 : 1976], [1969 : 1992] and [1988 : 2002], which correspond to the periods led by three different leaders in China. These local periods are overlapping, which is designed to maintain the strength of similarity (formula 6) between topics in two successive local periods.

RPD contains news and reports in newspaper every year, whereas RGPC only contains reports in particular dates, and is a small data-set, so word embeddings are trained from RPD. RGPC is used to estimate joint probability distribution of frequent noun terms, which is made use of to calculate PMI and coherence of models on RGPC. Note that besides corpus from the global period, corpus from three local periods are also pre-processed and used to learn word embeddings respectively.

We use python tool Jieba [15] to implement Chinese word segmentation, meanwhile a stopword list is used to avoid meaningless terms and a new words list is used to update new terms for term library of Jieba [15]. The stopword list is provided by human experts in social science area. Following NAM mentioned in Sect. 3.2, we choose 1000 most frequent noun terms of RGPC (1950-2003), utilize the similarity mentioned in Sect. 3.3 to build and visualize the global semantic networks and use louvain algorithm [3] to induce topics. In detail, we rely open source tools Gensim[3] to train word embeddings and Networkx[4] to implement filtering algorithm. Visualization and community detection are finished by Gephi [1]. Following reference [14], the threshold increment of edge strength, i.e., Δv is set as a fixed constant equal to 0.01.

5   Experimental Results

5.1   Global Semantic Networks

Three global networks are shown in Fig. 1 and 2. We can observe that vertices representing "cities and provinces" are away from the main component in sub-picture (b) and Fig. 2, whereas related vertices in sub-picture (a) are not. Because networks in sub-picture (b) and Fig. 2 are based on the same word embeddings learned from RPD, whereas the network in sub-picture (a) is based on RGPC, the difference of the degree that "cities and provinces" group is away from the whole network actually reflects the semantic difference between different datasets. Notably, the layout algorithm only serves for visualization and the spatial distance in these visual pictures doesn't strictly reflect the real semantic differences.

We can also observe that except the sub components corresponding to "cities and provinces", the structures of the rest graph are also similar, which indicates that the substitution of s_{SCS} for s_{PMI} should be feasible, although some semantic bias may also be introduced.

5.2   Topic Coherence

Reference [13] concludes various coherence metrics by a unified framework. We choose four metrics to measure coherence of topics induced by three different similarity. They are C_{UMass} [11], C_V [13], C_{UCI} and C_{NPMI} [4].

---

[3] https://radimrehurek.com/gensim/.
[4] https://networkx.org/.

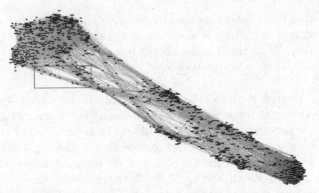

(a) Directed graph G_p^T based on s_{PMI}, with 956 vertices and 25192 edges. 7 communities are detected.

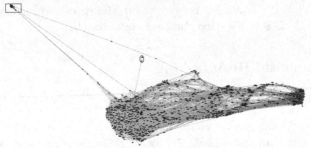

(b) Directed graph G_p^T based on s_{SCS}, with 994 vertices and 23587 edges. The threshold θ is 0.2. 7 communities are detected.

Fig. 1. The networks based on three different similarity, visualized by Gephi [1], with their communities detected by louvain algorithm [3]. Vertices of different communities are dyed in different colors and the size of vertices is positively correlated with term's frequency in corpus. Every vertices is labeled by corresponding Chinese terms. Built-in "Force Atlas" layout algorithm is used. Components in three red square frames are noun terms of "cities and provinces", meanwhile a small component in the small red ellipsoidal frame in picture (b) only contains three semantically similar terms.

We choose 10 top words to represent topics, and words are sorted by contribution score $v(w, \phi_p^l)$ (see Sect. 3.2). Top words are used to calculate coherence scores. Topics produced by NAM may contain a few even only two words, and these small topics can bring about the fluctuation in coherence scores, therefore if the number of words in a topic is less than 10, we don't take this topic into the calculation of coherence scores. It needs to be pointed out that once the frequent noun term set W_p, the similarity $s(i,j)$, and the parameters θ and Δv are given, basing on filtering process, the structure of final network G_p^T is uniquely determined. However, louvain algorithm [3] built in Gephi [1] has certain degrees of randomness and this algorithm is used to recognize communities. Therefore,

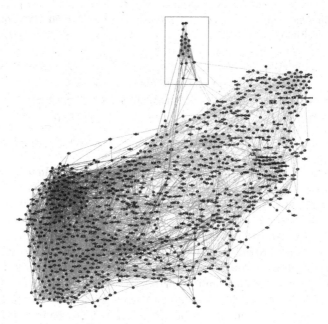

Fig. 2. Undirected graph G_p^T based on TRc, with 977 vertices and only 6314 edges. For visualization, this graph is not shown in Fig. 1.

communities detected after different runs of louvain algorithm [3] are slightly different, and it may also lead to the fluctuation in coherence scores.

We exam coherence of topics on RGPC, which is also used to induce s_{PMI}. Coherence results of networks shown in Fig. 1 are shown in Table 1. And a higher value of coherence means a better performance. As we can see in Table 1, the scores of TRc in all four metrics are lower than those of the other two similarity, which proves the compatibility between formula 2 and NAM. More importantly, s_{SCS} ($\theta = 0.2$) is slightly inferior to s_{PMI} in C_{UMass} and C_V, but outperforms s_{PMI} in C_{UCI} and C_{NPMI}. Notably, shared the same frequent noun term set W_p extracted from $RGPC$, measured by topic coherence on $RGPC$, s_{SCS} ($\theta = 0.2$) induced by word embeddings learned from RPD can achieve comparable performance with s_{PMI} induced by PMI based on $RGPC$. The results of coherence can quantitatively prove the effectiveness of s_{SCS} depending on an appropriate parameter θ.

5.3 Dynamic Topic Model

The processes of inducing topic models on the corpus produced in global period are also implemented on the corpus produced in three local periods, i.e., [1950 : 1976], [1969 : 1992] and [1988 : 2002]. Corresponding parameter θ is also adjusted respectively based on the topic coherence on RGPC in these local periods. Moreover, similarity (formula 6) between topics in successive periods is calculated and used as edge weights.

Table 1. Coherence table: four coherence scores on RGPC (1950–2003) of three networks shown in Fig. 1.

Similarity	C_{UMass}	C_V	C_{UCI}	C_{NPMI}
s_{PMI}	−2.44117	0.77497	−2.53045	0.01203
s_{SCS} ($\theta = 0.2$)	−3.64082	0.61669	−1.87727	0.02881
TRc	−5.72246	0.48859	−4.52341	−0.08177

Dynamic topics are shown in Fig. 3. For visualization, top three words ordered by contribution score $v(w, \phi_p^l)$ are applied to represent the topic. Tiny topics only containing two words are also shown, such as "Russia and the October Revolution" in 1950–1976 and "Library and Museum" in 1969–1992.

As we can see, some flows in both sub-figures are still similar, although the representative words are different. For example, the flows at the top of Fig. 3a and 3b are the stable flows of topics involving diplomacy, i.e., there are no confluence of topics involving diplomacy and other topics. Semantic difference introduced by word embeddings learned from different corpus source (see Sect. 5.1), e.g., the stable flow representing "cities and provinces" in the middle of Sub-Fig. 3b can also be observed.

However, notably, in Fig. 3b, the change from "State-owned enterprise" in 1950–1976 to "Enterprise" in 1969–1992, captures the change in the economic system in China to some extent, reflecting the different political concepts and attitudes about economic between two successive leaders. And "Macro-control" in 1988-2002 should correspond to the financial policy adjustment and control implemented by the government in 1998 in response to the Southeast Asian financial crisis since 1997.

The second flow from top to bottom in the Fig. 3b contains "socialism" and "party" in all three periods, which are the themes of "invariability" of national governance discourse. These two themes mean the consolidation of the ruling party and China's stable social system and ideology, showing strong stability and continuity. This flow contains "revolution", "party" and "socialism" in the first period 1959–1976, of which "revolution" shows the revolutionary spirit of the first leader and may be related to individual historical events. In next second time period 1969–1992, "constitution" replaces "revolution" in the flow, which shows China gradually attaches importance to the construction of the rule of law. In 1988-2002, the flow divides into two new branches, one of which contains "Laws and regulations", "administrative supervision", etc., showing the refinement of the law and the determination to promote the rule of law, the other one still contains "party" and "socialism", inheriting the political theme of flow.

These results also show the ability of s_{SCS} to capture key semantics, even in the dynamic topic model. Dynamic topic determined by s_{SCS} can actually capture the evolution of political themes of China.

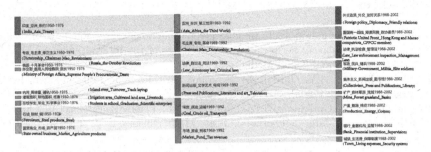

(a) Dynamic topics determined by s_{PMI}.

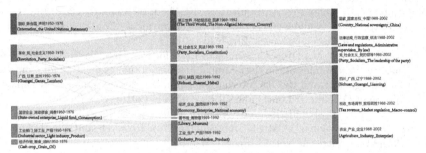

(b) Dynamic topics determined by s_{SCS}.

Fig. 3. The dynamic evolution of topics are shown in this figure. Picture (a) and (b) are dynamic topics corresponding to s_{PMI} and s_{SCS} respectively. Blocks in the same column are the communities, i.e., topics in the same local period. The further to the right, the later the corresponding period. The belts connecting different blocks are in fact the proximity measures of topics. A wider belt represents a stronger connection between two topics, meaning the main direction of topic evolution. The English in the figure is added later to facilitate readers' understanding.

6   Conclusions

In this paper, we propose to use shifted cosine values between word embeddings to replace point-wise mutual information (PMI) as the weight function to calculate distributional similarity between words. A shift parameter θ is crucial for performance of the topic model and needs to be adjusted by experiments. The experiments qualitatively show that in terms of the final semantic network and the dynamic evolution of topics, our proposed similarity s_{SCS} is able to capture semantic between words. And given topic coherence on RGPC and the difference between data-sets RGPC and RPD, the effectiveness of s_{SCS} is also quantitatively proven.

References

1. Bastian, M., Heymann, S., Jacomy, M.: Gephi: an open source software for exploring and manipulating networks. In: Proceedings of the International AAAI Conference on Web and Social Media, vol. 3 (2009)
2. Blei, D.M., Ng, A.Y., Jordan, M.I.: Latent Dirichlet allocation. J. Mach. Learn. Res. **3**, 993–1022 (2003)
3. Blondel, V.D., Guillaume, J.L., Lambiotte, R., Lefebvre, E.: Fast unfolding of communities in large networks. J. Stat. Mech: Theory Exp. **2008**(10), P10008 (2008)
4. Bouma, G.: Normalized (pointwise) mutual information in collocation extraction. In: Proceedings of GSCL, pp. 31–40 (2009)
5. Cointet, J.P., Mogoutov, A., Bourret, P., El Abed, R., Cambrosio, A.: Les réseaux de l'expression génique-émergence et développement d'un domaine clé de la génomique. médecine/sciences, **28**, 7–13 (2012)
6. Das, R., Zaheer, M., Dyer, C.: Gaussian LDA for topic models with word embeddings. In: Proceedings of the 53rd Annual Meeting of the Association for Computational Linguistics and the 7th International Joint Conference on Natural Language Processing (Volume 1: Long Papers), pp. 795–804 (2015)
7. Hofmann, T.: Probabilistic latent semantic indexing. In: Proceedings of the 22nd Annual International ACM SIGIR Conference on Research and Development in Information Retrieval, vol. 51, pp. 50–57 (1999)
8. Li, C., Wang, H., Zhang, Z., Sun, A., Ma, Z.: Topic modeling for short texts with auxiliary word embeddings. In: Proceedings of the 39th International ACM SIGIR Conference on Research and Development in Information Retrieval, pp. 165–174 (2016)
9. Li, D., et al.: Adding community and dynamic to topic models. J. Informet. **6**(2), 237–253 (2012)
10. Mikolov, T., Sutskever, I., Chen, K., Corrado, G.S., Dean, J.: Distributed representations of words and phrases and their compositionality. In: Advances in Neural Information Processing Systems, vol. 26, pp. 3111–3119 (2013)
11. Mimno, D., Wallach, H., Talley, E., Leenders, M., McCallum, A.: Optimizing semantic coherence in topic models. In: Proceedings of the 2011 Conference on Empirical Methods in Natural Language Processing, pp. 262–272 (2011)
12. Newman, M.E.: Modularity and community structure in networks. Proc. Natl. Acad. Sci. **103**(23), 8577–8582 (2006)
13. Röder, M., Both, A., Hinneburg, A.: Exploring the space of topic coherence measures. In: Proceedings of the Eighth ACM International Conference on Web Search and Data Mining, pp. 399–408 (2015)
14. Rule, A., Cointet, J.P., Bearman, P.S.: Lexical shifts, substantive changes, and continuity in state of the union discourse, 1790–2014. Proc. Natl. Acad. Sci. **112**(35), 10837–10844 (2015)
15. Sun, J.: Jieba Chinese word segmentation tool (2012)
16. Weeds, J., Weir, D.: Co-occurrence retrieval: a flexible framework for lexical distributional similarity. Comput. Linguist. **31**(4), 439–475 (2005)

A Genetic Algorithm Based Artificial Neural Network for Production Rescheduling Problem

Pakkaporn Saophan[1] and Warut Pannakkong[2(✉)]

[1] School of Knowledge Science, Japan Advanced Institute of Science and Technology,
Ishikawa, Japan
s2020020@jaist.ac.jp
[2] School of Manufacturing Systems and Mechanical Engineering, Sirindhorn
International Institute of Technology, Thammasat University, Bangkok, Thailand
warut@siit.tu.ac.th

Abstract. Production rescheduling plays an essential role in endorsing the effectiveness of a dynamic manufacturing environment. When the significant disruptive changes invalidate the original schedules, the rescheduling system should be adopted by responding quickly to lessen the effects on the performance of the production. Among the fourth industrial revolution, digital technologies (e.g., Internet of Things or IoT) and machine learning are creating new opportunities to execute production rescheduling. This paper presents a rescheduling approach based on a genetic algorithm (GA) and artificial neural network (ANN) to address the problem of flow shop scheduling with machine disruption. The objective is to find a new sequence or schedule of jobs that minimize makespan in satisfactorily computational time. This study first generates simulated scenarios of the interruptions. Then, we propose GA for solving each scenario. Secondly, we apply ANN to store the knowledge from simulated scenarios that can provide initial solutions for novel GA. It is found that the GA-based knowledge from ANN renders the new schedule 35.8% faster than the standard GA. Through observing the results, the proposed rescheduling methodology for flexible manufacturing not only has a productive performance in handling machine disruption in a scheduling problem but also contributes a faster new schedule to fill the gaps in state-of-the-art heuristic approaches whose computational time is inapplicable in implementation.

Keywords: Flow shop scheduling · Rescheduling · Machine disruption · Genetic algorithm · Artificial neural network

1 Introduction

Production scheduling is an essential process in manufacturing systems, where all production complex activities are controlled on a timescale, including the allocation of resources in the performance with maximum productivity.

© Springer Nature Switzerland AG 2022
K. Honda et al. (Eds.): IUKM 2022, LNAI 13199, pp. 279–290, 2022.
https://doi.org/10.1007/978-3-030-98018-4_23

In a dynamic manufacturing environment, the significant disruptions convey previous schedules to unacceptable performance, which enforces the rescheduling process to moderate the effect of such disturbances while obtaining optimum performance [1]. Many types of significant disturbances in performance can lead to rescheduling as the previous schedules are no longer feasible. Therefore, these disruptive events are called rescheduling factors [2], such as machine failure, rush order arrival, and order cancellation. Rescheduling is also known as real-time scheduling, relates to adjusting pre-decided schedules. It needs to generate high-quality and react in a reasonable amount of time to respond the disruptive events or other changes.

Intelligent industries coordinate different manufacturing resources (e.g., the machine's capacity for production and raw materials) based on the industry 4.0 technologies (e.g., Internet of Things or IoT), and integrate these resources by scheduling and rescheduling approaches. The internet and sensors provide the statuses of resources in real-time. The advent of the industry 4.0 revolution has brought great opportunities to improve the manufacturing industries. For this reason, tremendous perspectives and challenges for production rescheduling arise from two directions, including competency redesign and nearly optimal scheduling [3].

This research proposes a methodology for implementing production rescheduling knowledge, which can be integrated into the current schedule when unplanned situations have appeared. The methodology framework consists of four phases: Firstly, before the operation is actually performed, various simulated machine disruption scenarios are generated. Then, the conventional genetic algorithm (GA) is implemented for rescheduling in each scenario to generate rescheduling results. Secondly, an artificial neural network (ANN) is trained to capture the knowledge in the rescheduling results from GA. When the operation is performing, the trained ANN can make real-time predictions for a new schedule according to a given machine disruption scenario. In case that the predicted schedule from ANN is infeasible, the GA is applied by using the predicted schedule from ANN as the initial solution to determine a feasible schedule. Finally, the results from our proposed methodology are compared to the conventional GA.

The rest of this paper is structured as follows. Section 2 presents the contributions in the literature that is relevant to the production scheduling and rescheduling area. Section 3 offers the model development that aims to generate rapid rescheduling to minimize the effect of disruptions in the performance under a dynamic manufacturing system. Then, the results from the proposed methodology are compared with the results from the traditional approach, and presented in Sect. 4. The last section, Sect. 5, offers our final remarks about directions of future advances.

2   Literature Review

This section focuses on reviewing former researches of scientific literature related to production rescheduling problems to determine the research gaps. It is appropriate to confirm that the previous reviews specify the need for content deliberation and positions issues.

2.1 Production Flow Shop Scheduling

Scheduling is defined as the problem of arranging, managing, and optimizing a production process. The jobs have received a series of operations and sequence processes on several machines in an optimal objective function [4]. Scheduling procedure also depends on a well-studied machine environment which is the single machine, the parallel machine, the flow shop, and the job shop [5].

The production system proposed in this study belongs to the flow shop scheduling. The manufacturing process of the flow shop follows a fixed linear structure that the operations have to be completed on all jobs in the same sequence, and the jobs must pass through all the machines in the same order. The previous literature review has reported that sufficient and appropriate approaches for flow shop operation require further investigations in the field of rescheduling problems [6]. The flow shop scheduling has been an active area since Johnson [7] proposed a simple algorithm for flow shop production with two machines and unique three machines problem. Then many researchers have been exhaustively explored algorithms for developing the production sequencing problem. Stafford and Tseng [8–10] widely presented the permutation flow shop problem using Mixed Integer Linear Programming (MILP) models. However, a major problem with this application is that the mathematical model does not formulate the NP-hard problem [11]. The complexness of production scheduling problems renders exact solver algorithms incapable of high-quality solutions for large-size problems in a suitable amount of time. Consequently, many researchers have turned to heuristics that search for near-optimal solutions in a shorter computational time to solve scheduling problems [12].

A heuristic is designed to search large spaces of candidate solutions to find optimal or near-optimal solutions in a more potent fashion than traditional methods [13]. Heuristic algorithms are developed by Palmer [14] that established approaches for solving the scheduling with minimum total time. It has been widely used in solving flexible scheduling problems with makespan as the criterion. Solimanpur et al. [15] presented EXTS, which is the algorithm from neural networks and tabu search method, proposed for the flow shop scheduling. The EXTS gets an initial permutation from constructive algorithms and exploits a neuro-dynamical system to enhance the initial permutation. Etiler et al. [16] developed a GA-based heuristic, which is easily implementable and performs efficiently, for the flow shop scheduling problem with the objective as minimum makespan. Nevertheless, heuristic algorithms for solving NP-hard problems still take remarkably long computational time [17] depending on the number of generations. For this reason, common heuristics are still unable to contain the requirement of rapid rescheduling.

The algorithm introduced in this paper involves GA based on ANN for the rapid production rescheduling problem. The basic concept of GA is a method based on Charles Darwin's theory of biological evolution and natural selection.

GA is commonly used to generate search problems by representing them in biologically inspired operators: crossover, mutation, and selection. GA is a class of heuristic methods that are effective and popular in solving scheduling problems such as examination and course timetabling [18], maintenance scheduling [19], and diver scheduling [20]. Moreover, GA is also developed to address the production scheduling problem [21,22], and usually obtain satisfying results.

2.2 Production Rescheduling

Real-world scheduling problems are dynamic industrial environments with various unforeseen events that may invalidate the original schedules [23]. The procedure of modifying the production schedule is called production rescheduling. The conventional scheduling approaches take extra time to reform the primal schedules and reschedule the unprocessed work orders to achieve a new optimal solution [24]. The rapid updating of the previous schedule should be analyzed to mitigate the effects of unexpected situations. Mason et al. [25] presented rescheduling outperforms that used modified shifting bottleneck method to minimize total weighted tardiness. Proactive scheduling was proposed by Sabuncuoglu and Goren [26]. It focused on generating schedules that are capable of absorbing disturbance with rescheduling policies. Knowledge-based methods such as expert systems [27], a primitive form of artificial intelligence (AI), are presented to solve fuzzy and random disturbances in production problems. Dong and Jang [28] developed two heuristic algorithms based on a generation procedure of an active schedule and the well-known algorithm, Wilkerson-Irwin, for minimizing mean tardiness of production rescheduling due to machine breakdown. Kundakci and Kulak [29] built hybrid GA approaches for fluctuating job shop scheduling problems with machine breakdown, new order arrival, and change in processing time. Li et al. [30] introduced the artificial bee colony algorithms and Tabu Search algorithm for solving the flexible job-shop scheduling problems and addressed dynamic events with three rescheduling strategies. Nonetheless, solving production rescheduling problems in stochastic environments published in journal papers is challenging to implement in real-world industries [3].

Along with the fourth industrial revolution (Industry 4.0), different appliances from optimization, data analytics, IoT, and AI are stimulating opportunities in production systems. In the early years, research on AI has received more attention. Li et al. [31] modeled flexible job shop scheduling and solved it through a hybrid metaheuristic algorithm. Then, they train the classification model, which integrates machine learning (ML) for identifying rescheduling patterns. Zhou et al. [32] presented intelligent factories using novel cyber-physical integration for online scheduling low-volume-high-mix orders. They used IoT technologies to interconnect multi-agent systems and proposed an AI scheduler to schedule dynamic production with real-time sensor data. However, state of the art from Cadavid et al. [33] inform that 75% of the possible research domain in ML with production planning and control in industry 4.0 are hardly analyzed or not addressed. Although, the field of most cited research published in prominent journals [33,34] suggests that AI is used to determine to overcome the limitations of dynamic manufacturing environments.

ANN, which is suited for time series forecast [35], is used in this paper. Their process can be trained to allow computer programs to recognize patterns between input and target. In order to encourage to the literature, this study presents a dynamic flow shop scheduling problem in which machine disruption occurs and also proposes GA-based knowledge from ANN to generate rapid rescheduling.

3 Methodology

This section explains the proposed methodology for the rapid rescheduling processes to relieve the impact of unexpected events that able to occur anytime in the flexible manufacturing system. Fig. 1 demonstrates the proposed general methodology in a workflow diagram. It contains two main parts:

1. The knowledge of the rescheduling process is generated before starting the production. The optimal sequence of flow shop scheduling with disruptive events (DE_t) is stored in the ANN using simulated scenarios. The considering disruptive events that are implemented in this paper is machine disruption. In the simulation environment, schedule scenarios are modified by GA, which executes the sequence with minimizing idle time of the last machine operation, also known as minimizing tardiness.
2. The rescheduling process is performed when IoT sensors derive real-time disturbing machine data, the applying knowledge from the ANN as the initial solutions are placed into our proposed GA. Therefore, the novel GA-based rescheduling knowledge is able to provide a rapid new sequence.

3.1 Problem Description

The rescheduling problem addressed in this paper is the flow shop production with disruptive situations by considering the following assumptions. The manufacturer has n jobs that have to be scheduled on a queue of production with m machines, denoted by $Jobs = \{J_1, J_2, ..., J_n\}$ and $Machines = \{M_1, M_2, ..., M_m\}$. All the jobs and machines are available at time zero. Each machine has exactly the same sequence of jobs and can process one operation at a time. Each job requires p_{ij} that represents a processing time on a specific machine, where (i, j) refers to the operation of job j on machine i $(i \in 1, ..., m; j \in 1, ..., n)$. Setup times for the operations are included in processing times, and each operation cannot be interrupted.

The problem is to decide an optimal sequence $S = \{s_1, s_2, ..., s_n\}$ of the n jobs. Let I_{iq} indicates the idle time on machine i between the operation of the jobs in the qth position and $(q + 1)$th position. The objective of the problem is to minimize the makespan that is equitable to minimizing the total idle time on the last machine, machine m [11]:

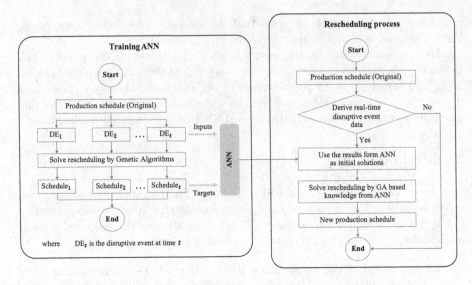

Fig. 1. Workflow diagram of proposed rescheduling methodology

$$min(\sum_{i=1}^{m-1} p_{i,1} + \sum_{j=1}^{n-1} I_{mj}),$$

which is summation between the idle time before starting the first job on the last machine and the idle time between the jobs on the last machine.

Moreover, unexpected situations frequently occur during production in the present factory. For adaptation to the changes, the operation needs to promptly reschedule production sequences to engender new optimal sequences into the schedules to maintain the quality of production performance.

3.2   Proposed Rescheduling Methodology

We proposed the methodology for the rescheduling problem by improving GA with initial solutions from ANN. The ANN is trained with optimal or near-optimal solutions of known cases to return excellent solutions for new cases, which are then given to GA. Accordingly, the GA is applied in two parts of the rescheduling process. First, before launch production, it is developed for searching a near-optimal sequence of each scenario. Then, after receiving the initial solutions from ANN, when actual disruptive events occur, GA is again used to generate a new sequence by using the knowledge from the simulation designs to execute in real-time or near real-time.

Genetic Algorithm. To get a suitable sequence in the ordering problem, GA first needs to create feasible solutions produced by binary code or actual code.

Our solutions are coded by the characteristics of a random permutation sequence of jobs. The individual characteristics are called chromosomes and represent the initial population. Thereupon, GA operators are established to create new solutions - crossover, mutation, and selection.

This work uses the crossover operator as a two-point crossover, randomly picking two crossover points from the parent chromosomes and swapping these points. A mutation procedure is applied by altering a gene in a chromosome to maintain genetic diversity for better solutions. Hence, offspring individuals are produced, and the populations are expanded. Afterward, the fitness value of the chromosomes, including parents and offspring, is calculated by minimizing the idle time at the last machine as criteria. The selection operator is the process of the chosen individual chromosome that has the best fitness values.

Artificial Neural Network. The ANN for rescheduling progress in this paper is presented to store knowledge and effectively provide initial solutions. The networks are trained by simulated scenarios consisting of inputs as the cases of changing processing time because of machine disruption and the targets as an optimal production sequence for each case that results from GA. Moreover, to facilitation the procedure of the ANN, all inputs are assigned into actual processing time (p_{ij}) of job i on machine j and targets are normalized into $[0, 1]$. After going through the learning process, the ANN is able to deliver the output as the new optimal sequence when real disruption appears. Note that the ANN will provide an excellent solution if the training instances demonstrate the disturbed production well. The proposed ANN is illustrated in Fig. 2.

GA Based Knowledge from ANN. The relevant paper publication mentioned that the computational time of optimization and heuristic approaches are unaffordable in practice. Thus, to conquer the limitation of previous publications in the literature, GA-based knowledge from ANN is being used. When disruptive situations arise, the population of GA in the first phase is expanded by the initial sequences, which might be the new optimal sequence provided by the ANN above. By doing this, improved GA contributes the faster searching the best solution with the guideline from the knowledge.

4   Experiments

To demonstrate the rescheduling process in dynamic production, we present the procedure for generating the instance of schedule scenarios, the implementation details, and then elucidate the results from our experiments.

4.1   Generating Simulated Scenarios

The scenarios were composed by using a general way to build. An industrial case study has been considered to illustrate the proposed methodology for receiving

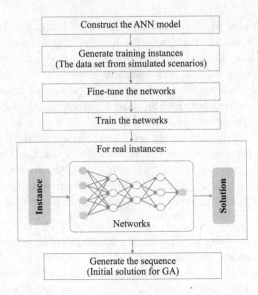

Fig. 2. The architecture of the ANN for the proposed rescheduling methodology

rescheduling knowledge. The manufactory possesses large-size flow shop production of 20 jobs and 10 machines. Each machine's setup time of a particular job is not determined since setup time depends solely on the job to be processed, regardless of its preceding job.

The proposed unexpected situation in this experiment is machine disruption. Consequently, in the scenarios, we begin with a base instance that the processing time of all jobs is randomly varied from 10 to 50 time units. Then, we create ten scenarios of increasing double the job's processing time on the disordered machine due to machine interruption.

4.2   Implementation Details

For our numerical parameters used in the proposed GA-based knowledge on the ANN are shown in Table 1. The GA and ANN are programmed on Python language in *Jupiter notebook* 6.3. Moreover, the experiments are performed on 2.40 GHz *Intel Core(TM)* i5 with 16 GB of RAM.

4.3   Numerical Results and Discussions

The experiments aim to alleviate the consequence of machine disruption that leads the primary sequence to intolerable execution. The proposed methodology attempts to acquire the new sequence in rapid computational time. Furthermore, the criteria for maximizing production scheduling and rescheduling performance are minimizing idle time at the last machine.

To evaluate the capability of the proposed methodology for solving the rescheduling problem, we test the performance by comparing the results between

Table 1. The parameters used in GA and ANN

Models	Parameters	Values
GA	Population size	300
	Crossover rate	0.8
	Mutation rate	0.1
	Number of iterations	250
ANN	Input layer	ReLU activation
	Hidden layers	7 layers, 15 units, and ReLU activation
	Output layer	Sigmoid activation
	Learning rate	0.001
	Epochs	1000

our rescheduling methodology and the ordinary GA, as demonstrated in Fig. 3. After the above scenarios train the ANN, we compose a new instance that differs from our scenarios into the ANN. The new instance represents a disruptive event that possibility emerges while performing the production. The trained ANN takes the new instance as a new input. Then, the solution from the output of ANN is applied as an initial population in GA. We also set the same instance into standard GA to observe the different results from both methodologies.

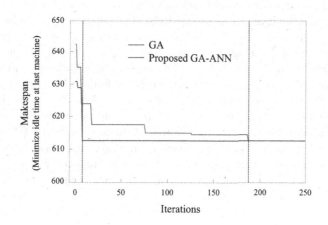

Fig. 3. Comparison between standard GA and proposed GA-based knowledge form ANN

We attempt the number of iterations into 100, 200, 500, and 1000 iterations. Then, we ascertained that the iterations after 200th are invariable. Therefore, we set 250 iterations for our final experiment. As Fig. 3 illustrates, the GA-based knowledge on ANN obtains the solution faster than the standard GA. At 1st iteration, our suggested GA that has the initial solution from ANN grants better fitness value (minimize makespan). Moreover, the proposed GA is able to reach minimum idle time (615 time units) since 9th iteration. On the other hand, the GA without the knowledge from the ANN reaches the fittest at 188th iteration. In summary, the GA based on the initial solution from the ANN completes for the best sequence with 35.8% faster than the ordinary GA.

5   Conclusions and Future Work

Most publications in the rescheduling area are computationally arduous for optimization solvers, and heuristics also generally take tens of minutes to hours to obtain a gratifying solution. For this reason, our proposed methodology, GA-based improvement from ANN, is presented to contribute fast and effective reschedules for large-size dynamic flow shop scheduling with machine disruption problems. In this way, our proposed methodology delivers preeminent solutions for the dynamic flow shop scheduling problem with permissible computational time for rescheduling faster than the conventional approach.

We propose that ANN provide the initial rescheduling production with faster computational time. We acknowledge that the potentiality of ANN can be built from the training data, which are inputs and targets. A lack of extrapolation property or destitute prior knowledge to train the networks may be cemented in a local minimum solution. The ANN may never be constituted to enhance its preciseness over a certain threshold. Therefore, the critical limitation of ANN is that the trained networks should have been well-variant instances.

In our study, the limitations are that the new instance needs to be similar to the provided scenarios, otherwise, the ANN cannot grant the well-initial solutions to GA. Consequently, using ANN based on reinforcement learning [36] is an appealing architecture to collect a plenty amount of knowledge and then store it in Q-learning for the instances that were never found in the simulations.

Moreover, further research will focus on more experiments with different types of disruptive events, such as rush orders and the arrival of new orders, scale up the simulated scenarios that represent unexpected real-world situations. Additionally, the results from our proposed methodology should be compared with the results from other approaches to ensure the precision of our experiments. However, the limitations of this work are that the new instance needs to be similar to the provided scenarios. Consequently, using ANN based on reinforcement learning [36] is an appealing architecture to collect a plenty amount of knowledge and then store it in Q-learning for the instances that were never found in the simulations.

References

1. Vieira, G.E., Herrmann, J.W., Lin, E.: Rescheduling manufacturing systems: a framework of strategies, policies, and methods. J. Sched. **6**(1), 39–62 (2003). https://doi.org/10.1023/A:1022235519958

2. Dutta, A.: Reacting to scheduling exceptions in FMS environments. IIE Trans. **22**(4), 300–314 (1990)
3. Uhlmann, I.R., Frazzon, E.M.: Production rescheduling review: opportunities for industrial integration and practical applications. J. Manuf. Syst. **49**, 186–193 (2018)
4. Johnson, L.A., Montgomery, D.C., Montgomery, D.C.: Operations Research in Production Planning, Scheduling, and Inventory Control. Wiley, Hoboken (1974)
5. Pinedo, M., Hadavi, K.: Scheduling: theory, algorithms and systems development. In: Gaul, W., Bachem, A., Habenicht, W., Runge, W., Stahl, W.W. (eds.) ORP, vol. 1991, pp. 35–42. Springer, Heidelberg (1992). https://doi.org/10.1007/978-3-642-46773-8_5
6. Khodke, P., Bhongade, A.: Real-time scheduling in manufacturing system with machining and assembly operations: a state of art. Int. J. Prod. Res. **51**(16), 4966–4978 (2013)
7. Johnson, S.M.: Optimal two-and three-stage production schedules with setup times included. Naval Res. Logist. Q. **1**(1), 61–68 (1954)
8. Stafford, E.F., Jr., Tseng, F.T.: On the Srikar-Ghosh MILP model for the IVX M SDST flowshop problem. Int. J. Prod. Res. **28**(10), 1817–1830 (1990)
9. Stafford, E.F., Jr., Tseng, F.T., Gupta, J.N.: Comparative evaluation of MILP Flowshop models. J. Oper. Res. Soc. **56**(1), 88–101 (2005). https://doi.org/10.1057/palgrave.jors.2601805
10. Tseng, F.T., Stafford, E.F., Jr.: New MILP models for the permutation flowshop problem. J. Oper. Res. Soc. **59**(10), 1373–1386 (2008)
11. Pinedo, M.: Scheduling, vol. 29. Springer, Heidelberg (2012). https://doi.org/10.1007/978-1-4614-2361-4
12. Branke, J., Nguyen, S., Pickardt, C.W., Zhang, M.: Automated design of production scheduling heuristics: a review. IEEE Trans. Evol. Comput. **20**(1), 110–124 (2015)
13. Cook, S.A.: An overview of computational complexity. ACM Turing Award Lect. (2007)
14. Palmer, D.: Sequencing jobs through a multi-stage process in the minimum total time-a quick method of obtaining a near optimum. J. Oper. Res. Soc. **16**(1), 101–107 (1965)
15. Solimanpur, M., Vrat, P., Shankar, R.: A neuro-Tabu search heuristic for the flow shop scheduling problem. Comput. Oper. Res. **31**(13), 2151–2164 (2004)
16. Etiler, O., Toklu, B., Atak, M., Wilson, J.: A genetic algorithm for flow shop scheduling problems. J. Oper. Res. Soc. **55**(8), 830–835 (2004). https://doi.org/10.1057/palgrave.jors.2601766
17. Wu, C.X., Liao, M.H., Karatas, M., Chen, S.Y., Zheng, Y.J.: Real-time neural network scheduling of emergency medical mask production during COVID-19. Appl. Soft Comput. **97**, 106790 (2020)
18. Rezaeipanah, A., Matoori, S.S., Ahmadi, G.: A hybrid algorithm for the university course timetabling problem using the improved parallel genetic algorithm and local search. Appl. Intell. **51**(1), 467–492 (2020). https://doi.org/10.1007/s10489-020-01833-x
19. Javanmard, H., Koraeizadeh, A.A.W.: Optimizing the preventive maintenance scheduling by genetic algorithm based on cost and reliability in national Iranian drilling company. J. Ind. Eng. Int. **12**(4), 509–516 (2016)
20. Li, J., Kwan, R.S.: A fuzzy genetic algorithm for driver scheduling. Eur. J. Oper. Res. **147**(2), 334–344 (2003)

21. Guo, K., Yang, M., Zhu, H.: Application research of improved genetic algorithm based on machine learning in production scheduling. Neural Comput. Appl. **32**(7), 1857–1868 (2019). https://doi.org/10.1007/s00521-019-04571-5

22. Chen, R., Yang, B., Li, S., Wang, S.: A self-learning genetic algorithm based on reinforcement learning for flexible job-shop scheduling problem. Comput. Ind. Eng. **149**, 106778 (2020)

23. Salido, M.A., Escamilla, J., Barber, F., Giret, A.: Rescheduling in job-shop problems for sustainable manufacturing systems. J. Clean. Prod. **162**, S121–S132 (2017)

24. Abumaizar, R.J., Svestka, J.A.: Rescheduling job shops under random disruptions. Int. J. Prod. Res. **35**(7), 2065–2082 (1997)

25. Mason, S., Jin, S., Wessels, C.: Rescheduling strategies for minimizing total weighted tardiness in complex job shops. Int. J. Prod. Res. **42**(3), 613–628 (2004)

26. Sabuncuoglu, I., Goren, S.: Hedging production schedules against uncertainty in manufacturing environment with a review of robustness and stability research. Int. J. Comput. Integr. Manuf. **22**(2), 138–157 (2009)

27. Li, H., Li, Z., Li, L.X., Hu, B.: A production rescheduling expert simulation system. Eur. J. Oper. Res. **124**(2), 283–293 (2000)

28. Dong, Y.H., Jang, J.: Production rescheduling for machine breakdown at a job shop. Int. J. Prod. Res. **50**(10), 2681–2691 (2012)

29. Kundakcı, N., Kulak, O.: Hybrid genetic algorithms for minimizing makespan in dynamic job shop scheduling problem. Comput. Ind. Eng. **96**, 31–51 (2016)

30. Li, X., Peng, Z., Du, B., Guo, J., Xu, W., Zhuang, K.: Hybrid artificial bee colony algorithm with a rescheduling strategy for solving flexible job shop scheduling problems. Comput. Ind. Eng. **113**, 10–26 (2017)

31. Li, Y., Carabelli, S., Fadda, E., Manerba, D., Tadei, R., Terzo, O.: Machine learning and optimization for production rescheduling in Industry 4.0. Int. J. Adv. Manuf. Technol. **110**(9), 2445–2463 (2020)

32. Zhou, T., Tang, D., Zhu, H., Zhang, Z.: Multi-agent reinforcement learning for online scheduling in smart factories. Robot. Comput.-Integr. Manuf. **72**, 102202 (2021)

33. Usuga Cadavid, J.P., Lamouri, S., Grabot, B., Pellerin, R., Fortin, A.: Machine learning applied in production planning and control: a state-of-the-art in the era of Industry 4.0. J. Intell. Manuf. **31**(6), 1531–1558 (2020). https://doi.org/10.1007/s10845-019-01531-7

34. Çaliş, B., Bulkan, S.: A research survey: review of AI solution strategies of job shop scheduling problem. J. Intell. Manuf. **26**(5), 961–973 (2015)

35. Feindt, M., Kerzel, U.: The NeuroBayes neural network package. Nucl. Instrum. Methods Phys. Res. Sect. A **559**(1), 190–194 (2006)

36. Sutton, R.S., Barto, A.G.: Reinforcement Learning: An Introduction. MIT Press, Cambridge (2018)

Transductive Learning Based on Low-Rank Representation with Convex Constraints

Yoshifumi Kusunoki[1]([✉]) [iD], Katsuhiko Kojima[2], and Keiji Tatsumi[2] [iD]

[1] Graduate School of Humanities and Sustainable System Sciences,
Osaka Prefecture Univeristy, 1-1 Gakuen-cho, Naka-ku, Sakai, Osaka 599-8531, Japan
yoshifumi.kusunoki@kis.osakafu-u.ac.jp
[2] Graduate School of Engineering,
Osaka University, 2-1 Yamada-oka, Suita, Osaka 565-0871, Japan
kojima@se.eei.eng.osaka-u.ac.jp, tatsumi@eei.eng.osaka-u.ac.jp

Abstract. Transductive learning is a problem to predict labels of unlabeled data exploiting both of labeled and unlabeled data. There are various methods for transductive learning, which are often variants of existing machine learning methods. In this paper, we use one of the existing unsupervised methods, row-rank representation (LRR), for transductive learning. The proposed method consists of two phases: clustering and classification. In the clustering phase, we apply a revised LRR to the data set including both of labeled and unlabeled data. Then, we obtain a modification of the data set, which reflects cluster structure behind the data set. In the classification phase, we classify unlabeled data by using the modified data set obtained in the clustering phase. We use a classification method which is inspired by LRR. That is, for each class, we approximate each unlabeled data point by the labeled data set of the class, then classify the point to the class with the smallest approximation error. Finally, we examine performance of the proposed method by numerical experiments.

Keywords: Transductive learning · Low-rank representation · Subspace clustering

1 Introduction

One of the tasks of the machine learning is to classify given data into a number of groups. Learning problems related to classification task are divided into two kinds. One is supervised learning, in which a data set with class labels is given, and a classifier is learned by exploiting the labeled data set. The other is unsupervised learning, in which a data set without class labels is given, and a cluster structure is learned which intrinsically exists in the data set. Moreover, semi-supervised learning is a combination of problems, in which a classifier is learned by exploiting not only labeled data but also unlabeled data. In general,

© Springer Nature Switzerland AG 2022
K. Honda et al. (Eds.): IUKM 2022, LNAI 13199, pp. 291–301, 2022.
https://doi.org/10.1007/978-3-030-98018-4_24

we can easily obtain a large amount of data, but assigning labels to data is expensive. Hence, it is useful that performance of a classifier can be improved by using unlabeled data.

Semi-supervised learning problems are divided into inductive and transductive. Inductive learning is a task to build a classifier which works for unseen data. On the other hand, transductive learning is a task to assign labels to unlabeled data which are provided before learning. In other words, in the transductive learning, classification of unseen data is not required. Hence, transductive learning is a easier problem than inductive learning.

In this paper, we propose a transductive learning method based on low-rank representation (LRR) [4], which is an unsupervised method. LRR is a method to induce a low-rank model from data and divide the data into multiple subspaces. Such a subspace clustering method is useful for image classification problems. Additionally, LRR robustly estimates intrinsic subspaces despite noise in data. By using LRR, we expect that cluster structure of data is detected, which is useful for classification task.

The proposed method consists of clustering and classification phases. In the clustering phase, we apply a revised LRR to the data set including both of labeled and unlabeled data. In the original LRR, each data point is represented (approximated) by a linear combination of a (selected) data set. Additionally, the representation is regularized by minimizing the nuclear norm of the matrix of coefficients. On the other hand, the LRR of the proposed method uses convex combinations instead of linear combinations. In the classification phase, we classify unlabeled data by using the data set approximated by the process of the clustering phase. We use a classification method which is inspired by LRR. That is, for each class, we approximate each unlabeled data point by the labeled data set of the class, then classify the point to the class with the smallest approximation error. Finally, we examine performance of the proposed method by numerical experiments.

2 Low-Rank Representation

We introduce Low-Rank Representation (LRR) [4] which forms the basis of our proposed method. LRR is a clustering method in which an ideal data set is assumed to be distributed in a union of subspaces $S_1, \ldots, S_q$. Additionally, we assume that these subspaces are independent, that is each subspace is a complement of the subspace spanned by the other subspaces. For example, Fig. 1(a) shows a data set consisting of 3-dimensional vectors distributed in the union of a plane (2-dimensional subspace) and a line (1-dimensional subspace). LRR can divide the data points into two groups corresponding to the subspaces.

Let $x_1, x_2, \ldots, x_m \in \mathbf{R}^n$ be given data points (n-dimensional vectors). We define $X = [x_1 \ x_2 \ \cdots \ x_m]$, i.e., the matrix whose columns corresponding to the data points. Additionally, we prepare "dictionary" $a_1, a_2, \ldots, a_k \in \mathbf{R}^n$, and define $A = [a_1 \ a_2 \ \cdots \ a_k]$. LRR achieves clustering using the following optimization problem.

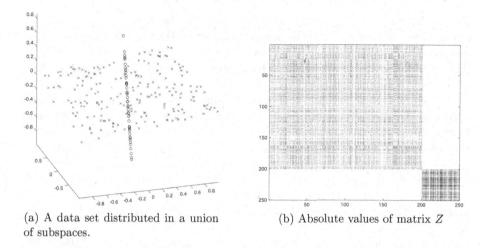

(a) A data set distributed in a union of subspaces.

(b) Absolute values of matrix Z

Fig. 1. An example of low-rank representation

$$\underset{Z,E}{\text{minimize}} \quad \|Z\|_* + \mu\|E\|_\ell \tag{1}$$
$$\text{subject to} \quad X = AZ + E.$$

The decision variables are $Z \in \mathbf{R}^{k \times m}$ and $E \in \mathbf{R}^{n \times m}$. Symbol $\|\cdot\|_*$ means the nuclear norm, i.e., $\|Z\|_*$ is the sum of the singular values of Z. Symbol $\|\cdot\|_\ell$ means an arbitrary matrix norm. In the original paper [4], the $\ell_{2,1}$ norm $\|\cdot\|_{2,1}$ is used, i.e., $\|E\|_{2,1}$ is the sum of the ℓ_2 norms of the columns of E. The weight $\mu > 0$ is a hyperparameter to control the minimizations of $\|Z\|_*$ and $\|E\|_\ell$.

This optimization problem tries to represent each data point x_i by a linear combination of the column vectors of the dictionary A. The i-th column $z_i = [z_{1i} \ z_{2i} \ \cdots \ z_{ki}]^\top$ of Z is the coefficients of this linear combination. The i-th column $e_i = [e_{1i} \ e_{2i} \ \cdots \ e_{ni}]^\top$ is the residual for the linear combination. Therefore, we minimize the norm of e_i to obtain a good approximation.

On the other hand, minimization of the nuclear norm $\|Z\|_*$ aims to obtain a low-rank representation of X. By low-rank representation and appropriately selecting a dictionary A, we can find simple clustered structure in the data X, that reflects the union of subspaces where the data intrinsically locate. That is, we expect that data points in the same subspace form a cluster and the clusters are found by LRR.

Figure 1(b) shows the result of LRR for the data X of the left figure. We use X as the dictionary, i.e., $A = X$. This figure shows the absolute values of elements of the solution Z. The horizontal and vertical axes are the column and row indices of Z. The elements of Z are shown in gray scale, and white regions indicate elements are zero. We can find the clusters of the plane and the line by using the two blocks in the diagonal of Z.

We mention the analytic solution of LRR when the column space of A contains $x_1, \ldots, x_m$ and E is fixed to 0. In that case, the optimum Z for LRR is

$Z = A^\dagger X$, where $A^\dagger$ is the pseudoinverse of A. Furthermore, when $A = X$, we obtain $Z = \tilde{V}\tilde{V}^\top$, where V is the matrix comes from the "skinny" singular value decomposition (SVD) $X = \tilde{U}\tilde{\Sigma}\tilde{V}^\top$. The skinny SVD is the reduced form of SVD $X = U\Sigma V^\top$, where zero singular values and the corresponding column vectors of U and V are removed.

We explain why the solution Z of (1) becomes block diagonal. As mentioned above, in the special case, we have $Z = A^\dagger X$. We assume $AA^\top$ is nonsingular. Hence, $A^\dagger = A^\top(AA^\top)^{-1}$. Let x be a column of X and $z = A^\dagger x$. Moreover, the columns of A are divided into two groups A_1 and A_2, namely $A = [A_1 \ A_2]$. The columns A_1 and A_2 are included in subspaces $\mathcal{S}_1$ and $\mathcal{S}_2$, respectively. The vector z is also divided into $z = [z_1^\top \ z_2^\top]^\top$, according to the column indices of A_1 and A_2. Here, we show that $z_2 = 0$ when x is included in $\mathcal{S}_1$. Let $y = (AA^\top)^{-1}x$, and $z = A^\top y$. Then, we have $A_1 A_1^\top y + A_2 A_2^\top y = x$ and $z_1 = A_1^\top y$, $z_2 = A_2^\top y$. Since x is included in only $\mathcal{S}_1$, we have $A_2 A_2^\top y = 0$. Therefore, $z_2 = A_2^\top y$ is included in the image of $A_2^\top$ and the kernel of A_2. Since these two subspaces are orthogonal, we obtain $z_2 = 0$.

3   Transductive Learning Based on Low-rank Representation with Convex Constraints

3.1   Framework

In this paper, we propose a transductive learning method by using LRR. First, we introduce transductive learning. Let $C = \{1, 2, \ldots, c\}$ be a set of class label. Let $(x_1, y_1), \ldots, (x_m, y_m)$ be a given data set, where $x_i \in \mathbf{R}^n$ is an input vector and $y_i \in C \cup \{0\}$ is a class label for x_i. A vector x_i or an index i is called labeled if $y_i \neq 0$ ($y_i \in C$), and it is called unlabeled if $y_i = 0$. The task of transductive learning is to assign labels to unlabeled data by exploiting both of the labeled and unlabeled data.

The proposed method consists of two phases. First, we apply a modified version of LRR to the data set $X = [x_1 \ x_2 \ \cdots \ x_m]$. We expect that the modified LRR induces clustered structure of X by the solution Z, and we can remove noise and/or data-point-specific information from the data set by replacing X with $\tilde{X} = XZ$. We call this phase "clustering phase".

The second is "classification phase". There are many classification algorithms or supervised learning methods, however in this paper, we use a reconstruction classifier (RC), which is a classifier based on the same idea of LRR. RC classifies an unlabeled data point by approximating them by linear combinations (or convex combinations) of labeled data set. For each data point x and each class k, we compute the error of the best approximation of x by the data in the class k. Then, we classify x to the class k^* that its reconstruction error is the smallest.

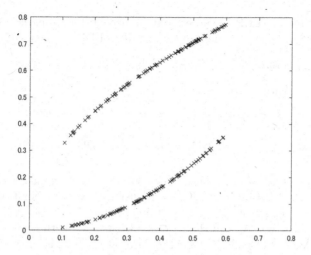

Fig. 2. A data set with nonlinearity.

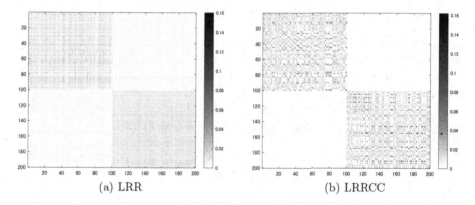

(a) LRR   (b) LRRCC

Fig. 3. Solutions Z for Fig. 2.

3.2   LRR with Convex Constraints

First, we mention a motivation to use the modified LRR in our method. Figure 2 shows a data set consisting of points generated from two curves. The first 100 data points form the lower curve, and the next 100 data points form the upper curve. Applying LRR (1) (setting $\|E\|_\ell$ by the $\ell_{2,1}$ norm and $\mu = 10^3$) to this data set, we obtain the solution Z that is shown in Fig. 3(a). We can see that there exist many nonzero elements in the off-diagonal part, namely the part consisting of pairs of indices (data points) in the different curves. This is because that the data set includes nonlinearity, but LRR is intended to group data by using subspaces (lines).

Hence, we modify LRR to successfully induce clusters from data sets such as Fig. 2. The optimization problem of the modified LRR is given as follows:

$$\underset{Z,E}{\text{minimize}} \quad \|Z\|_* + \mu\|E\|_\ell$$

$$\text{subject to} \quad X = AZ + E,$$

$$z_{ij} \geq 0, \ i = 1,\ldots,k, j = 1,\ldots,m, \tag{2}$$

$$\sum_{i=1}^{k} z_{ij} = 1, \ j = 1,\ldots,m.$$

The additional constraints ensure that each x_i is approximated by a convex combination of the dictionary $a_1,\ldots,a_k$. By using convex combination, we intend that each x_i is reconstructed by a local line segment, and the global nonlinear structure is preserved. We call the modified LRR, LRR with convex constraints (LRRCC). The result of LRRCC for the data set of Fig. 2 is shown in Fig. 3(b). We can see that almost the elements in the off-diagonal part are zero.

3.3   Clustering Phase

In this subsection, we explain the clustering phase of our proposed method. We use LRRCC in the clustering phase. Additionally, we combine LRRCC with the label information of the give data set. We define the following set L of index paris.

$$L = \{(i,j) \mid y_i = y_j \text{ or } y_i = 0 \text{ or } y_j = 0, \ i,j = 1,\ldots,m\}. \tag{3}$$

Then, we consider the following optimization problem:

$$\underset{Z}{\text{minimize}} \quad \frac{1}{2}\|Z\|_*^2 + \frac{\mu}{2}\sum_{i=1}^{m}\|x_i - Xz_i\|^2$$

$$\text{subject to} \quad z_{ij} \geq 0, \ (i,j) \in L,$$

$$z_{ij} = 0, \ (i,j) \notin L, \tag{4}$$

$$\sum_{i=1}^{m} z_{ij} = 1, \ j = 1,\ldots,m.$$

We use the data set X as the dictionary: $A = X$. Moreover, we note that $\|E\|_\ell$ is replaced with the square of the Frobenius norm.

Remark 1. To simplify the notation of this paper, we use the same symbol Z for the solutions of (2) and (4).

If x_i is a labeled data point, then the labeled data points in the classes other than y_i are not used for the reconstruction of x_i. On the other hand, if x_i is unlabeled, then no restriction is performed.

Another difference between (2) and (4) is $\|Z\|_*$ and $\|Z\|_*^2$. We square the nuclear norm to make the solution less sensitive to the hyperparameter μ. It is because the clustering method requires fine tuning of the hyperparameter to distinguish noise and cluster structure.

3.4   Algorithm

The optimization problem (4) is solved by the alternating direction method of multipliers (ADMM). We introduce an additional variable $W \in \mathbf{R}^{m \times m}$, and reformulate (4) as follows.

$$\underset{Z,W}{\text{minimize}} \quad f(W) + g(Z)$$
$$\text{subject to} \quad Z - W = 0, \tag{5}$$

where

$$f(W) = \frac{1}{2}\|W\|_*^2, \tag{6}$$

and

$$g(Z) = \begin{cases} \dfrac{\mu}{2}\sum_{i=1}^{m}\|x_i - Xz_i\|^2 & \text{if } Z \text{ satisfies the constraint of (4),} \\ \infty & \text{otherwise.} \end{cases} \tag{7}$$

We apply ADMM to (5). ADMM is an iterative method. Let (Z^k, W^k) be the k-th solution in iteration. Additionally, we consider a Lagrange multiplier Λ^k in each computation of iteration. The variables Z^k, W^k and Λ^k are updated by the following computation.

$$Z^{k+1} = \underset{Z}{\text{argmin}}\{g(Z) + \frac{\beta}{2}\|Z - (W^k - \beta^{-1}\Lambda^k)\|^2\},$$
$$W^{k+1} = \underset{W}{\text{argmin}}\{f(W) + \frac{\beta}{2}\|W - (Z^{k+1} + \beta^{-1}\Lambda^k)\|\}, \tag{8}$$
$$\Lambda^{k+1} = \Lambda^k + \beta(Z^{k+1} - W^{k+1}).$$

Positive value β is a parameter of ADMM, which affects speed of convergence. The iteration is terminated if $\|Z^k - W^k\|$ is sufficiently small. It is ensured that the sequence of solutions $\{(Z^k, W^k)\}$ converge to the optimum of (5) [1].

We explain how to compute Z^{k+1} and W^{k+1} of (8). The optimization problem to obtain Z^{k+1} is expressed as follows:

$$\underset{Z}{\text{minimize}} \quad \frac{\mu}{2}\sum_{i=1}^{m}\|Xz_i - x_i\|^2 + \frac{\beta}{2}\sum_{i=1}^{m}\|z_i - (w_i^k - \beta^{-1}\lambda_i^k)\|^2$$
$$\text{subject to} \quad z_{ij} \geq 0, \ (i,j) \in L,$$
$$z_{ij} = 0, \ (i,j) \notin L, \tag{9}$$
$$\sum_{i=1}^{m} z_{ij} = 1, \ j = 1, \dots, m,$$

where z_i, w_i^k and λ_i^k are the k-th column vectors of Z, W^k and Λ^k, respectively. This is a convex quadratic optimization problem. Let z_i This problem can be divided into the subproblems corresponding to $z_1, \dots, z_m$, respectively.

The optimization problem to obtain W^{k+1} is expressed as follows:

$$\underset{s,W}{\text{minimize}} \quad \frac{1}{2}s^2 + \frac{\beta}{2}\|W - (Z^{k+1} + \beta^{-1}\Lambda^k)\|^2$$

$$\|W\|_* = s \tag{10}$$

Consider the Lagrange function:

$$\mathcal{L}(s,W,\nu) = \frac{1}{2}s^2 + \frac{\beta}{2}\|W - (Z^{k+1} + \beta^{-1}\Lambda)\|^2 + \nu(\|W\|_* - s), \tag{11}$$

where ν be the multiplier. By the method of Lagrange multipliers, the optimality condition of this problem is given as follows: (W,s) is the optimum iff the following conditions hold.

$$W = \underset{W'}{\text{argmin}}\left\{\nu\|W'\|_* + \frac{\beta}{2}\|W - (Z^{k+1} + \beta^{-1}\Lambda)\|^2\right\} = \text{SVT}_{\nu/\beta}(Z + \beta^{-1}\Lambda), \tag{12}$$

and $s = \nu$, $\|W\|_* = s$. The function SVT, which is called singular value thresholding [2], is defined as follows: for a matrix A and a positive α,

$$\text{SVT}_\alpha(A) = U\text{diag}(\{\max\{0, \sigma_i - \alpha\}\})V^\top, \tag{13}$$

where $\sigma_1, \ldots, \sigma_m$ are the singular values of A and U and V are the left and right orthogonal matrices of the singular value decomposition, namely $A = U\text{diag}(\{\sigma_i\})V^\top$.

. Let $Z + \beta^{-1}\Lambda = U\text{diag}(\{\sigma_i\})V^\top$ be the singular value decomposition. From the optimality condition, variable s is uniquely determined by the following equation:

$$\sum_{i=1}^{m} \max\{0, \sigma_i - s/\beta\} = s. \tag{14}$$

Note that the nuclear norm is the sum of the singular values of a given matrix. Then, W is computed as follows:

$$W = U\text{diag}(\{\max\{0, \sigma_i - s/\beta\}\})V^\top. \tag{15}$$

3.5   Classification Phase

By the clustering phase, we obtain Z by the optimization problem (4). Then, the data set X (both of labeled and unlabeled) is replaced with XZ. Using the replaced X, we classify the unlabeled data points.

Following the idea of LRR, we use RC (reconstruction classifier). For each class label, each unlabeled data point is approximated by a convex combination of the data set in the class. Then, the unlabeled data point is classified by the class with the smallest approximation error.

Let X_k be the matrix of the labeled data points of class label k. Additionally, let M_k be the index set of class label k. The optimization problem to reconstruct a data point x is defined as follows.

$$\underset{z}{\text{minimize}} \quad \|x - X_k z\|^2$$

$$\text{subject to} \quad z_i \geq 0, \; i \in M_k,$$

$$\sum_{i \in M_k} z_i = 1. \tag{16}$$

After solving the optimization problem of class k and obtaining the optimum z^*, we compute the function $F_k(x)$ whose value is the approximation error for data point x.

$$F_k(x) = \|x - X_k z^*\|. \tag{17}$$

Finally, we classify x to the class k with the smallest error.

$$y(x) = \underset{k}{\operatorname{argmin}} \, F_k(x). \tag{18}$$

4   Numerical Experiments

By numerical experiments, we examine performance and characteristic of the proposed method. We use `optdigits` data set obtained from UCI machine learning repository [3], which is a collection of images of hand-written digits 0, 1 ,..., 9, i.e., 10 class labels.

First, we evaluate classification errors of the proposed method. We compare results of classification with and without the clustering phase. Moreover, we compare those with the reconstruction classifier and the nearest neighbor classifier. Table 1 shows the results. The columns of NN (resp. NNC) and RC (resp. RCC) in the table are the results of the nearest neighbor classifier and the reconstruction classifier without (resp. with) clustering phase, respectively. The first column shows the sizes of labeled and unlabeled data, which are randomly drawn from the `optdigits` data set. Additionally, each class has the same size. The values of the table for each method and each size are the average and the standard deviation of 10 misclassification rates for unlabeled data sets. The parameter of the clustering phase μ is set to 0.1.

From the table, we can see that RC has better classification performance than NN. It may implies that classification by reconstruction is better than that by distance for image classification task. On the other hand, RCC is better than RC, in particular data sets with the smaller sizes of labeled data (100/400 and 100/900). That is, the clustering phase effectively works to induce cluster structure consistent with class separation.

Figures 4, 5 and 6 show the effect of the parameter μ. We compare the results of the clustering phase for $\mu = 0.1$ and $\mu = 1$. Moreover, we use a random sample from the `optdigits` data set, whose sizes of labeled and unlabeled data are 100 and 100.

Table 1. Comparison of classification errors.

Labeled/Unlabeled	NN	NNC	RC	RCC
100/400	9.10 ± 1.85	8.15 ± 1.51	7.33 ± 1.55	6.78 ± 1.79
200/300	5.77 ± 1.79	5.50 ± 1.68	4.57 ± 1.20	4.40 ± 1.31
100/900	9.74 ± 1.33	8.30 ± 1.20	8.18 ± 1.61	7.32 ± 1.33
200/800	7.06 ± 0.75	6.08 ± 0.73	5.40 ± 0.72	4.95 ± 0.81

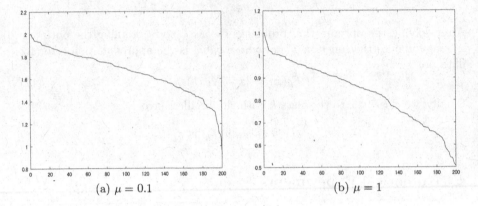

(a) $\mu = 0.1$　　　　　　　　　　　(b) $\mu = 1$

Fig. 4. Effect of μ for the approximation error E.

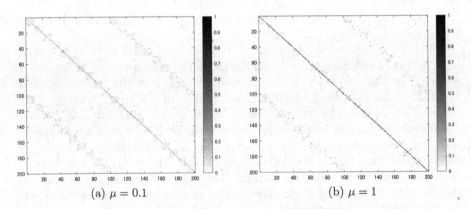

(a) $\mu = 0.1$　　　　　　　　　　　(b) $\mu = 1$

Fig. 5. Effect of μ for the values of Z.

Figure 4 shows approximation errors $\|x_i - X z_i\|/\sqrt{n}$ for data points x_i. They are sorted in descending order. As expected, the errors are decreasing when μ is increasing.

Figure 5 shows values of the solution Z (200×200 matrix) of the problem (4). The first 100 indices are labeled data, and the following 100 indices are unlabeled data. We can see that the block diagonal, which a reflection of cluster structure

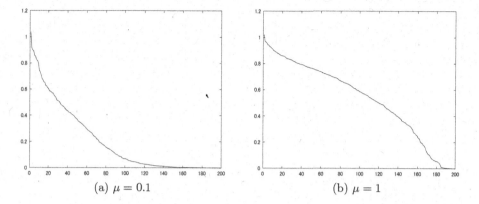

Fig. 6. Effect of μ for the singular values of Z.

that is consistent with the class labels. The upper-right and lower-left block lines express associations between labeled and unlabeled data in the same classes.

Figure 6 shows the singular values of Z. In contrast to approximation errors, they are decreasing when μ is decreasing.

5 Conclusion

In this paper, we have proposed a transductive learning method based on the low-rank representation (LRR). The numerical experiments confirm that the clustering phase of the proposed method can improve classification accuracy, especially for data sets with small sizes of labeled data. In the future work, we will apply the proposed method to other image data sets and examine its performance. Additionally, detailed investigation on the advantage of the convex constraint and/or the nonnegative constraint in the LRR model is also a task that should be addressed.

Acknowledgements. This work was supported by JSPS KAKENHI Grant Number JP21K12062.

References

1. Bertsekas, D.P.: Convex Optimization Algorithms. Athena Scientific, Belmont (2015)
2. Cai, J.F., Candès, E.J., Shen, Z.: A singular value thresholding algorithm for matrix completion. SIAM J. Optim. **20**(4), 1956–1982 (2010)
3. Dua, D., Graff, C.: UCI machine learning repository (2017). http://archive.ics.uci.edu/ml
4. Liu, G., Lin, Z., Yan, S., Sun, J., Yu, Y., Ma, Y.: Robust recovery of subspace structures by low-rank representation. IEEE Trans. Pattern Anal. Mach. Intell. **35**(1), 171–184 (2013)

Economic Applications

Hedging Agriculture Commodities Futures with Histogram Data Based on Conditional Copula-GJR-GARCH

Roengchai Tansuchat[1,2] and Pichayakone Rakpho[1(✉)]

[1] Center of Excellence in Econometrics, Chiang Mai University,
Chiang Mai 50200, Thailand
[2] Faculty of Economics, Chiang Mai University, Chiang Mai 50200, Thailand
pichayakone@gmail.com

Abstract. This paper aims to suggest the optimal hedge ratio for agriculture commodities using copula based GJR-GARCH models, including the conventional static and dynamic conditional copulas. High frequency data are also considered as the information for constructing the hedge ratio. To find the best fit hedging model, we use the AIC and BIC to compare the performance of the models. In order to obtain the reliable frequency data, we use the hedging effectiveness for evaluating the variance reduction of the portfolio. Our results show that dynamic Student-t, static Student-t, and static Gumbel copulas are utilized to capture the dependence structure between spot and futures of wheat. We also find that 1-h frequency provide the best information for reducing the risk of the portfolio.

Keywords: Hedging strategy · Histogram data · GJR-GARCH model · Time varying · Copula

1 Introduction

Future contract is a tool for assisting investors and risk managers to reduce the risk of the agricultural spot. For example, farmers can prevent the risk of agriculture prices by using the agriculture futures, investors can use it to minimize or offset the chance that their assets will lose value in the future. Although, this hedging strategy is working well in practice, it is quite difficult to obtain the optimal hedging strategy. This issue has become an important issue in the field of risk management. In this study, we thus consider the most recent method, namely copula based GARCH approach.

In a hedging strategy and a safe haven have been investigated in the literature and most of them consider to use the close price data as the information for obtaining the optimal hedging strategy which may lead to unreliable risk management and might not reflect the real behavior of data set (See [1,11]). Arroyo and Maté [2] mentioned that this traditional data type does not faithfully describe the whole phenomena where a set of realizations of the observed

© Springer Nature Switzerland AG 2022
K. Honda et al. (Eds.): IUKM 2022, LNAI 13199, pp. 305–316, 2022.
https://doi.org/10.1007/978-3-030-98018-4_25

variable is available for each time point. They suggested a Symbolic data like histogram-value data in order to gain more information of the prices. This type of data is new and has been receiving more and more attention because they are able to summarize huge sets of data [7]. Therefore, in this study, we will apply the histogram data,e.g. with a hourly, 30-min frequency and 5-min frequency data. In the methodology literature, the hedge ratio can be constructed form the variance and covariance of the spot and future. In the variance part, Engle [5] and Bollerslev [3] proposed conditional volatility model namely, GARCH, which is taking into account the conditional heteroscedasticity inherent in time series. GARCH model has become popular econometric tools to study the volatility of time series. However, Li [8] presented the main problem of GARCH model. They stated that the GARCH process is a symmetric variance process and it ignores the sign of the disturbance. This is to say, it assumes that the positive or negative information have the same impact on the volatility. This may not be true in the financial time series as negative impact is more valuable than positive in the reality. To deal with this problem of paper, we consider GJR-GARCH model. Li [8] confirm that GJR-GARCH with skewed-t distribution provide the best fit specification for modelling the volatility of the financial assets.

In the covariance part, the most recent and acceptable method for capturing the correlation is Copulas approach which was introduced by Sklar [12]. This method is widely used in joining random variables and it does not require the assumption of multivariate normal. This model can join all the possible distribution of the random variables [8]. However, this traditional copula is assumed to be constant overtime and it fails to exhibit the sensitivity of price changes and neglect the time factor in the correlation patterns. Patton [10] proposed that the interdependence of financial asset returns is time-varying, and asset prices in different financial markets also tend to have dynamic tail dependence. Thus, this lead us to consider time-varying copula function of this study. We make the empirical contribution to the hedging strategy literature by further applying our model to forecast hedge ratio in the agriculture commodity portfolio which consists of spot and futures. We use Time-varying copula based on GJR-GARCH model to capture time-varying dependence. Four classes of copula, i.e., Gaussian copula, Student-t copula, Clayton copula and Gumbel copula are employed in this study. Furthermore, we also contribute the literature by using a high frequency of spot and futures data to build the hedging models. We believe that our proposed method will become more realistic and flexible reflecting the variety of agriculture commodities characteristics. Besides, this innovation provides an ideal alternative model for the construction of hedging portfolios.

The remainder of the paper is organized as follows. In Sect. 2 we briefly describe GJR-GARCH(1,1) and Time-varying copula function consisting four classes of copula, i.e., Gaussian copula, Student-t copula, Clayton copula and Gumbel copula. We also present the hedging portfolios and hedging effectiveness analysis under the histogram value data context. In Sect. 3 the data and the descriptive statistics are presented. Section 4 analyses the empirical estimation results, and Sect. 5 presents the conclusions.

2   Methodology

2.1   Histogram Value

Let we briefly review the histogram-valued data concept. Note that, high-frequency data, i.e. 5 min, 30 min and 1 h are considered. Let $Y_t = \{y(1),\ldots,y(T)\}$ be a histogram-valued variable at time t, the histogram-valued variables can be expressed as follows:

$$H_{y(t)} = \left\{ \left[\underline{I}_{y(t)1},\overline{I}_{y(t)1}\right],\omega_{t1}; \left[\underline{I}_{y(t)2},\overline{I}_{y(t)2}\right],\omega_{t2};\ldots \left[\underline{I}_{y(t)n_t},\overline{I}_{y(t)n_t}\right],\omega_{tn_t} \right\}, \tag{1}$$

where $\underline{I}_{y(t)i}$ and $\overline{I}_{y(t)i}$ denote the lower and upper boundary of the interval $i \in \{1,2,\ldots n_t\}$ in the histogram. n_t is the number of subintervals for the in the histogram t. Note that $\underline{I}_{y(t)i} \leq \overline{I}_{y(t)i}$ and $\underline{I}_{y(t)i+1} \geq \overline{I}_{y(t)i}$. ω_{it} is a nonnegative frequency associated to the subinterval $\left[\underline{I}_{y(t)i},\overline{I}_{y(t)i}\right]$ and $\sum_{i=1}^{n_t} \omega_{it} = 1$. Then, according to Dias and Brito [4] we can transform the histogram $y(t)$ using the empirical quantile function, $\Omega_{y(t)}^{-1}$

$$\Omega_{y(t)}^{-1} = \begin{cases} \underline{I}_{y(j)1} + \frac{k}{w_{t1}}a_{y(t)1} & if \ \ 0 \leq k < w_{t1} \\ \underline{I}_{y(t)2} + \frac{k-w_{t1}}{w_{t2}-w_{t1}}a_{y(t)2} & if \ w_{t1} \leq k < w_{t2}) \\ \vdots \\ \underline{I}_{y(t)n_t} + \frac{k-w_{tn_{t-1}}}{1-w_{tn_{t-1}}}a_{y(t)n_t} & if \ w_{tn_{j-1}} \leq k \leq 1 \end{cases} \tag{2}$$

where $a_{y(t)i} = \overline{I}_{y(t)i} - \underline{I}_{y(t)i}$ with $i \in \{1,2,\ldots,n_t\}$ and $w_{t,0} = 0$ and $w_{tl} = \sum_{h=1}^{n_t} \omega_{th}$. Note that when we work with histogram-valued variables, the frequency associated to the subinterval ω_{ih} and the number of subintervals in the histograms η_t may be different, the subintervals of histograms.

2.2   GJR-GARCH (p, q) Model

In this paper, we choose GJR-GARCH model of Glosten et al. [6] to model the volatility of the spot and futures. The model is defined as

$$\varepsilon_t = \sigma_t z_t, \tag{3}$$

$$\sigma_t^2 = \omega + \sum_{i=1}^{p}(\alpha_i + \gamma_i I_{t-i})\varepsilon_{t-i}^2 + \sum_{j=1}^{q}\beta_j\sigma_{t-j}^2, \tag{4}$$

where

$$1_{t-i} = \begin{cases} 0 \ if \ \varepsilon_{t-i} \geq 0, \\ 1 \ if \ \varepsilon_{t-i} < 0. \end{cases} \tag{5}$$

γ is leverage effect, $\varepsilon_{it} = \sigma_{it}\eta_{it}$ and η_{it} i.i.d. of standard innovation. $\omega,\alpha,\beta,\gamma \geq 0$,, stationary condition are $\alpha + \gamma \leq 0$ and $z_t \sim F(\cdot)$ is sequence of independent random variable or innovation.

2.3   Copula and Dependence

Copula is defined as functions that join more than two random variables through the multivariate distribution functions [12]. Sklar's theorem states that any multivariate distribution can be factored into the marginal cumulative distributions. Let $C(u_1, \ldots, u_d)$ is a cumulative distribution function (cdf) with uniform marginals on the unit interval. If $F_i(y_i)$ is the CDF of a univariate continuous random variable y_i. Then $C(F_1(y_1), \ldots, F_d(y_d))$ is a d-variate distribution for $Y = (y_1, \ldots, y_d)$ with marginal distributions $F_i; i = 1, \ldots, d$. Conversely, if H is a continuous d-variate cdf with univariate marginal cdfs $F_1, \ldots, F_d$, then there exists a unique d-variate copula C such that:

$$F(Y) = C(F_1(y_1)), \ldots, F_d(y_d)), \forall Y = (y_1, \ldots, y_d). \tag{6}$$

The corresponding density is:

$$c(F_1, (y_1), \ldots, F_d(y_d)) = \frac{h(F_1^{(-1)}(u_1), \ldots, F_d^{(-1)}(u_d))}{\prod_{i=1}^{d} f_i(F_i^{(-1)}(u_i))}, \tag{7}$$

where h is the density function associated to H, f_i is the density function of each marginal distribution and C is the copula density.

2.4   Time-Varying Copulas

In applying the conditional copula with a time-varying dependence structure, we assume that the dependence parameter is determined by past information and follows GARCH(p,q) process.

Time-Varying Gaussian Copula. The first considered is the time-varying in Gaussian copula, which can be following;

$$\rho_t = A\left(\omega_\rho + \beta_{\rho 1} \cdot \rho_{t-1} + \ldots + \beta_{\rho p} \cdot \rho_{t-p} + \alpha_\rho \cdot \frac{1}{q}\sum_{j=1}^{q} |u_{t-j} - v_{t-j}|\right), \tag{8}$$

where $A(x)$ is the logistic transformation which defined as $A(x) = (1 - e^{-x})(1 - e^{-x})^{-1}$, ρ_t is the correlation coefficient at time t which is the modified logistic transformation needed to keep as $\rho_t \in (-1, 1)$.

Time-Varying Student-t Copula. The second is time-varying in the student-t copula, which can be following;

$$\rho_t = A\left(\omega_\rho + \beta_{\rho 1} \cdot \rho_{t-1} + \ldots + \beta_{\rho p} \cdot \rho_{t-p} + \alpha_\rho \cdot \frac{1}{q}\sum_{j=1}^{q} |u_{t-j} - v_{t-j}; v_t|\right), \tag{9}$$

where $A(x)$ is the logistic transformation, ρ_t and v_t are the Pearson correlation coefficient and the degree of freedom, respectively. In addition, it is not only the degree of freedom but also the correlation can change by the time.

Time-Varying Clayton Copula. The third is time-varying in the conditional Clayton copula, which can be following;

$$\delta_t = \omega_U + \beta_{U1} \cdot \delta_{t-1} + \ldots + \beta_{Up} \cdot \delta_{t-p} + \alpha_U \cdot \frac{1}{q}\sum_{j=1}^{q} |u_{t-j} - v_{t-j}|, \qquad (10)$$

where $\delta_t \in [0, \infty)$ is the degree of dependence between u_t and v_t. δ_t is the rank correlation Kendall's tau. it is similar with the correlation parameter in the time-varying Gaussian copula.

Time-Varying Gumbel Copula. The finally is time-varying in the conditional Gumbel copula, which can be following;

$$\theta_t = \omega_L + \beta_{L1} \cdot \theta_{t-1} + \ldots + \beta_{Lp} \cdot \theta_{t-p} + \alpha_L \cdot \frac{1}{q}\sum_{j=1}^{q} |u_{t-j} - v_{t-j}|. \qquad (11)$$

where $\theta_t \in [1, \infty)$ is the degree of dependence between u_t and v_t.

2.5 The Hedging Ratio

The optimal hedge ratio is defined as the holdings of futures which minimize the risk of the hedging portfolio and number of futures contracts ($\Psi^{-1}_{yFutures,t}$) held to hedge against spot position ($\Psi^{-1}_{ySpot,t}$). Malliaris and Urrutia [9] determined the risk as the variance between the returns in the portfolio. If the joint distribution of the spot and futures returns remains the same over time, the conventional risk-minimizing hedge ratio δ^* will be

$$\delta^* = \frac{Cov\left(\Delta\Psi^{-1}_{ySpot,t}, \Delta\Psi^{-1}_{yFutures,t}\right)}{Var\left(\Delta\Psi^{-1}_{yFutures,t}\right)} = \frac{Cov\left(r_{\Psi^{-1}_{ySpot,t}}, r_{\Psi^{-1}_{yFutures,t}}\right)}{Var\left(\Delta r_{\Psi^{-1}_{yFutures,t}}\right)}. \qquad (12)$$

Let $\Delta\Psi^{-1}_{ySpot,t}$ and $\Delta\Psi^{-1}_{yFutures,t}$ be the respective changes in the spot and futures prices, δ^* serves to estimate the number of futures contracts used to mitigate the risk of changes in the price of spot.

$$\delta^* = \frac{Cov\left(r_{\Psi^{-1}_{ySpot,t}}, r_{\Psi^{-1}_{yFutures,t}}\right)}{Var\left(r_{\Psi^{-1}_{yFutures,t}}\right)} = \frac{\rho_{\Psi^{-1}_{ySpot,t},\Psi^{-1}_{yFutures,t}} * h_{\Psi^{-1}_{ySpot,t}} * h_{\Psi^{-1}_{yFutures,t}}}{h^2_{\Psi^{-1}_{yFutures,t}}}, \qquad (13)$$

$$= \rho_{\Psi^{-1}_{ySpot,t},\Psi^{-1}_{yFutures,t}}\left[\frac{h_{\Psi^{-1}_{ySpot,t}}}{h_{\Psi^{-1}_{yFutures,t}}} = \frac{\rho_{\Psi^{-1}_{ySpot,t},\Psi^{-1}_{yFutures,t}}\sqrt{h_{\Psi^{-1}_{ySpot,t}}}}{\sqrt{h_{\Psi^{-1}_{yFutures,t}}}}\right]. \qquad (14)$$

2.6   Hedging Effectiveness

In order to test the efficiency of which models and different frequency ranges that are able to prevent the most risks. Calculation of variances in the case of unhedged (U) and hedged (δ^*)

$$Var(U) = h^2_{\Psi^{-1}_{ySpot,t}} \tag{15}$$

$$Var(\delta^*) = h^2_{\Psi^{-1}_{ySpot,t}} + \delta^{*2} h^2_{\Psi^{-1}_{yFuture,t}} - 2\delta^* h_{\Psi^{-1}_{ySpot,t} \Psi^{-1}_{yFuture,t}} \tag{16}$$

Measuring hedging effectiveness is measured in terms of percentages. Reduction of the variance of hedged ports compared to ports that are not hedged is given by

$$E = \frac{Var\,(U) - Var(\delta^*)}{Var(\delta^*)} \times 100. \tag{17}$$

3   Data Description

Table 1. Descriptive statistics

	Wheat-s	Wheat-f
Mean	−5.21E−07	−5.20E−07
Median	0.0000	0.0000
Maximum	0.0155	0.0782
Minimum	−0.0214	−0.0734
Std. Dev.	0.001084	0.0011
Skewness	0.069527	1.7713
Kurtosis	27.51	629.9
Jarque-Bera	1519482	9.94E+08
MBF of Jarque-Bera	0.0000	0.0000
MBF of Unit root test	0.0000	0.0000

Note: MBF is Minimum Bayes factor.

In this study, we illustrate our model using wheat spot (wheat-s) and wheat futures (wheat-f). The data are high frequency 5-min, 30-min, 1-h and daily time series for the period from July 2017 to July 2018. All data for this study are collected from Bloomberg database. Additionally, we transform these time series variables into return rate before estimation. Table 1 shows the summary statistics for the wheat spot and futures returns, the Jarque-Bera normality test and Augmented Dickey-Fuller test (ADF) unit root test. The two data series exhibit a negative average return rate. The skewness of these series shows a positive value and the kurtosis values are all higher than 3, indicating a non-normal characteristic of the series. The Jarque-Bera normality test also confirms this non-normality pattern. The result shows that the Minimum Bayes factor of Jarque-Bera are equal to zero, indicating the decisive evidence for the Null hypothesis, that all returns are not normally distributed. In addition, the stationary test (Augmented Dickey-Fuller) is provided and we observe a MBF of our series are zero, indicating the decisive evidence for the Null hypothesis, that all returns are stationary.

4 Empirical Results

4.1 Model Selection

Table 2. AIC and BIC for model selection

Time	Family	Static Copula-GJR-GARCH		Conditional Copula-GJR-GARCH	
		AIC	BIC	AIC	BIC
5-min	Gaussian	−140.67	−137.15	−141.75	−123.82
	Student-t	**−156.39**	**−149.35**	−148.59	−127.46
	Clayton	−121.9	−118.38	−136.2	−125.64
	Gumbel	−125.18	−121.66	−99.42	−88.86
30-min	Gaussian	−274.74	−271.22	−310.51	−299.94
	Student-t	−304.98	−297.93	**−303.85**	**−282.73**
	Clayton	−191.64	−188.12	−196.84	−186.28
	Gumbel	−303.18	−299.66	−76.23	−65.67
1-h	Gaussian	−265.54	−262.02	−300.17	−289.6
	Student-t	−293.71	−286.66	−298.74	−277.61
	Clayton	−172	−168.47	−176.16	−165.6
	Gumbel	**−307.74**	**−304.22**	−75.92	−65.36
Daily	Gaussian	−351.84	−348.32	−356.2	−345.63
	Student-t	**−386.82**	**−379.78**	−382.97	−361.84
	Clayton	−261.7	−258.18	−263.97	−253.4
	Gumbel	−375.7	−372.18	−82.69	−72.13

In this study, we used static copula and Time-varying copula for four classes of copula, i.e., Gaussian copula, Student-t copula, Clayton copula and Gumbel copula based on GJR-GARCH(1,1) under the histogram value data context, used 5-min, 30-min, 1-h and daily data, are considered here for modeling the dependency or relationship between wheat spot (wheat-s) and wheat futures (wheat-f). The information criteria, namely AIC and BIC statistics are adopted to select the most suitable models. The estimated values of models are presented in Table 2. The results find that the Time-varying student-t copula is the best-fit for 30-min histogram data as AIC(−303.85) and BIC(−282.73), static Student-t copula is the best-fit for 5-min histogram data as AIC(−156.39) and BIC(−149.35) and daily data as AIC(−386.82) and BIC(−386.82) and the static Gumbel copula is the best-fit for 1-h histogram data as AIC(−307.74) and BIC(304.22). Note that we look at minimum AIC and BIC.

4.2   Estimates for the Copula Function of Wheat Return

This section discusses the result of copula function under the histogram value data context, used 5-min, 30-min, 1-h and daily data. The results of the estimated parameters are shown in Table 3, that each period gives different results. Therefore, investors should focus on the frequency of trading. Fig. 1. presents the comparison of the static copulas and the correlation of the time-varying student-t copula. The correlation of the time-varying student-t copula with 30-min data (black dotted lines) found that the lowest time vary correlation value of Kendall tau is close to 0.5 while the highest value is approximately 0.87. This indicates a time varying correlation of spot and futures along our sample. The correlations of the static student-t copulas of 5-min data (red line) and the static student-t copulas with daily data (green line) are 0.4 and 0.68 respectively. The correlations of the static Gumbel copula with 1-h (blue line) the value equal 0.61.

Table 3. Estimates for the copula models

Time	Parameter	Coef.	S.E.	Stat	MBF
Student-t copula					
5-min	ρ	0.5894	0.0373	15.8195	0.0000
	ν	6.2595	1.8876	3.3161	0.0041
Daily	ρ	0.8795	0.0138	63.9516	0.0000
	ν	4.5084	1.1846	3.8059	0.0007
Gumbel copula					
1-h	δ	2.5637	0.131	19.5681	0.0000
Time varying student-t copula					
30-min	ω_U	2.3155	8.4825	0.273	0.7849
	ρ_U	0.6583	9.1017	0.0723	0.9423
	ν_U	−0.5896	1.1325	−0.5206	0.6026
	ω_L	−0.5308	0.0145	−36.707	0.0000
	ρ_L	−0.9394	0.0145	−64.6057	0.0000
	ν_L	−0.5158	0.0011	−462.5397	0.0000

Note: MBF is Minimum Bayes factor.

4.3   Hedge Ratios

Hedging application, the results of hedge ratio and portfolio weight are plotted in Figs. 2, 3, respectively. The time-varying hedge ratios clearly change when new information arrives in the market. We find that the time-varying hedging ratio seem to response to the change of real economic situation. The hedging ratio is volatile over time.

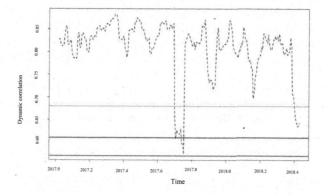

Fig. 1. The comparison of the conditional correlations of copulas and the correlation of the time-varying student-t copula (Color figure online)

We observe that the highest average hedge ratio of 5-min data is 0.47 meaning that, if investors buy spot 100 contract, they will long (buy) futures 47 contracts to prevent the risk. In part of 30-min data of the highest average hedge ratio is 0.87 meaning that, if investors buy spot 100 contract, they will long (buy) futures 87 contracts to prevent the risk. In part of 1-h data of the highest average hedge ratio is 0.65 meaning that, if investors buy spot 100 contract, they will long (buy) futures 65 contracts to prevent the risk. Finally, daily data, the highest average hedge ratio is 0.72 meaning that, if investors buy spot 100 contract, they will long (buy) futures 72 contracts to prevent the risk. As we can observe, the time varying hedge ratios obtained from different frequencies show a different pattern, the question is which frequency provide the best suggestion for the investors making the hedging strategy? Therefore, we further examine the impact of frequency specification to fit the model using the hedging effectiveness approach.

4.4 Hedging Effectiveness

The hedging effectiveness is used to check the efficiency of each frequency data. The results is shown in be Figs. 4, 5. The results suggest that the hedging effectiveness are quite different compared to the spot and futures prices for different frequencies.

It is observed that the hedging effectiveness obtained form 1-h and daily frequencies data illustrate a similar pattern. However, we compare the values of hedging effectiveness, we can find that 1-h frequency hedging model is better than daily frequency hedging model and other two frequency models (5-min and 30 min). This indicates that 1-h frequency data is the best information for constructing the hedging model as it could substantially reduce the risk of the portfolio. Thus, the investor may consider 1-h data to find the hedging strategy in spot-futures portfolio.

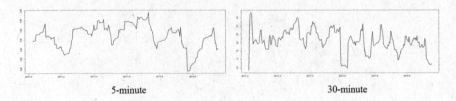

5-minute   30-minute

Fig. 2. Hedge ratio of 5-min and 30-min

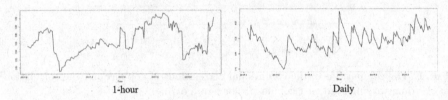

1-hour   Daily

Fig. 3. Hedge ratio of 1-h and daily

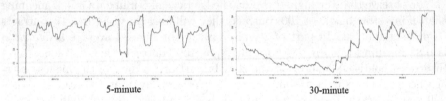

5-minute   30-minute

Fig. 4. Hedging effectiveness of 5-min and 30-min

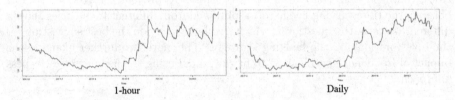

1-hour   Daily

Fig. 5. Hedging effectiveness of 1-h and daily

5   Conclusion

One of the main functions of the future market is to provide a hedging mechanism. It is also a well-documented claim in the future market literature that the
optimal hedge ratio should be time-varying and not constant. An optimal hedge
ratio is the proportion of a cash position that should be covered with an opposite
position on a future market to estimate the time-varying hedge ratio. To do this,
we consider various dynamic conditional copula families to find the time-varying
hedge ratio. We note that both static and dynamic copulas are employed in this
study. In addition, we also use the histogram value data which contains a tone
of information to build our models. Therefore, 5-min, 30-min, 1-h and daily data

are considered in this study. This paper applies our models to examine the wheat market (spot and futures prices). We model the volatility of the spot and futures prices using the GJR-GARCH(1,1) model. In addition, we apply our method to quantify the hedge ratios and hedging effectiveness of wheat. The empirical evidence shows that the time-varying student-t copula is the best-fit for 30-min histogram data, static Student-t copula is the best-fit for 5-min histogram data and daily data. Gumbel copula is the best-fit for 1-h histogram data. We find that the time-varying hedging ratios seem to response to the change over time and response to the change of economy. Our findings are important for portfolio managers, especially during periods of market stress, since they can use this information to further improve their hedging performance.

However, this study focused on minute, hourly, and daily data, which required a significant amount of time and resources to collect. As a result, this study was only captured one variable as wheat. In future studies, it should take into account other financial variables, such as commodities such as oil and gold, as well as cryptocurrency assets, which are very popular today. They are traded throughout the day and have a wide range of frequency trading data, which is a very interesting topic.

References

1. Arroyo, J., González-Rivera, G., Maté, C.: Forecasting with interval and histogram data. Some financial applications. In: Handbook of Empirical Economics and Finance, pp. 247–280 (2010)
2. Arroyo, J., Maté, C.: Forecasting histogram time series with k-nearest neighbours methods. Int. J. Forecast. **25**(1), 192–207 (2009)
3. Bollerslev, T.: Generalized autoregressive conditional heteroskedasticity. J. Econ. **31**(3), 307–327 (1986)
4. Dias, S., Brito, P.: A new linear regression model for histogram-valued variables. In: 58th ISI World Statistics Congress, Dublin, Ireland (2011). http://isi2011.congressplanner.eu/pdfs/950662.pdf
5. Engle, R.F.: Autoregressive conditional heteroscedasticity with estimates of the variance of United Kingdom inflation. Econometrica: J. Econ. Soc. 987–1007 (1982)
6. Glosten, L.R., Jagannathan, R., Runkle, D.E.: On the relation between the expected value and the volatility of the nominal excess return on stocks. J. Financ. **48**(5), 1779–1801 (1993)
7. Irpino, A., Verde, R.: A new Wasserstein based distance for the hierarchical clustering of histogram symbolic data. In: Batagelj, V., Bock, H.H., Ferligoj, A., Ziberna, A. (eds.) Data science and classification, pp. 185–192. Springer, Heidelberg (2006). https://doi.org/10.1007/3-540-34416-0_20
8. Li, Y.: GARCH-copula approach to estimation of value at risk for portfolios (2012)
9. Malliaris, A.G., Urrutia, J.L.: The impact of the lengths of estimation periods and hedging horizons on the effectiveness of a hedge: evidence from foreign currency futures. J. Futur. Mark. **11**(3), 271–289 (1991)
10. Patton, M.Q.: Qualitative Research and Evaluation Methods, Thousand Oaks (2002)

11. Rakpho, P., Yamaka, W., Tansuchat, R.: Risk valuation of precious metal returns by histogram valued time series. In: Kreinovich, V., Sriboonchitta, S., Chakpitak, N. (eds.) TES 2018. SCI, vol. 753, pp. 549–562. Springer, Cham (2018). https://doi.org/10.1007/978-3-319-70942-0_39
12. Sklar, M.: Fonctions de repartition an dimensions et leurs marges. Publ. Inst. Statist. Univ. Paris **8**, 229–231 (1959)

Predicting Energy Price Volatility Using Hybrid Artificial Neural Networks with GARCH-Type Models

Pichayakone Rakpho[1], Woraphon Yamaka[1,2(✉)], and Rungrapee Phadkantha[1]

[1] Center of Excellence in Econometrics, Chiang Mai University,
Chiang Mai 50200, Thailand
[2] Faculty of Economics, Chiang Mai University, Chiang Mai 50200, Thailand
woraphon.yamaka@cmu.ac.th

Abstract. This paper aims at analyzing the energy price volatility forecasts for crude oil, ethanol, and natural gas. Several hybrid Artificial Neural Networks (ANN)-GARCH models consisting of ANN-GARCH, ANN-EGARCH, and ANN-GJR-GARCH models are introduced. However, a challenge in the ANN design is the selection of activation function. Thus, various forms of activation function, namely logistic, Gompertz, tanh, ReLU and leakyReLU are also considered to analyze the increase in the hybrid models' predictive power. In our investigation, both in-sample and out-of-sample analysis are used and the results provide the strong evidence of the higher performance of the hybrid-ANN-GARCH compared to the single GARCH-type models. However, when five activation functions are applied over the parameters, the results tend to be similar, indicating the robustness of our forecasting results.

Keywords: ANN-GARCH-type models · Activation functions · Predicting volatility · Energy

1 Introduction

With the important role of energy in the world economy, the increase in energy price volatility has caused a great concern among investors and businesses. Volatility refers to the spread of unlikely outcomes which results in the uncertainty in the world economy. Hence, market participants utilize different methods for forecasting the volatility of energy prices to manage their financial risks. The methods and models for analyzing and predicting energy market and price volatility have been discussed and investigated in many studies ([3,12,14]). The most widely used are the generalized autoregressive conditional heteroskedasticity (GARCH) model of Bollerslev [1] and its variants. However, the literature has suggested that energy price volatility may exhibit the asymmetric nature, thus, the original GARCH may fail to capture this feature and result in an incorrect inference [4]. To deal with this problem, Nelson [10] and Glosten et

© Springer Nature Switzerland AG 2022
K. Honda et al. (Eds.): IUKM 2022, LNAI 13199, pp. 317–328, 2022.
https://doi.org/10.1007/978-3-030-98018-4_26

al. [5] proposed the EGARCH and GJR-GARCH models, respectively. These models are able to capture asymmetry, which refers to the negative unexpected returns affecting future volatility more than the positive unexpected returns, and leverage, which refers to the negative correlation between the shocks and the subsequent shocks to volatility [9]. The review related to the volatility forecasting of energy prices can be found in Wei et al. [14].

Despite the higher performance of the GARCH-type models in forecasting the volatility of energy prices, the errors in the prediction using these approaches are often quite high [13]. The recent studies of Kristjanpoller and Minutolo [13], Lu et al. [7], and Liao et al. [6] revealed that the forecasting accuracy of the GARCH-type models can be improved by applying the Artificial Intelligence techniques, in particular, the Artificial Neural Networks (ANN). The ANN model is one of the most popular methods in machine learning (ML) as it has an ability to learn and model non-linear and complex structure of the volatility. Moreover, Donaldson and Kamstra [4], Kristjanpoller and Minutolo [13], Lu et al. [7] and Liao et al. [6] have shown that ANN can model better heteroskedasticity, i.e., volatility clustering, nonlinear behavior and non-constant variance, as it is able to learn the hidden relationships between the input and output without imposing any fixed relationships.

Given lessons learned from the existing literature, the purpose of this study is twofold. Our first objective is to improve the hybrid ANN-GARCH type models for the conditional energy price volatility which can capture important asymmetric and nonlinear effects that the existing models do not capture. In particular, we extend the research streams that forecast energy price volatility using various ANN-GARCH-type models to demonstrate improvements in precision over the hybrid classical-forecasting models and the conventional GARCH-type models. Specifically, this study aims at forecasting the volatility of energy prices consisting of crude oil, natural gas, and ethanol by applying the ANN approach to the GARCH-type models consisting of GARCH, EGARCH, and GJR GARCH. We extend previous research in the hybrid modeling domain to the area of crude oil, natural gas, and ethanol volatility. Our second objective is a response to the observation that the appropriate activation function for the ANN-GARCH-type models has not been investigated in the literature yet. To this end, we introduce several activation functions to the ANN- GARCH-type models and compare the forecasting performance of each activation function. Specifically, we introduce several activation functions namely logistic, tanh, Gompertz, Rectified Linear Unit (RLU) and Rectified Exponential Linear Unit (ReLU). It is necessary to know which activation function performs the best fit for the ANN-GARCH-type models. To the best of our knowledge, this is the first attempt to investigate the activation function selection for the ANN-GARCH-type models. Selecting an appropriate activation function is a challenging pursuit, as it affects the accuracy and the complexity of the given ANN-GARCH-type models.

The rest of this paper is organized as follows. Section 2 presents our methodology and discusses the hybridization of the ANN approach with the GARCH-type models (GARCH, EGARCH, and GJR GARCH). Section 3 presents descriptive data. Section 4 shows the empirical results of in-sample and out-of-sample volatility forecasting. Finally, the conclusions are made in Section 5.

2 Methodology

To forecast the energy price volatility, various hybrid ANNs-GARCH-type models are employed. In this section, we briefly present the conventional GARCH-type models, namely GARCH, EGARCH and GJR-GARCH, and ANN model. In the last sub-section, we introduce our hybrid-ANN-GARCH type models.

2.1 GARCH-Type Models

GARCH Model. The GARCH model was proposed by Bollerslev [1] and it is used to forecast the conditional variance as well as capture the time-dependent heteroskedasticity of financial data. In this study, GARCH(1,1) is considered and can be expressed as follows:

$$
\begin{aligned}
y_t &= \mu + \varepsilon_t \\
\varepsilon_t &= \sigma_t \nu_t \\
\sigma_t^2 &= \omega + \alpha \varepsilon_{t-1}^2 + \beta \sigma_{t-1}^2
\end{aligned}
\tag{1}
$$

where $\omega, \alpha, \beta \geq 0$ are unknown parameters with parameter restrictions $\alpha + \beta \leq 1$. ε_t is the uncorrelated random variable with mean zero and variance σ_t^2. ν_t is a standardized residual which is assumed to have the skewed student-t distribution.

EGARCH Model. EGARCH is a nonlinear GARCH model which was proposed by Nelson [10] to capture the long-memory and short-memory volatility effects, and asymmetric leverage effects of financial variables. The specification for the conditional variance of the EGARCH(1,1) model is

$$
\ln(\sigma_t^2) = \omega + \gamma \left| \frac{\varepsilon_{t-1}}{\sqrt{\sigma_{t-1}^2}} \right| + \alpha \left(\frac{\varepsilon_{t-1}}{\sqrt{h_{t-1}}} \right) + \beta \ln \left(\sigma_{t-1}^2 \right),
\tag{2}
$$

where γ is the asymmetric leverage coefficient to describe the volatility leverage effect.

GJR-GARCH Model. Another popular nonlinear GARCH model is GJR-GARCH of Glosten et al. [5]. It is introduced to capture the potential larger impact of negative shocks on volatility of the data (asymmetric leverage volatility effect). The GJR-GARCH(p,q) model is defined as

$$
\sigma_t^2 = \omega + (\alpha + \gamma I_{t-1}) \varepsilon_{t-1}^2 + \beta \sigma_{t-1}^2
\tag{3}
$$

where

$$
I_{t-1} = \begin{cases} 0, & \text{if } \varepsilon_{t-1} \geq 0 \\ 1, & \text{if } \varepsilon_{t-1} < 1 \end{cases}
\tag{4}
$$

and again γ is an asymmetric leverage effect.

2.2 Artificial Neural Network (ANN)

ANN is a network of artificial neurons which presents the connection between input and output signals similar to the human brain. In this study, we consider the multilayer perceptron, with only 1 input layer, 1 hidden layer, and 1 output layer, in order to simplify the estimation. The input layer is represented by vector $d = (x_1, x_2, \ldots, x_d)'$ and the output layer is a vector $c = (y_1, y_2, \ldots y_c)'$. The ANN model can be presented by

$$c = \tilde{\phi}(\hat{\phi}(dw^I + b^I)w^O + b^O), \tag{5}$$

where $\tilde{\phi}$ and $\hat{\phi}$ are the input and output activation functions, respectively. b^I and b^O are the bias term of input and output, respectively. w^I and w^O are the weight vector between the hidden layer and the output layer, respectively.

2.3 Hybrid Models

In this section, we extend the ANN model to the GARCH-type models. In this respect, the conditional variance processes of ANN-GARCH(1,1), ANN-EGARCH(1,1), and ANN-GJR-GARCH(1,1) augmented with ANN are defined as follows:

ANN-GARCH Model

$$\sigma_t^2 = \omega + \alpha \varepsilon_{t-1}^2 + \beta \sigma_{t-1}^2 + \theta \left(\psi \left(z_t \delta \right) \right) \tag{6}$$

ANN-EGARCH Model

$$\ln(\sigma_t^2) = \omega + \gamma \left| \frac{\varepsilon_{t-1}}{\sqrt{\sigma_{t-1}^2}} \right| + \alpha \left(\frac{\varepsilon_{t-1}}{\sqrt{h_{t-1}}} \right) + \beta \ln \left(\sigma_{t-1}^2 \right) + \theta \left(\psi \left(z_t \delta \right) \right), \tag{7}$$

ANN-GJR-GARCH Model

$$\sigma_t^2 = \omega + \left(\alpha + \gamma I_{t-1} \right) \varepsilon_{t-1}^2 + \beta \sigma_{t-1}^2 + \theta \left(\psi \left(z_t \delta \right) \right) \tag{8}$$

where θ is the additional output weight of the neural network part. $\psi \left(z_t \delta \right)$ is the sigmoid activation function of the output layer of the ANN part which can be written as

$$\psi \left(z_t \delta \right) = \left(1 + \exp \left(\delta z_{t-1} \right) \right)^{-1} \tag{9}$$

where $z_{t-1} = (\varepsilon_{t-1} - E(\varepsilon_t)) / \sqrt{E(\varepsilon_t^2)}$ is the standardized residual in the log-sigmoid activation function. We follow the computationally simple approach of first choosing the with a uniform random number generated between -1 and 1. In the estimation point of view, the maximum likelihood estimation is employed to estimate all unknown parameters.

2.4   Activation Function

The selection of activation function for linking the input, hidden, and output layers is important. In this study, five activation functions are considered and presented as follows

Logistic. The logistic function is known as the sigmoid function and can map the input signal between 0 and 1. This function is defined as

$$f(x) = \frac{1}{1 + e^{-x}} \tag{10}$$

Gompertz. The Gompertz curve or Gompertz function is similar to the logistic function but it exhibits a slower growth at the initial cultivation stage and at the end of stage. The function can be written as

$$N(t) = N(0)\exp(-c(\exp(-at))) \tag{11}$$

where $N(0)$ is the initial number of cells/organisms at time zero.

Tanh. Tanh or the hyperbolic tangent activation function is also similar to the logistic function, but it maps the resulting values in between 0 and 1. The formula is written as

$$f(x) = \tanh(x) = \frac{(e^x - e^{-x})}{(e^x + e^{-x})} \tag{12}$$

Rectified Linear Unit (ReLU). ReLU is a nonlinear activation function [2]. This function has been found to provide a better result in many different settings. This is due to the sparseness property which is useful in many contexts. Specifically, this function can map the negative argument values to zero

$$f(x) = \begin{cases} 0, & \text{if } x \leq 0 \\ 1, & \text{if } x > 1 \end{cases} \tag{13}$$

Leaky ReLU. Leaky ReLU function is an improved version of the ReLU activation function. As the gradient of ReLU is 0 for all the values of inputs that are less than zero, which would deactivate the neurons in the ANN. This activation function can be presented as

$$f(x) = \begin{cases} 0.01x, & \text{for } x < 0 \\ x, & \text{for } x \geq 1 \end{cases} \tag{14}$$

2.5   Forecasting Comparison Criteria

In this study, the Root Mean Square Error (RMSE) is used as a prediction error for our forecasting models. It measures how far it is the distance between the true and the forecasted volatilities. The formula is:

$$RMSE = \sqrt{\frac{\sum_{t=1}^{N}\left(\widehat{\sigma}_t^2 - \sigma_t^2\right)^2}{N}} \tag{15}$$

where $\widehat{\sigma}_t^2$ represent the forecasted and experimental values of energy price volatility, respectively and N is the number of forecasting data test data points. Note that the actual volatility σ_t^2 can be computed as the sample variance log returns in a 21 days' window to the future (approximately one month of transactions),

$$\sigma_t^2 = \frac{1}{21} \sum_{i=t+1}^{t+21} (r_i - \bar{r}_t)^2 \tag{16}$$

where n is the number of day, r_i is the return of energy price and $\bar{r}_t$ is the average energy price.

3   Data Description

This paper collects Brent crude oil, ethanol, and natural gas prices from the Bloomberg database, covering the period from January 2005 to December 2019. Table 1 presents some descriptive statistics for energy price returns (log-difference). For all cases, the mean is close to 0 and the standard deviation is around 0.03. These values are typical for stationary series as their averages are close to 0 with small variations. Thus, the Augmented Dickey Fuller test (ADF) is further conducted to examine the stationarity. The Minimum Bayes Factor(MBF) of ADF test is reported in the last row. We note that the MBF is the p-value calibration and the closer to zero, the higher chance to reject the non-stationarity [8]. The result shows that the MBF of ADF values are 0, indicating the returns are stationary

The skewness of crude oil and natural gas are positive indicating that the upper tails of the empirical distributions of returns are longer that the lower tails. All returns have kurtosis greater than 3, indicating that the return series have fat-tailed distributions. Moreover, the MBF-Jarque-Bera test results strongly reject the null hypothesis of normality in all returns series. Figure 1 presents the energy price returns. We can observe that the variance of energy returns is not stable and there exists the volatility clustering along the sample period.

Table 1. Data statistics

	Crude oil	Ethanol	Natural gas
Mean	5.65E−06	4.84E−05	−0.000309
Median	0.000677	0.000000	−0.00067
Maximum	0.164097	0.160343	0.267712
Minimum	−0.130654	−0.309978	−0.180545
Std. Dev.	0.023388	0.020834	0.031398
Skewness	0.125840	−2.43719	0.529386
Kurtosis	7.598936	34.03754	7.965910
Jarque-Bera	3223.577	149996.4	3917.677
MBF-Jarque-Bera	0.000000	0.000000	0.000000
Unit root test	−64.28157	−63.70705	−56.75844
MBF-Unit root	0.000000	0.000000	0.000000

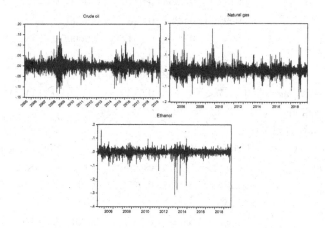

Fig. 1. The daily return of the baseload price for each series

4   Results

4.1   Estimation Results for Different Volatility Models

To find the best model for forecasting the energy price volatility, we conduct the in-sample and out-of-sample forecasts. The in-sample data covers a 13-year period from January 2005 to December 2017, and the out-of-sample data for model evaluation covers the period from January 2018 to December 2019. In this section, we present the in-sample estimation results of the hybrid ANN-GARCH-type models namely the ANN-GARCH(1,1), ANN-EGARCH(1,1), and ANN-GJR-GARCH(1,1) in Table 2. We note that the conventional hybrid GARCH models with logistic activation function is presented here as the example.

Table 2. Estimation results of the hybrid ANN-GARCH-type models

ANN-GARCH(1,1)

Parameter	Crude oil		Ethanol		Natural gas	
	Estimate	MBF	Estimate	MBF	Estimate	MBF
ω	0.0357	0.0082	0.2247	0.0232	0.1155	0.0002
α	0.0601	0.0000	0.1703	0.0000	0.0777	0.0000
β	0.9341	0.0000	0.7943	0.0000	0.9136	0.0000
θ	0.0866	0.0000	0.0554	0.0000	0.0013	0.0000
ARCH-LM lag[10]	0.6823		0.4631		0.3979	
Log-Likelihood	−7743.54		−7406.557		−9027.253	

ANN-EGARCH(1,1)

Parameter	Crude oil		Ethanol		Natural gas	
	Estimate	MBF	Estimate	MBF	Estimate	MBF
ω	0.0147	0.0055	0.0410	0.0000	0.0337	0.0000
α	0.0095	0.0297	0.0539	0.0000	0.0004	0.0099
β	0.9912	0.0000	0.9812	0.0000	0.9868	0.0000
γ	0.0925	0.0000	0.2018	0.0000	0.1717	0.0000
θ	0.0038	0.0000	0.0028	0.0000	0.0425	0.0000
ARCH-LM lag[10]	0.1017		0.5143		0.2154	
Log-Likelihood	−7702.935		−7364.264		−9027.701	

ANN-GJR-GARCH(1,1)

Parameter	Crude oil		Ethanol		Natural gas	
	estimate	MBF	estimate	MBF	estimate	MBF
ω	3.2258	0.0055	0.0277	0.0055	0.1152	0.0002
α	0.5515	0.0297	0.0147	0.0297	0.0749	0.0000
β	0.4893	0.0000	0.9446	0.0000	0.9131	0.0000
γ	6.9752	0.0000	0.0711	0.0000	0.0071	0.0083
θ	0.0013	0.0000	0.0037	0.0000	0.0037	0.0000
ARCH-LM lag[10]	0.3362		0.9556		0.4278	
Log-Likelihood	−7713.823		−7724.872		−9027.023	

First, the results reported in Table 2 show that the degree of volatility persistence of the hybrid GARCH-type models, which can be measured by the sum of ARCH and GARCH estimates, $(\alpha + \beta)$, is close to 1 with decisive evidence. This indicates a high degree of volatility in oil, ethanol, and natural gas returns. Second, the asymmetric leverage parameters γ of ANN-EGARCH (1,1) and ANN-GJR-GARCH(1,1) are significantly different from 0, implying the occurrence of leverage effects in all energy returns and the nonlinear EGARCH and GJR-GARCH may be the preferred specifications to model asymmetric effects in the oil, ethanol, and natural gas returns. Third, we find the output weight of neural network θ for all GARCH-type models to be significant, thereby confirming that the ANN term can be used to predict the conditional volatility. Finally, the ARCH-LM test is used to investigate the existence of ARCH effects in the models. We find no significant ARCH effect (ARCH-LM [8]) in the variance equations with decisive evidence for all the ANN-GARCH-type models.

4.2   Performance Results for the Forecasting Models

In this section, we investigate the model performance based on RMSE. Tables 3 and 4 present in-sample and out-of-sample forecasts comparison results.

Table 3. In-sample performance results for the conventional and hybrid GARCH-type models using different activation functions

Activation	Crude oil		
	GARCH	EGARCH	GJR-GARCH
	0.14092	0.1499	0.14083
	ANN-GARCH	ANN-EGARCH	ANN-GJR-GARCH
Logistic	0.12765	0.14847	0.1268
Gompertz	0.12765	0.14844	0.1268
Tanh	0.12682	0.14811	0.12666
ReLU	0.12765	0.14815	0.1268
LeakyReLU	0.1275	0.14803	0.12659
Activation	Natural gas		
	GARCH	EGARCH	GJR-GARCH
	0.26639	0.26639	0.26639
	ANN-GARCH	ANN-EGARCH	ANN-GJR-GARCH
Logistic	0.25391	0.24814	0.23011
Gompertz	0.2539	0.24824	0.23038
Tanh	0.25392	0.24877	0.23099
ReLU	0.25394	0.24834	0.2309
LeakyReLU	0.25392	0.24883	0.23078
Activation	Ethanol		
	GARCH	EGARCH	GJR-GARCH
	0.25936	0.25946	0.2599
	ANN-GARCH	ANN-EGARCH	ANN-GJR-GARCH
Logistic	0.25314	0.24399	0.252
Gompertz	0.25306	**0.24380**	0.2521
Tanh	0.25251	0.24381	0.25199
ReLU	0.25066	0.24392	0.25203
LeakyReLU	0.25065	0.24393	0.25202

Note: Bold number indicates the best forecasting model

Table 4. Out-of-sample performance results for the conventional and hybrid GARCH-type models using different activation functions

Activation	Crude oil		
	GARCH	EGARCH	GJR-GARCH
	0.16242	0.1614	0.16246
	ANN-GARCH	ANN-EGARCH	ANN-GJR-GARCH
Logistic	0.15941	0.16666	0.157
Gompertz	0.16081	0.16675	0.1569
Tanh	0.16094	0.16649	0.15706
ReLU	0.16028	0.16693	0.157
LeakyReLU	0.16053	0.16613	0.15683
Activation	Natural gas		
	GARCH	EGARCH	GJR-GARCH
	0.24651	0.24651	0.24651
	ANN-GARCH	ANN-EGARCH	ANN-GJR-GARCH
Logistic	0.25262	0.24833	0.23202
Gompertz	0.2521	0.24832	0.2421
Tanh	0.25208	0.24834	0.23208
ReLU	0.25462	0.24835	0.23462
LeakyReLU	0.25397	0.24889	0.23397
Activation	Ethanol		
	GARCH	EGARCH	GJR-GARCH
	0.06989	0.06989	0.06989
	ANN-GARCH	ANN-EGARCH	ANN-GJR-GARCH
Logistic	0.26989	0.25343	0.25402
Gompertz	0.26736	**0.25330**	0.25403
Tanh	0.26748	0.25344	0.25409
ReLU	0.26744	0.25344	0.25404
LeakyReLU	0.2673	0.25343	0.25404

Note: Bold number indicates the best forecasting model

Considering the in-sample-forecast in Table 3, we find that the volatility forecasting of the GARCH-type models can be improved when the ANN is augmented because the RMSEs of the hybrid-models are lower than the traditional GARCH models. For example, the GARCH, EGARCH, and GJR-GARH models applied to forecast the crude oil price volatility have the RMSE equal to 0.14092, 0.14990, and 0.14083, respectively, while the estimation by the three hybrid models with the logistic activation function has the RMSE equal to 0.12765, 0.14847, and 0.12680, respectively. Similar results are obtained from using other activation functions.

We then further analyze the model performance for different activation functions, we find that (i) the ANN-GJR-GARCH with LeakyReLU is superior to other models in forecasting oil price volatility, while the ANN-GJR-GARCH and ANN-EGARCH achieve the lowest RMSE in the cases of natural gas and ethanol, respectively. These results confirm the higher performance of the nonlinear ANN-GARCH-types models. In other words, the GARCH-types models applied to predict the energy price volatility produced an error in the forecasting which could potentially be reduced using the hybrid ANN-GARCH [13].

Regarding the out-of-sample forecast, which is reported in Table 4, we find that the proposed ANN-GARCH-type models still perform better than the conventional ones. In addition, the LeakyReLU, Logistic, and Gompertz functions remain the best fit activation to forecast through the hybrid ANN-GARCH-type models for oil, natural gas, and ethanol price volatilities, respectively.

5   Conclustions

Accurately estimating and forecasting volatility in energy returns is a crucial issue. To achieve this goal, this study proposed three hybrid GARCH-type models, namely the ANN-GARCH, ANN-EGARCH, and ANN-GJR-GARCH models. In addition, four activation functions, namely Gompertz, Tanh, ReLU and LeakyReLU, are also investigated in order to find the best transfer function for ANN structure. In this forecasting investigation, we compare the forecasting ability of each model using in and out-of-sample volatility forecasts. In this regard, we can obtain the best model for forecasting energy price volatility (Crude oil, Natural gas, Ethanol). In this comparison, the RMSE is used as the loss function and criterion. The results show that the hybrid models perform better than the conventional GARCH-type models in all cases. This indicates that the augmentation of ANN can potentially improve the forecasting precision of the GARCH-type models. Furthermore, we investigate the effect of activation function in our hybrid models. In this regard, our findings demonstrate the mixed results. As discussed by Yamaka et al. [15], it is not obvious which activation function is more appropriate for linking with the ANN, and different activation functions may play different roles in the practical applications In essence, our results suggest that we should not arbitrarily select a volatility forecasting model by referring to the literature. The reliable and accurate model depends on not only the given data but also the correspondence of the particular forecasting purpose with the GARCH specification and activation function considered. In the further research, the hybrid GARCH model can be extended the fuzzy-based approach to gain more accurate volatility forecast (see, [11]).

References

1. Bollerslev, T.: Generalized autoregressive conditional heteroskedasticity. J. Econometrics **31**(3), 307–327 (1986)
2. Clevert, D.A., Unterthiner, T., Hochreiter, S.: Fast and accurate deep network learning by exponential linear units (elus) (2015). arXiv preprint arXiv:1511.07289

3. Chang, C.L., McAleer, M., Tansuchat, R.: Analyzing and forecasting volatility spillovers, asymmetries and hedging in major oil markets. Energy Econ. **32**(6), 1445–1455 (2010)
4. Donaldson, R.G., Kamstra, M.: An artificial neural network-GARCH model for international stock return volatility. J. Empir. Financ. **4**(1), 17–46 (1997)
5. Glosten, L.R., Jagannathan, R., Runkle, D.E.: On the relation between the expected value and the volatility of the nominal excess return on stocks. J. Financ. **48**(5), 1779–1801 (1993)
6. Liao, R., Yamaka, W., Sriboonchitta, S.: Exchange rate volatility forecasting by hybrid neural network Markov switching Beta-t-EGARCH. IEEE Access **8**, 207563–207574 (2020)
7. Lu, X., Que, D., Cao, G.: Volatility forecast based on the hybrid artificial neural network and GARCH-type models. Procedia Comput. Sci. **91**, 1044–1049 (2016)
8. Maneejuk, P., Yamaka, W.: Significance test for linear regression: how to test without P-values? J. Appl. Stat. **48**(5), 827–845 (2021)
9. Martinet, G.G., McAleer, M.: On the invertibility of EGARCH (p, q). Economet. Rev. **37**(8), 824–849 (2018)
10. Nelson, D.B.: Conditional heteroskedasticity in asset returns: A new approach. Econometrica: J. Econometric Soc. 347–370 (1991)
11. Novák, V.: Fuzzy vs. probabilistic techniques in time series analysis. In: Anh, L.H., Dong, L.S., Kreinovich, V., Thach, N.N. (eds.) ECONVN 2018. SCI, vol. 760, pp. 213–234. Springer, Cham (2018). https://doi.org/10.1007/978-3-319-73150-6_17
12. Tarkhamtham, P., Yamaka, W., Maneejuk, P.: Forecasting volatility of oil prices via google trend: LASSO approach. In: Ngoc Thach, N., Kreinovich, V., Trung, N.D. (eds.) Data Science for Financial Econometrics. SCI, vol. 898, pp. 459–471. Springer, Cham (2021). https://doi.org/10.1007/978-3-030-48853-6_32
13. Kristjanpoller, W., Minutolo, M.C.: Gold price volatility: a forecasting approach using the artificial neural network-GARCH model. Expert Syst. Appl. **42**(20), 7245–7251 (2015)
14. Wei, Y., Wang, Y., Huang, D.: Forecasting crude oil market volatility: further evidence using GARCH-class models. Energy Econ. **32**(6), 1477–1484 (2010)
15. Yamaka, W., Phadkantha, R., Maneejuk, P.: A convex combination approach for artificial neural network of interval data. Appl. Sci. **11**(9), 3997 (2021)

Estimating Wind Speed by Using Confidence Intervals for the Median in a Three-Parameter Lognormal Model

Patcharee Maneerat[1], Sa-Aat Niwitpong[2], and Pisit Nakjai[3(✉)]

[1] Department of Applied Mathematics, Rajabhat Uttaradit University,
Uttaradit 53000, Thailand
m.patcharee@uru.ac.th
[2] Department of Applied Statistics, King Mongkut's University of Technology
North Bangkok, Bangkok 10800, Thailand
sa-aat.n@sci.kmutnb.ac.th
[3] Department of Computer Sciences, Rajabhat Uttaradit University,
Uttaradit 53000, Thailand
pisit.nak@uru.ac.th

Abstract. Low wind speed encourages $PM_{2.5}$ (fine particular matter ≤ 2.5 μm) accumulation primarily caused by agricultural burning during the transition from the winter to the summer season in northern Thailand. How to improve the accuracy of wind speed estimation is our motivation behind this study. Herein, the wind speed is estimated by using confidence intervals for the median of a three-parameter lognormal model based on bootstrap-t, percentile bootstrap, normal approximation, and the generalized pivotal quantity (GPQ). Monte Carlo simulation is used to compare our proposed methods in terms of their coverage probabilities and expected lengths. A numerical evaluation shows that the GPQ method performed quite well, even with small sample sizes. The efficacies of our proposed methods are illustrated by using daily wind speed data from Chiang Mai, northern Thailand.

Keywords: Daily wind speed data · Generalized pivotal quantity · Median · Three-parameter lognormal model.

1 Introduction

In meteorology, wind is a fundamental atmospheric characteristic caused by air moving from an area of high pressure toward an area of low pressure in any direction, and the difference between the air pressures and the temperature, among other factors, determines the wind speed [22]. Wind speed is usually affected by season temperature changes while wind direction is in response to the Earth's rotation [4]. Importantly, wind, along with temperature and humidity, are meteorological factors contributing to long-distance air pollutants in northern Thailand [2]. Low wind speed has an important impact on $PM_{2.5}$ (fine particular matter

© Springer Nature Switzerland AG 2022
K. Honda et al. (Eds.): IUKM 2022, LNAI 13199, pp. 329–341, 2022.
https://doi.org/10.1007/978-3-030-98018-4_27

$\leq 2.5 \ \mu m$) accumulation, which markedly increases during the transition from the winter to the summer season. Since the wind speed is primarily determined by the difference between the air pressures of two areas, a low pressure difference results in a low wind speed.

In early 2021, Thais became knowledgeable about the existence of $PM_{2.5}$, which can easily enter the human body and cause several health problems including respiratory illness, allergic symptoms affecting the eyes and nasal passage, etc. High $PM_{2.5}$ levels occur in most regions in Thailand but are especially prevalent in the northern region [23] and, unfortunately, are increasing each . year. Using results from Teerasuphaset and Culp [24], it can be seen that the upper region of Thailand has lower temperatures making cold in some areas in the end of winter, which this has been influenced the northeast monsoon season reducing to a clam wind. Temperature changes under the atmosphere could double the volume of dust floating which provides the gathering of dust, smog and smoke in the atmosphere. At the same time, the $PM_{2.5}$ level progressively occurs during winter to summer seasons as well. Evidence for in support of this position, can be found in Liu et al. [18]. These reasons lead to our motivation to estimate wind speed as it might provide important information on the current trend of $PM_{2.5}$ levels by assessing historical data. Importantly, daily wind speed data from Chiang Mai (the largest city in northern Thailand) follow a three-parameter lognormal (TPLN) model.

A TPLN distribution is a statistical model suitable for highly right-skewed data that cannot specifically be fitted to a lognormal distribution [1]. Its three parameters (scale, shape, and threshold) include the threshold parameter defined as the lower bound of the data, which is not in the lognormal model. Thus, the TPLN and lognormal models are the same when the threshold parameter value equals zero. Furthermore, the median of the TPLN model used to determine the central tendency of highly skewed data is a parameter of interest for probability and statistical inference.

Statistical inference includes point and interval estimations (the latter is also known as the confidence interval (CI)). Both point and interval estimates for the parameters of a TPLN model have been developed and discussed by a few researchers. Cohen et al. [9] modified moment estimates by replacing the function of the first-order statistic in the third moment. Singh et al. [21] conducted a performance evaluation of several methods: the method of moments (MMs), modified MMs, maximum likelihood estimation (MLE), modified MLE, and the entropy for the parameter and quantity of a TPLN model. Royston [20] constructed CIs for the reference range of random samples from a three-parameter lognormal distribution. Pang et al. [19] used a simulation-based approach to assess Bayesian CIs for the coefficient of variation of a TPLN distribution as one of their studied distributions of interest. Finally, Chen and Miao [7] presented exact CIs and upper CIs for the threshold parameter of a TPLN model.

However, interval estimates for the median of a TPLN model have not yet been formulated. Herein, we propose bootstrap-t, percentile bootstrap, normal approximation (NA), and generalized pivotal quantity (GPQ) methods for

constructing CIs for the median of a TPLN model. Furthermore, their efficacies are illustrated by estimating the wind speed during the transition from the
winter to the summer season in Chiang Mai, northern Thailand. The outline
of the article is as follows. The TPLN model and our proposed methods for
constructing interval estimates for the TPLN median are elaborated in Sect. 2.
The accuracies of the CI estimation methods numerically assessed via Monte
Carlo simulation are discussed in Section 3. Later, the efficacies of the proposed
method are illustrated by using real wind speed data from Chiang Mai, northern
Thailand, in Sect. 4. Some concluding remarks are provided in Sect. 5.

2   Model and Methods

2.1   Model

Assume that $W = (W_1, W_2, ..., W_n)$ be identically and independent random variable from a three-parameter lognormal (TPLN) distribution (model) with scale
parameter μ_W, shape parameter σ_W^2 and threshold parameter θ. The relationship between the random variables W and $X = \ln(W - \theta)$ is the random variable
W has a TPLN model, denoted by $W \sim TLN(\theta, \mu_W, \sigma_W^2)$ if X has a normal
model, denoted by $X = \ln(W - \theta) \sim N(\mu_X, \sigma_X^2)$. The probability distribution
function of W is given by

$$f(w; \theta, \mu_W, \sigma_W^2) = \left[(x - \theta)\sqrt{2\pi\sigma_W^2} \right]^{-1} \exp\left\{ -\left[\ln(w - \theta) - \mu_X\right]^2 / (2\sigma_X^2) \right\} \quad (1)$$

where $\mu_X = E(X)$ and $\sigma_X^2 = Var(X)$. Thus, the median of W is

$$M_W = \theta + \exp(\mu_X) \quad (2)$$

which is the parameter of interest in the present study. Accordance with Griffiths
[12], the idea is skewness as a point for consideration, while the random variable
$X = \ln(W - \theta)$ is a normal distribution (the skewness of X is equal to zero) if
θ is known. Thus, the zero-skewness estimate of θ is the value that satisfying

$$\frac{(n)^{-1} \sum_{i=1}^{n} [\ln(W_i - \hat{\theta}) - (n)^{-1} \sum_{i=1}^{n} \ln(W_i - \hat{\theta})]^3}{\left\{ (n)^{-1} \sum_{i=1}^{n} [\ln(W_i - \hat{\theta}) - (n)^{-1} \sum_{i=1}^{n} \ln(W_i - \hat{\theta})]^2 \right\}^{3/2}} = 0 \quad (3)$$

Also, this leads to obtain the zero-skewness estimates of (μ_X, σ_X^2) is

$$(\hat{\mu}_X, \hat{\sigma}_X^2) = \left((n)^{-1} \sum_{i=1}^{n} \ln(W_i - \hat{\theta}), (n - 1)^{-1} \sum_{i=1}^{n} [\ln(W_i - \hat{\theta}) - \hat{\mu}_X]^2 \right) \quad (4)$$

The interval estimations for M_W are established based on the different concepts,
detailed in the Sect. 2.2.

2.2   Methods

To estimate the median of a TPLN model, the methods are constructed the CIs based on different concepts as follows: bootstrap, normal approximation and generalized pivotal quantity methods.

Bootstrap Method. It is well-known that bootstrap method is one of the resampling techniques to estimate parameters on a distribution or statistics on a population by sampling with replacement. Efron and Tibshirani [11] was firstly introduced the bootstrap method, and after that it was recovered by Hall [13] It can be used to construct an asymmetric CI which does not depend on normal theory assumptions. There are the bootstrap intervals: bootstrap-t and percentile bootstrap based on calculating the probability distribution of the bootstrap replications. Then, both of bootstrap intervals are considered. First, let $W = (W_1, W_2, ..., W_n)$ be a random sample from a TPLN distribution function $F(W; \theta, \mu_X, \sigma_X^2)$, while the median $M_W = \theta + \exp(\mu_X)$ be a parameter of interest and σ_X^2 be a nuisance parameter. Given observed values $w = (w_1, w_2, ..., w_n)$, let $\widehat{M_X}$ be the estimate of M_W. In the boostrap world, the empirical distribution $\widehat{F}(W; \theta, \mu_X, \sigma_X^2)$ provides bootstrap samples with sample size n, that are

$$w_1^* = (w_{11}, w_{12}, ..., w_{1n}) \tag{5}$$
$$w_2^* = (w_{21}, w_{22}, ..., w_{2n})$$
$$...$$
$$w_b^* = (w_{b1}, w_{b2}, ..., w_{bn})$$
$$...$$
$$w_B^* = (w_{B1}, w_{B2}, ..., w_{Bn})$$

where $b = 1, 2, 3, ..., B$. A random variable is

$$t_b^* = \widehat{se}^{-1}(\widehat{M_{X,b}^*} - \widehat{M_W^*}) \sim t_{df} \tag{6}$$

which has a Student's t-distribution with $df = b - 1$ degree of freedom where

$$\widehat{M_{W,b}^*} = n^{-1} \sum_{i=1}^{n} w_{bi}^* \tag{7}$$

$$\widehat{se} = \left\{ (B-1)^{-1} \sum_{i=1}^{B} [\widehat{M_{W,b}^*} - \widehat{M_W^*}]^2 \right\}^{1/2} \tag{8}$$

which are the estimated median for the bth bootstrap replication and the standard error of $\widehat{M_{W,b}^*}$, respectively, where $\widehat{M_W^*} = B^{-1} \sum_{b=1}^{B} \widehat{M_{W,b}^*}$ be the estimated TPLN median based on bootstrap samples. Define the γth quantile (q_γ) as

$$\frac{\# \{t_b^* \leq q_\gamma\}}{B} = \gamma \tag{9}$$

The coverage probability is considered as

$$
\begin{aligned}
1 - \gamma &= P(q_{\gamma/2} < t_b^* < q_{1-\gamma/2}) \\
&\approx P(q_{\gamma/2} < t < q_{1-\gamma/2}) \\
&= P\left(q_{\gamma/2} < \widehat{se}^{-1}(\widehat{M}_X^* - M_W) < q_{1-\gamma/2}\right) \\
&= P\left(\widehat{M}_X^* - q_{1-\gamma/2}\widehat{se} < M_W < \widehat{M}_X^* - q_{\gamma/2}\widehat{se}\right)
\end{aligned}
\tag{10}
$$

The $100(1 - \gamma)\%$ bootstrap-t (BT) interval for M_W is

$$
[L_{BT}, U_{BT}] = \left[\widehat{M}_W^* - q_{1-\gamma/2}\widehat{se}_B, \widehat{M}_W^* - q_{\gamma/2}\widehat{se}_B\right]
\tag{11}
$$

Furthermore, the percentile bootstrap (PB) interval for M_W is

$$
[L_{PB}, U_{PB}] = \left[\widehat{M}_W^*(\gamma/2), \widehat{M}_W^*(1 - \gamma/2)\right]
\tag{12}
$$

where $\widehat{M}_W^*(\gamma)$ be the γth percentile of $\widehat{M}_W^*$.

Normal Approximation Method. In the present study, we are also interested in constructing a CI using a normal model to approximate the TPLN median M_W. The central limit theorem (CLT) is statistical tools that allows us to attempt an asymptomatic distribution of a random variable when a sample size was large. Using the normal approximation method, let $W = (W_1, W_2, ..., W_n)$ be an independent random variable of a TPLN distribution with the parameters $(\theta, \mu_W, \sigma_W^2)$. Using the logarithm transformation, a random variable $X = \ln(W - \theta)$ is a normal distribution with the parameter (μ_X, σ_X^2). Here $M_W = \theta + \exp(\mu_X)$ is a parameter of interest, more importantly it needs for statistical estimations. The target parameter is suggested plug-in its estimate as

$$
\widehat{M}_W = \hat{\theta} + \exp(\hat{\mu}_X)
\tag{13}
$$

where $(\hat{\theta}, \hat{\mu}_X)$ are the zero-skewness estimate. According to Hollander and Wolfe [16], the asymptotic variance of $\widehat{M}_W$ is defined a distribution-free of $Var(\widehat{M}_X)$ as

$$
Var_H(\widehat{M}_W) = \left[\frac{n^{3/10}}{max\{1, A\}}\right]
\tag{14}
$$

where $A = \sum_{i=1}^{n} a_i$;

$$
a_i = \begin{cases} 1, & if \quad \left\{X_i \middle| \left[\widehat{M}_W - n^{-1/5} < X_i < \widehat{M}_W + n^{-1/5}\right]\right\} \\ 0, & if \quad otherwise \end{cases}
\tag{15}
$$

From (14) and (15), the random variable $G = [Var_H(\widehat{M}_W)]^{-1/2}(\widehat{M}_W - M_W)$ has a standard normal distribution using CLT, then the coverage probability is considered as

$$1 - \gamma = P(g_{\gamma/2} < G < g_{1-\gamma/2}) \tag{16}$$
$$= P\left(g_{\gamma/2} < [Var_H(\widehat{M_W})]^{-1/2}(\widehat{M_W} - M_W) < g_{1-\gamma/2}\right)$$
$$= P\left(\widehat{M_W} - g_{1-\gamma/2}[Var_H(\widehat{M_W})]^{1/2} < M_W < \widehat{M_W} - g_{\gamma/2}[Var_H(\widehat{M_W})]^{1/2}\right)$$

Then, the normal approximation interval-based the zero-skewness estimate is

$$[L_N, U_N] = \left[\widehat{M} - g_{1-\gamma/2}[Var_H(\widehat{M_W})]^{1/2}, \widehat{M}_X - g_{\gamma/2}[Var_H(\widehat{M_W})]^{1/2}\right] \tag{17}$$

where g_γ denotes the γth percentile of a standard normal distribution.

Generalized Pivotal Quantity Method. A pivotal quantity is defined as a random variable whose distribution does not depend on the parameter of interest, and it can be expressed in terms of a quantity. This quantity is also called as "pivots". Barnard [5] used a pivotal quantity for constructing a confidence interval, called as "pivotal inference". A pivotal quantity is defined by Casella and Berger [6] in Definition 1.

Definition 1. *A random variable* $Q(\boldsymbol{X}, \tau) = Q(X_1, X_2, ..., X_n, \tau)$ *is a pivot quantity (or pivot) if the distribution of* $Q(\boldsymbol{X}, \tau)$ *is independent of all parameters. That is, if* $\boldsymbol{X} \sim F(x|\tau)$, *then* $Q(\boldsymbol{X}, \tau)$ *has the same distribution for all values of* τ.

Later, the pivotal quantity was generalized by Weerahandi [25] based on a pivot inference and defined in Definition 2. This was proven to be an important tools in many practical problems, and importantly its simulation studies could provide the coverage probability of such intervals close to a target values as well, evidence in Hannig [14].

Definition 2. *Let* $\boldsymbol{S} \in \boldsymbol{R}^k$ *denote an observable random vector whose distribution is indexed by a (possibly vector) parameter* $\xi \in \boldsymbol{R}^p$. *Suppose that* $\tau = \pi(\xi) \in \boldsymbol{R}^q$ *be a parameter of interest in making inference;* $q \geq 1$. *Let* $\boldsymbol{S}^*$ *represent an independent copy of* $\boldsymbol{S}$. *Also, let* $\boldsymbol{s}$ *and* $\boldsymbol{s}^*$ *denote realized values of* $\boldsymbol{S}$ *and* $\boldsymbol{S}^*$. *A GPQ for* τ, *denoted by* $R_\tau(\boldsymbol{S}, \boldsymbol{S}^*, \xi)$ *(or simply* R_τ *or* R *when there is no ambiguity) is a function of* $(\boldsymbol{S}, \boldsymbol{S}^*, \xi)$ *with the following properties:*

(Property 1) Conditional on $\boldsymbol{S} = \boldsymbol{s}$, *the conditional distribution* $R_\tau(\boldsymbol{S}, \boldsymbol{S}^*, \xi)$ *is free of* ξ.

(Property 2) For every allowable $\boldsymbol{s} \in \boldsymbol{R}^k$, $R_\tau(\boldsymbol{s}, \boldsymbol{s}^*, \xi)$ *depends on* ξ *only through* τ.

If a GPQ satisfies the properties (1) and (2), then the CI-based GPQ is $[L, U] = [R_{\tau, \gamma/2}(\boldsymbol{S}, \boldsymbol{S}^*, \xi), R_{\tau, 1-\gamma/2}(\boldsymbol{S}, \boldsymbol{S}^*, \xi)]$ where $R_{\tau, \gamma}(\boldsymbol{S}, \boldsymbol{S}^*, \xi)$ denotes the γth percentile of $R_\tau(\boldsymbol{S}, \boldsymbol{S}^*, \xi)$. Here $M_W = \theta + \exp(\mu_X)$ is interested in constructing CI-based GPQ. It can be seen that there are two unknown parameters: θ and μ_X, thus their GPQs are considered. Cohen and Whitten [8] produced the point estimates of $(\theta, \mu_X, \sigma_X^2)$ using the modified method of moments, denoted by $(\hat{\theta}^{(C)}, \hat{\mu}_X^{(C)}, \hat{\sigma}_X^{2(C)})$. The asymptotic variance of $\hat{\theta}^{(C)}$ is given by

$$Var(\hat{\theta}^{(C)}) = (\sigma_X^2/n)\left[\exp(\sigma_X^2)(1+\sigma_X^2) - (1+2\sigma_X^2)\right]^{-1}\left[\exp(\mu_X + \sigma_X^2/2)\right] \quad (18)$$

where $\hat{\theta}^{(C)}$ has a property of random variable, while $\hat{\mu}_X^{(C)} = (n)^{-1}\sum_{i=1}^{n}\ln(W_i - \hat{\theta}^{(C)})$ and $\hat{\sigma}_X^{2(C)} = (n-1)^{-1}\sum_{i=1}^{n}[\ln(W_i - \hat{\theta}^{(C)}) - \hat{\mu}_X^{(C)}]^2$ are obtained. Replacing $(\hat{\theta}^{(C)}, \hat{\mu}_X^{(C)}, \hat{\sigma}_X^{2(C)})$ from the sample, the $\widehat{Var}(\hat{\theta}^{(C)})$ is obtained. Using the CLT, the random variable is $V^{(C)} = [\widehat{Var}(\hat{\theta}^{(C)})]^{-1/2}(\hat{\theta}^{(C)} - \theta) \sim N(0,1)$ leading to obtain the GPQ of θ is

$$R_\theta(\mathbf{W}, \mathbf{W}^*, \mu_X, \sigma_X^2) = \hat{\theta}^{(C)} - V^{(C)}[\widehat{Var}(\hat{\theta}^{(C)})]^{1/2} \quad (19)$$

Next, the GPQ of μ_X is obtained from the random variable $T = \sqrt{n}(\hat{\mu}_X^{(C)} - \mu_X)/\sigma_X \sim N(0,1)$ using the CLT, then

$$R_{\mu_X}(\mathbf{W}, \mathbf{W}^*, \theta, \sigma_X^2) = \hat{\mu}_X^{(C)} - T(\sigma_X/\sqrt{n}) \quad (20)$$

where $\sigma_X = [(n-1)\hat{\sigma}_X^{2(C)}/U]^{1/2}$; $U = [(n-1)\hat{\sigma}_X^{2(C)}/\sigma^2]$ is a chi-square distribution with $n-1$ degrees of freedom. From the two pivots (19) and (20), the quantity of M_W in term of GPQ is $R_{M_W}(\mathbf{W}, \mathbf{W}^*, \theta, \mu_X, \sigma_X^2) = R_\theta(\mathbf{W}, \mathbf{W}^*, \mu_X, \sigma_X^2) + \exp[R_{\mu_X}(\mathbf{W}, \mathbf{W}^*, \theta, \sigma_X^2)]$, and then

$$[L_G, U_G] = \left[R_{M_W, \gamma/2}(\mathbf{W}, \mathbf{W}^*, \theta, \mu_X, \sigma_X^2), R_{M_W, 1-\gamma/2}(\mathbf{W}, \mathbf{W}^*, \theta, \mu_X, \sigma_X^2)\right] \quad (21)$$

which is the $100(1 - \gamma)\%$ CI-based GPQ for M_W; the $R_{M_W, \gamma}(\mathbf{W}, \mathbf{W}^*, \theta, \mu_X)$ stands for the γth percentile of $R_{M_W}(\mathbf{W}, \mathbf{W}^*, \theta, \mu_X)$.

3   Details and Results of the Monte Carlo Simulation Study

To measure the performance of the proposed methods derived in the previous section, their coverage probabilities (CPs) and expected lengths (ELs) are calculated for several realistic values of the parameters: threshold ($\theta = 3, 6, 10$); mean ($\mu_X = 1, 2$); variance ($\sigma_X^2 = 0.5, 1, 2$); and and small-to-large sample size ($n = 30, 60, 100, 200$). In practical applications, the values of σ_X^2 are usually small, and so realistically small values were used in the simulation study. Note that σ_X^2 is a statistical measure of the dispersion of the data.

In the Monte Carlo simulation, repeated random sampling fixed at 5000 iterative simulations was used to estimate the CPs and ELs of the four CI estimation methods. In addition, the number of bootstrap samples was set as 1000 ($B = 1000$), while the GPQs were set at as 2500 ($m = 2500$). For each parameter combination in Tables 1 and 2, the best-performing method is the one that provided a CP close or greater than the nominal confidence level $100(1 - \gamma) = 0.95$ with the narrowest EL.

The median CI estimates were evaluated to determine which gave suitable CPs with the most efficient (narrowest) interval width. The results of the simulation study provide insight into the sampling behavior of the CIs as follows. Although the median CI estimates based on GPQ and NA fulfilled the CP criteria, the GPQ-based CI method produced the shortest interval estimates for a small variance and small-to-moderate sample sizes (Table 1). When the sample size was increased (Table 2), the NA-based median CI estimate gave suitable CPs with the narrowest ELs. On the other hand, both the bootstrap-t and percentile bootstrap intervals revealed poor CPs.

Table 1. Performance measures of the CIs for $M_W : n = 30, 60$

	(σ_X^2, μ_X)	θ	CPs				ELs			
			GPQ	NA	BT	BP	GPQ	NA	BT	BP
30	(1,1)	3	0.956	0.940	0.359	0.374	**2.611**	3.122	0.569	0.598
		6	0.965	0.941	0.355	0.366	**2.598**	3.070	0.564	0.573
		10	0.958	0.932	0.357	0.363	**2.640**	3.091	0.550	0.565
	(1,2)	3	0.966	0.914	0.361	0.382	**7.099**	7.606	1.604	1.923
		6	0.966	0.920	0.356	0.375	**7.070**	7.580	1.611	1.919
		10	0.963	0.920	0.353	0.364	**7.164**	7.572	1.599	1.907
	(2,1)	3	0.743	0.927	0.373	0.392	3.456	4.560	0.770	0.809
		6	0.758	0.931	0.370	0.386	3.428	4.521	0.768	0.806
		10	0.751	0.933	0.372	0.389	3.465	4.564	0.772	0.811
	(2,2)	3	0.758	0.888	0.387	0.411	9.313	8.581	2.075	2.179
		6	0.755	0.895	0.377	0.397	9.331	8.635	2.086	2.191
		10	0.758	0.888	0.379	0.399	9.328	8.655	2.076	2.180
60	(1,1)	3	0.935	0.949	0.495	0.511	**1.639**	1.975	0.518	0.536
		6	0.938	0.952	0.503	0.519	**1.648**	1.988	0.523	0.542
		10	0.939	0.952	0.509	0.522	**1.643**	1.978	0.519	0.538
	(1,2)	3	0.940	0.936	0.510	0.526	**4.469**	6.322	1.408	1.458
		6	0.941	0.940	0.512	0.527	**4.455**	6.260	1.410	1.461
		10	0.941	0.933	0.514	0.528	**4.467**	6.312	1.418	1.469
	(2,1)	3	0.652	0.949	0.509	0.531	2.264	3.110	0.737	0.773
		6	0.664	0.952	0.511	0.515	2.272	3.077	0.734	0.768
		10	0.651	0.947	0.513	0.527	2.265	3.053	0.729	0.759
	(2,2)	3	0.656	0.937	0.510	0.534	6.170	8.349	2.006	2.105
		6	0.658	0.920	0.513	0.537	6.140	8.299	2.002	2.096
		10	0.660	0.928	0.515	0.540	6.128	8.285	1.990	2.050

Remark: Bold denotes the best-performing method.

Table 2. Performance measures of the CIs for M_W : $n = 100, 200$

n	(σ_X^2, μ_X)	θ	CPs				ELs			
			GPQ	NA	BT	BP	GPQ	NA	BT	BP
100	(0.5,1)	3	0.988	0.969	0.631	0.643	1.251	**0.994**	0.364	0.373
		6	0.986	0.965	0.619	0.627	1.256	**0.994**	0.363	0.372
		10	0.987	0.967	0.607	0.620	1.259	**0.993**	0.364	0.373
	(0.5,2)	3	0.987	0.954	0.621	0.631	3.383	**3.159**	0.986	1.011
		6	0.988	0.953	0.621	0.633	3.401	**3.150**	0.990	1.014
		10	0.986	0.956	0.610	0.622	3.432	**3.159**	0.986	1.010
	(1,1)	3	0.902	0.962	0.608	0.624	1.225	**1.461**	0.514	0.532
		6	0.904	0.956	0.619	0.634	1.221	**1.448**	0.512	0.530
		10	0.913	0.962	0.624	0.639	1.224	**1.460**	0.514	0.532
	(1,2)	3	0.916	0.944	0.621	0.637	3.326	**4.800**	1.395	1.444
		6	0.902	0.946	0.618	0.634	3.317	**4.901**	1.393	1.442
		10	0.916	0.945	0.618	0.635	3.322	**4.730**	1.394	1.443
	(2,1)	3	0.594	0.958	0.603	0.639	1.691	**2.182**	0.709	0.745
		6	0.604	0.959	0.615	0.643	1.686	**2.212**	0.711	0.743
		10	0.582	0.961	0.628	0.649	1.694	**2.197**	0.708	0.742
	(2,2)	3	0.604	0.939	0.610	0.633	4.599	**7.040**	1.969	2.064
		6	0.593	0.943	0.608	0.635	4.598	**7.111**	1.966	2.062
		10	0.593	0.938	0.605	0.638	4.606	**6.996**	1.963	2.060
200	(0.5,1)	3	0.982	0.974	0.778	0.791	0.842	**0.688**	0.360	0.369
		6	0.984	0.973	0.778	0.788	0.849	**0.687**	0.361	0.370
		10	0.984	0.969	0.784	0.796	0.842	**0.684**	0.360	0.369
	(0.5,2)	3	0.981	0.959	0.786	0.798	2.298	**2.011**	0.981	1.005
		6	0.982	0.963	0.791	0.804	2.295	**2.001**	0.981	1.005
		10	0.981	0.962	0.782	0.792	2.287	**2.001**	0.978	1.002
	(1,1)	3	0.868	0.969	0.781	0.795	0.837	**0.991**	0.507	0.525
		6	0.873	0.967	0.776	0.795	0.837	**0.990**	0.507	0.524
		10	0.875	0.969	0.789	0.805	0.837	**0.988**	0.507	0.525
	(1,2)	3	0.864	0.960	0.788	0.804	2.274	**3.017**	1.379	1.427
		6	0.875	0.962	0.783	0.802	2.278	**3.026**	1.382	1.430
		10	0.871	0.958	0.787	0.802	2.271	**3.038**	1.378	1.426
	(2,1)	3	0.500	0.966	0.785	0.797	1.156	**1.433**	0.712	0.747
		6	0.510	0.967	0.787	0.802	1.155	**1.426**	0.711	0.745
		10	0.503	0.969	0.781	0.813	1.159	**1.443**	0.708	0.741
	(2,2)	3	0.512	0.954	0.770	0.798	3.138	**4.795**	1.937	2.031
		6	0.513	0.953	0.782	0.806	3.137	**4.743**	1.930	2.026
		10	0.508	0.957	0.787	0.803	3.146	**4.718**	1.929	2.024

Remark: Bold denotes the best-performing method.

4   Application of the Methods to Wind Speed Data

$PM_{2.5}$ is the most common air pollutant in northern Thailand that significantly affects human health. Wind speed is an important determinant of $PM_{2.5}$ level along with rainfall and relative humidity in the winter and the summer season [15]. Amnuaylojaroen et al. [3] argued that monsoons characterize the climate of northern Thailand, and air pollutants are transported by wind from the surrounding area into northern Thailand during the transition period between the Northeast and Southwest monsoons.

Chiang Mai, the second-largest city in Thailand, is in the upper northern region in which $PM_{2.5}$ emitted from agriculture burning in the early year is particularly high. Hence, estimating wind speed data together with $PM_{2.5}$ concentration measurements is needed to aid decision-making by the Thai authorities. The daily wind speed data in Chiang Mai used in the study were recorded during January–March 2021 [10]. It can be concluded that the wind speed data fit a TPLN model, as evidenced by the probability density plot in Fig. 1 and the minimum Akaike and Bayesian information criteria (AIC and BIC) data in Table 3. Note that the wind speed data can be lognormally distributed if they are subtracted by $\hat{\theta}$, the evidence for which is provided in Sect. 2.

The basic statistics calculated for the daily wind speed data log-transformed measurements are $n = 93$, $\hat{\theta} = 3.057$, $\hat{\mu}_X = 2.567$. Thus, the zero-skewness estimate in terms of the median is 16.0867 km/hr, and the 95% CIs for this based on the bootstrap-t, percentile bootstrap, NA, and GPQ methods were computed via Eqs. (11), (12), (17) and (21), respectively (Table 4). It can be interpreted that the daily wind speed was quite low in Chiang Mai during January–March 2021, which resulted in very high $PM_{2.5}$ concentrations. This was due to agricultural burning being carried out earlier than usual because farmers wanted to plant their crops earlier to avoid drought conditions as a result of climate change [17]. Notably, these results follow the outcomes of the simulation study in the preceding section.

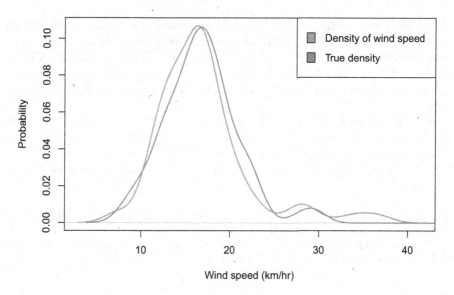

Fig. 1. Probability density plot between true (a three-parameter lognormal) and wind speed densities.

Table 3. Results of AIC and BIC for the wind speed data subtracted by $\hat{\theta}$.

Criteria	Models					
	Exponential	Weibull	Lognormal	Logistic	Normal	Cauchy
AIC	691.4882	575.0861	**549.5256**	558.7382	575.1305	560.3573
BIC	694.0208	580.1513	**554.5908**	563.8034	580.1957	565.4225

Note: a bold indicates an suitable model for the data.

Table 4. The 95% CIs for the speed wind data during January–March 2021 in Chiang Mai, northern Thailand.

CIs	CI-based GPQ	Normal approximation	Percentile bootstrap	Bootstrap-t
Lower	6.402	8.452	15.729	15.717
Upper	25.803	23.721	16.462	16.462
Lengths	19.401	15.269	0.733	0.745

5   Concluding Remarks

Four methods for CI estimation for the median of a TPLN model are proposed in this article. Two are based on bootstrap sampling (the bootstrap-t and percentile boostrap CIs), one is an approximation of the normal model, and the last one is the GPQ proposed by Weerahandi [25]. The bootstrap methods failed to produce accurate results, and so the NA and GPQ methods should be used to estimate the median of a TPLN model. In conclusion, the results indicate that the CI based on GPQ is the best-performing method for a small variance and small-to-moderate sample sizes. For a large sample size, the best-performing CI was based on NA for estimating the TPLN median. The latter also has the advantage that it can be computed in a straightforward manner using the R programming language.

Finally, it should be noted that none of the methods can handle extreme situations such as a large variance, as evidenced by the simulation results. In future research, we might explore new interval estimates of the median in a TPLN model that provides good performance for a population with a large variation.

Acknowledgements. The authors are grateful to the reviewers for their valuable comments and suggestions which help to improve this manuscript. This work was financially supported by Office of the Permanent Secretary, Ministry of Higher Education, Science, Research and Innovation. Grant No. RGNS 64–196.

References

1. Aitchison, J., Brown, J.A.: The Lognormal Distribution: With Special Reference to Its Uses in Economics. Cambridge University Press, Cambridge (1963)
2. Amnuaylojaroen, T.: Investigation of fine and coarse particulate matter from burning areas in Chiang Mai, Thailand using the WRF/CALPUFF. Chiang Mai J. Sci. **39**(2), 311–326 (2012)
3. Amnuaylojaroen, T., Inkom, J., Janta, R., Surapipith, V.: Long range transport of Southeast Asian PM2.5 pollution to Northern Thailand during high biomass burning episodes. Sustainability **12**(23), 10049 (2020)
4. Barber, D.: The Four Forces That Influence Wind Speed & Wind Direction (2019). https://sciencing.com/list-7651707-four-wind-speed-wind-direction.html
5. Barnard, G.A.: Pivotal inference and the Bayesian controversy. Trabajos de Estadistica Y de Investigacion Operativa **31**(1), 295–318 (1980)
6. Casella, G., Berger, R.L.: Statistical Inference, 2nd edn. Duxbury, Pacific Grove (2002)
7. Chen, Z., Miao, F.: Interval and point estimators for the location parameter of the three-parameter lognormal distribution. Int. J. Qual. Stat. Reliab. **2012**, 1–6 (2012)
8. Cohen, A.C., Whitten, B.J.: Estimation in the three-parameter lognormal distribution. J. Am. Stat. Assoc. **75**(370), 399–404 (1980)
9. Cohen, A.C., Whitten, B.J., Ding, Y.: Modified moment estimation for the three-parameter lognormal distribution. J. Qual. Technol. **17**(2), 92–99 (1985)
10. Thai Meteorological Department: Daily Weather Summary (2021). https://www.tmd.go.th/en/climate.php?FileID=1
11. Efron, B., Tibshirani, R.J.: An Introduction to the Bootstrap. Chapman & Hall/CRC, Boca Raton (1993)
12. Griffiths, D.A.: Interval estimation for the three-parameter lognormal distribution via the likelihood function. Appl. Stat. **29**(1), 58 (1980). https://doi.org/10.2307/2346411

13. Hall, P.: On symmetric bootstrap confidence intervals. J. R. Stat. Soc. Ser. B (Methodol.) **50**(1), 35–45 (1988)
14. Hannig, J., Iyer, H., Patterson, P.: Fiducial generalized confidence intervals. J. Am. Stat. Assoc. **101**(473), 254–269 (2006)
15. Hien, P.D., Bac, V.T., Tham, H.C., Nhan, D.D., Vinh, L.D.: Influence of meteo-rological conditions on PM2.5 and PM2.5-10 concentrations during the monsoon season in Hanoi, Vietnam. Atmos. Environ. **36**(21), 3473–3484 (2002)
16. Hollander, M., Wolfe, D.A.: Introduction to Nonparametric Statistics. Wiley, Hoboken (1973)
17. IQAir: New data exposes Thailand's 2021 "burning season" (2021). https://www.iqair.com/blog/air-quality/thailand-2021-burning-season
18. Liu, Z., Shen, L., Yan, C., Du, J., Li, Y., Zhao, H.: Analysis of the influence of precipitation and wind on PM2.5 and PM10 in the atmosphere. Adv. Meteorol. **2020**, 1–13 . https://doi.org/10.1155/2020/5039613. https://www.hindawi.com/journals/amete/2020/5039613/
19. Pang, W.K., Leung, P.K., Huang, W.K., Liu, W.: On interval estimation of the coefficient of variation for the three-parameter Weibull, lognormal and gamma distribution: a simulation-based approach. Eur. J. Oper. Res. **164**(2), 367–377 (2005)
20. Royston, P.: Estimation, reference ranges and goodness of fit for the three-parameter log-normal distribution. Stat. Med. **11**(7), 897–912 (1992)
21. Singh, V., Cruise, J., Ma, M.: A comparative evaluation of the estimators of the three-parameter lognormal distribution by Monte Carlo simulation. Comput. Stat. Data Anal. **10**(1), 71–85 (1990)
22. National Geographic Society: Wind (2012). http://www.nationalgeographic.org/encyclopedia/wind/
23. Tanraksa, P., Kendall, D.: Air quality in North still 'critical'. Bangkok Post (2020). https://www.bangkokpost.com/thailand/general/1888915/air-quality-in-north-critical
24. Teerasuphaset, T., Culp, J.: Stay Safe in the PM 2.5. Chulalongkorn University (2020). https://www.chula.ac.th/en/news/26593/
25. Weerahandi, S.: Generalized confidence intervals. J. Am. Stat. Assoc. **88**(423), 899–905 (1993)

Trust Uncertainty Modeling in Agri-Food Logistic Decision Making

Rindra Yusianto[1] (ID), Suprihatin Suprihatin[2] (ID), Hartrisari Hardjomidjojo[2] (ID),
and Marimin Marimin[2(✉)] (ID)

[1] Department of Industrial Engineering, Faculty of Engineering, Dian Nuswantoro University, Semarang 53192, Indonesia
[2] Department of Agro-Industrial Technology, Faculty of Agricultural Engineering and Technology, IPB University (Bogor Agricultural University), Bogor 16002, Indonesia
marimin@ipb.ac.id

Abstract. In agri-food logistics, group decision-makers are faced with high risk and uncertainty involving many criteria. They must have a good level of trust with an amount of risk consistent with their level of risk acceptance. In this study, we propose a new model of trust uncertainty built from the trust component based on the principles of uncertainty theory. We developed a trust level scale by calculating each trust component to calculate the trust level. A new method was developed by advancing ME-MCDM by measuring the trust of experts as decision-makers to measure the reliability of group decision results. We adopt a non-numeric approach and propose four alternative logistics routes with five conflicting criteria, i.e., distance, utilization, logistic cost, transportation condition, and traffic. The results showed that this new method obtained the most optimal route with a high aggregation value. The trust uncertainty modeling proposed was successfully used to measure the trust level of group decision-makers with a trust value of 1.7, which indicates reliability. Finally, by comparing this new model with commonly used approaches, it is shown that this new model has advantages over AHP.

Keywords: Agri-food logistic · Decision making · Trust uncertainty modeling

1 Introduction

An essential element of agri-food logistics is the trust between all members and behaviors in all fields of business [1]. Trust is an essential element for agri-food logistics because it provides a necessary basis for work distribution activities and the establishment of new methods for directing the work of organizational members. Trust is an essential aspect of any relationship that can improve the quality of a connection [2]. Reducing transaction costs, reducing the need to write complicated and difficult to enforce contracts between organizations, and increase agri-food logistics performance [3].

The definition of trust, in general, is a condition related to vulnerability and risk which consist of an invention to accept exposure based on the desires and expectations of the behaviors of others [4, 5]. Consumer trust needs to be made, and it has become an

© Springer Nature Switzerland AG 2022
K. Honda et al. (Eds.): IUKM 2022, LNAI 13199, pp. 342–354, 2022.
https://doi.org/10.1007/978-3-030-98018-4_28

important goal. They have efforts to protect trust in agri-food logistics and regard trust as a problem in informing risk. Agri-food logistics should verify the trustworthiness of a particular identity and decide how much trust it will place on the verification. Trust in agr0-food logistics is the willingness of one party to be vulnerable to the behaviors and activities of other parties.

Trust and trustworthiness in agri-food logistics are divided into two levels [6]. First, studies related to horizontal collaboration are reviewed and discussed, focusing on the objectives to be achieved and the existing trust model. Agri-food logistics performance may be unique and usually differ for each organization, reflecting the objectives and surrounding environment [7]. Second, the organizational theory that investigates inter-actions between companies is analyzed to determine how the key can contribute to the topic of choice. The agri-food logistics model consists of many layers. Agri-food logistics obtain raw materials from farmers or collectors. The agri-food logistics demand at the Distribution Center (DC) node usually follows a normal distribution [8]. In agri-food logistics, farmers will send goods to the agroindustry nearer, where farmers have agro-industrial preferences or interests. The trust model measured expert reliability to increase trust in expert decisions. The trust concept has received significant attention in the technical research community because trust is the basis for decision-making in many contexts. The trust model is used to measure the trust level of the experts who make uncertainty decisions. In logistics, the decision-makers' trust plays an essential role [9]. Decision results will be a reference if the decision-maker can be trusted [10].

Previous studies have discussed the multi-expert multi-criteria decision making (ME-MCDM) method and fuzzy logic. In making multi-criteria decisions regarding agri-food logistics, they have used the non-numeric preferences of many experts. However, they did not measure experts' trust as group decision-makers, so they could not measure the reliability of their decision results. In this study, we propose a new model of trust uncertainty built from the trust component and based on the principles of uncertainty theory by measuring the level of expert trust as decision-makers. To calculate the trust level, we developed a trust level scale by calculating each trust component. We can measure the reliability of the results of group decisions. We synergize the experts as group decision-makers using the non-numeric ME-MCDM. Many previous studies used this method, but they did not discuss the uncertainty trust level, so they could not measure the reliability level of the decision. We develop uncertainty trust levels by categorizing them into trust, moderate, and distrust.

2 Background

This section provides an overview of the related concepts of the multi-criteria, trust model, and uncertainty theory.

2.1 Definition 1: Multi-criteria

ME-MCDM can determine the best alternative results that are relatively complex and uncertainty. MCDM is a part of operations research that refers to decision-making for several uncertainty and conflicting criteria. Zadeh introduced fuzzy sets that paved the way for new models to solve challenges with standard MCDM techniques [7]. Correspondingly, group decision-making for many alternatives with many criteria is solved using a non-numeric approach. Experts independently evaluated each option on each standard. Therefore, ME-MCDM can be used [11, 12]. A non-numeric approach can evaluate using multi-criteria preferences. Each decision-maker (E_j) $(j = 1, 2, ..., m)$ can assess each alternative (A_i) $(I = 1, 2, ..., n)$ in each criterion (C_k) $(k = 1, 2, ..., j)$ freely.

Assessments are determined using qualitative labels and can be clarified by assuming that V represents the value of a set of $X = \{X_1, X_2, ..., X_n\}$ where X_n is the score in the qualitative symbol. The aggregation of multiple criteria and multiple experts determines the weight value of each expert (Q_k) and alternative values such as Eq. (1)

$$V_{ji} = f(V_i) = \max[Qj \wedge bj] \tag{1}$$

V_{ji}, aggregation on multiple experts; Q_j, weight values of each expert; bj, rank order of the most significant expert rank; j, number of experts.

The Aggregation result for Alternative 1 (A_1) is $V_1 = \max [Q_1 \wedge b_1]$, for A_2 is $V_2 = \max [Q_2 \wedge b_2]$, for A_{z-1} is $V_{z-1} = \max[Q_{(Z-1)} \wedge b_{(Z-1)}]$, and for A_Z is $V_z = \max[Q_{(z)} \wedge b_{(z)}]$.

2.2 Definition 2: Trust Model

The trust model measured expert reliability to increase trust in expert decisions. The trust concept has received significant attention in the technical research community because trust is the basis for decision-making in many contexts. In agri-food logistics, the trust of decision-makers plays an important role. Decision results will be a reference if the decision-maker can be trusted [13]. Reliability and trust in the expert's ability to determine the best alternative to many criteria is important.

Therefore, in this study, expert reliability was measured using a trust model. Hossain and Ouzrout (2012) introduced the trust model; they defined multiple criteria consisting of honesty, credibility, competence, goodwill, predictability, transparency, commitment, respect, and communication skills [14]. Uncertainty trust behavior (UC) is based on trust criteria such as in Eq. (2).

$$UC = \frac{(CT_1 + CT_2 + \cdots + CT_j)}{j} \tag{2}$$

UC, uncertainty trust behavior; CT, criteria of trust; j, number of trust criteria.

Trust is the weighted average of all the trust behavior criteria components $(UC_1, UC_2, ..., UC_r)$. The UC_{avg} is entered on the trust scale to obtain the expert's trust level. The expert's decision results are declared to be reliable if the average value (CT_{avg}) is above 1.5. We state the results are trusted or reliable if the mean is equal to or above 2.0, moderate if the average value is 0.5–1.5, and distrust if the average value is 0– 0.5.

2.3 Definition 3: Uncertainty Theory

The uncertainty model was developed based on the model structure and model parameterization. Uncertainty is fundamentally propagated in data models that represent real-world phenomena. If X denotes a non-empty set and Y represents the subset of X, then

(X, Y) is named a measurable space. Every element R in Y is called an uncertain event. The uncertain measure M is defined over Y. M{R} is the representation of the trust level that R is going to occur. At the same time, the distance between two uncertain variables V_1 and V_2 was calculated based on Eq. (3).

$$d(V_1, V_2) = E[|V_1 - V_2|] \tag{3}$$

Many studies have addressed the uncertainties that arise through the various components of the system. Uncertainty is inherent in data, where users can quickly ascertain the uncertainty of a data set, especially in terms of metadata.

3 The Propose Trust Uncertainty Model

The proposed trust uncertainty model contains a trust calculation function and a trust decision-making function. In this function, trust is not considered a single concept. Instead, it is viewed as an uncertain vector containing nine uncertain variables. These variables are as follows: (1) honesty, (2) credibility, (3) competence, (4) goodwill, (5) predictability, (6) transparency, (7) commitment, (8) resections, and (9) communication skills. These parameters were selected by reviewing the existing literature on the essential factors viewed as the primary trustworthiness constructs.

The decision-making in the proposed model depends not only on the trust level but also on the risk level of trusting the experts. Thus, an expert is selected to have the maximum trust level and the minimum associated risk level. The primary mechanism for selecting the best expert is uncertain using ME-MCDM. This approach allows us to consider the target levels corresponding to the trustworthiness variables.

3.1 Trust Calculation Function

Developing complex and uncertain decision-making trends integrates competence from various disciplines. Reliability and trust in the expert's ability to determine the best alternative to many criteria is essential. Therefore, in this study, expert reliability was measured using a trust model. Trust is the average trust behavior of all predefined uncertainty criteria such as in Eq. (4).

$$UC = \frac{(\alpha_1 CT_1 + \alpha_2 CT_2 + \alpha_3 CT_3 + \alpha_4 CT_4 + \alpha_5 CT_5 + \alpha_6 CT_6 + \alpha_7 CT_7 + \alpha_8 CT_8 + \alpha_9 CT_9)}{(\alpha_1 + \alpha_2 + \alpha_3 + \alpha_4 + \alpha_5 + \alpha_6 + \alpha_7 + \alpha_8 + \alpha_9)} \tag{4}$$

UC, uncertainty trust behavior; CT_1, honesty; CT_2, credibility; CT_3, competence; CT_4, goodwill; CT_5, predictability; CT_6, transparency; CT_7, commitment; CT_8, respections; CT_9, communication skills; α, constant.

Trust is a condition that psychologically consists of finding and logically accepting a vulnerability based on the expectations and desires of others. We explain the criteria for trust. Next, we formulate uncertainty trust behavior (UC) based on trust criteria. Trust is the weighted average of all the trust behavior criteria components (UC_1, UC_2, ..., UC_r), as shown in Eq. (5):

$$UC_{avg} = \frac{(UC_1 + UC_2 + \cdots + UC_r)}{r} \tag{5}$$

UC_{avg}, average of uncertainty trust behavior; r, number of experts.
The UC_{avg} is entered on the trust scale to obtain the expert's trust level (Eq. (6)).

$$TR = 0 < \text{distrust} \leq 0.5 < \text{moderate} \leq 1.5 < \text{trust} \leq 2.0 \tag{6}$$

The expert's decision results are declared to be reliable if the average value (CT_{avg}) is above 1.5.

3.2 Trust Decision Making

In complex and uncertain decision-making, many experts' participation is required. Decision-making involving many experts is a class of important and uncertain problems. Usually, there are conflicting criteria. Yager developed a multi-criteria ME-MCDM, and Kumar et al. (2017) expanded their research. They state that a non-numeric approach can evaluate multi-criteria preferences.

Marimin et al. stated that each decision maker (E_j) (j = 1, 2, ..., m) can assess each alternative (A_i) (I = 1, 2, ..., n) in each criterion (C_k) (k = 1, 2, ..., j). Assessments are determined using qualitative labels and can be clarified by assuming that V represents the value of a set of X = {X_1, X_2, ..., X_n} where X_n is the score in the qualitative symbol [15]. Aggregation conflict in ME-MCDM is an important task for experts to make effective decisions [16–18].

In this study, ME-MCDM steps start identifying experts (E_j) (j = 1, 2, ..., r). Then, some criteria and selection criteria related to the transportation route (C_k) (k = 1, 2, ..., j). The alternative (A_r) (r = 1, 2, ..., i) chosen was determined based on the current conditions. Next, we aggregated the multi-expert and multi-criteria. Multi-expert aggregation is used to obtain alternative values, whereas multi-criteria aggregation determines the evaluation of each criterion. After obtaining the multi-criteria aggregation, Alternatif 1 (A_1) based on criteria is $X_1 = (V_{11}, V_{21}, ..., V_{r1})$, for A_2 is $X_2 = (V_{12}, V_{22}, ..., V_{r2})$, for A_{z-1} is $X_{z-1} = (V_{1(z-1)}, V_{2(z-1)}, ..., V_{r(z-1)})$, and A_z is $X_z = (V_{1(z)}, V_{2(z)}, ..., V_{r(z)})$; then, we weighted the values for each expert (Q_k) (k = 1, 2, ..., m) using Eq. (7).

$$Q_k = \text{Int}[1 + \left(k * \frac{q-1}{r} \right)] \tag{7}$$

Q_k: weight values of each expert; k: index; q: rating scale number; r: number of experts.

After aggregating on multi-criteria (V_{ij}) and determining the weight value of each expert (Q_k), we aggregate on multiple experts (V_i) to obtain alternative values. The Aggregation result for Alternative 1 (A_1) is $V_1 = \text{Max} [Q_1 \wedge b_1]$, for A_2 is $V_2 = \text{Max} [Q_2 \wedge b_2]$, for A_{z-1} is $V_{z-1} = \text{Max} [Q_{(Z-1)} \wedge b_{(Z-1)}]$, and for A_Z is $V_z = \text{Max} [Q_{(z)} \wedge b_{(z)}]$. The final decision-making result from transportation route selection is the most optimal alternative with the highest aggregation (V_{max}) as follows:

$$V_{max} = f(V_i, V_{i+1}, ..., V_p) \tag{8}$$

V_{max}: the most optimal alternative with the highest aggregation; V: aggregation result; i: number of alternatives.

3.3   Trust Uncertainty Model Stages

The process begins with identifying experts and determining the criteria and alternatives. We extend the AHP by adding the unique characteristics of agri-food logistics, and propose four alternative logistics routes with five conflicting criteria. The following process negates the level of criteria importance and matrix to the most optimal alternative results. We advance the ME-MCDM by measuring the trust level (Fig. 1).

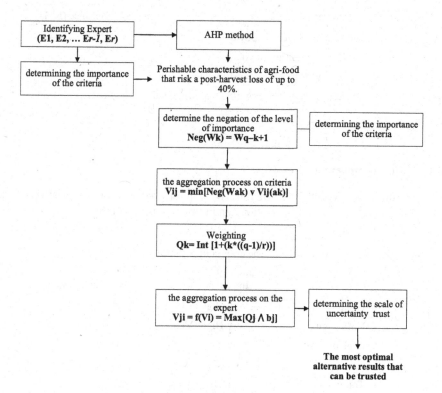

Fig. 1. The stages in the trust uncertainty model.

We propose a new method that combines the non-numeric ME-MCDM by measuring the expert's trust level using the uncertainty trust model by knowing the expert's trust level. We propose that if more than 50% of the average expert's trust level is trust (scale 1.5–2.0), one expert has a moderate level of trust. In contrast, the general average is trust; then, the decision category is reliable.

4   Illustrative Example

Food security has become a global issue in various scientific discussions. However, world food needs are expected to increase by 60% by 2030. Potatoes have been recommended as agri-food that supports food security by the United Nations. Potatoes have perishable characteristics that risk a post-harvest loss of up to 40% [19]. Thus, this study discusses potato logistics as an interesting agri-food logistics example. In this section, the starting point used is the DC in Kejajar, Indonesia to the Boyolali, Indonesia. We considered all route, and there were four alternatives (Table 1).

Table 1. Alternatives routes in agri-food logistic.

Alternative	Routes	Distance (KM)
A_1	Kejajar-Magelang-Boyolali	108
A_2	Kejajar-Salatiga-Boyolali	140
A_3	Kejajar-Kab. Magelang-Boyolali	140
A_4	Kejajar-Kab. Semarang-Salatiga-Boyolali	123

4.1 Criteria Importance Level

We considered five conflicting criteria. The criteria used were as follows: distance, utilization, logistics cost, transportation conditions, and traffic (Table 2).

Table 2. The criteria importance level.

Code	Criteria	Importance level	Conversion value
C_1	Distance	Very High (VH)	1
C_2	Utilization	Very High (VH)	1
C_3	Logistic cost	Low (L)	4
C_4	Transportation	Low (L)	4
C_5	Traffic	High (H)	2

The experts (E_1, E_2, ..., E_5) consisted of an expert from the Wonosobo Horticulture Department (E_1), an expert from Adhiguna Laboratory, Wonosobo with more than ten years of experience (E_2), an expert from IPB University, Bogor (E_3), an expert from the Wonosobo Transportation Agency (E_4), and expert from potato farmers in Garung with more than 15 years of experience (E_5).

4.2 The Aggregation Process

Five experts in groups assessed alternative 1 (A_1), so the results of the expert assessment of E_j for A_1 were as follows (Table 3):

The following process is an expert aggregation: Expert aggregation starts from weighting using Eq. (7) as follows:

Table 3. Expert assessment for alternative 1 (A_1)

Expert	Alternative	Criteria 1	Criteria 2	Criteria 3	Criteria 4	Criteria 5
E_1	A_1	VH	H	L	L	H
E_2	A_1	H	VH	L	M	H
E_3	A_1	VH	H	M	L	H
E_4	A_1	H	H	M	VL	M
E_5	A_1	H	H	M	VL	M

$$Q_1 = Int\left[1 + \left(1*\tfrac{5-1}{5}\right)\right] = Int[1.8] = 2 = L$$

$$Q_2 = Int\left[1 + \left(2*\tfrac{5-1}{5}\right)\right] = Int[2.6] = 3 = M$$

$$Q_3 = Int\left[1 + \left(3*\tfrac{5-1}{5}\right)\right] = Int[3.4] = 3 = M$$

$$Q_4 = Int\left[1 + \left(4*\tfrac{5-1}{5}\right)\right] = Int[4.2] = 4 = H$$

$$Q_5 = Int\left[1 + \left(5*\tfrac{5-1}{5}\right)\right] = Int[5] = 5 = VH$$

Values of $Q = (Q_1, Q_2, Q_3, Q_4, Q_5) = $ L, M, M, H, VH.

After weighting, to obtain an alternative value of A_1 (X_1), the expert aggregation results are collected with the criteria aggregation results using Eq. (8):

$X_1 = $ H, H, H, M, M; So that $b_1 = $ H, H, H, M, M

$V_1 = Max[L \wedge H, M \wedge H, M \wedge H, H \wedge M, VH \wedge M]$

$V_1 = Max\,[L, M, M, M, M]$

$V_1 = M$

The final result of Alternative1 (A_1) is M(Medium).

The final results for A_2, A_3, and A_4 are as shown in Table 4.

Table 4. The final result of agri-food logistic route alternatives.

Alternative	Aggregation result	Final result	Alternative ranking
A4	H, H, H, H, M	H	1
A1	H, H, H, M, M	M	2
A2	VL, L, L, M, M	M	2
A3	L, M, L, L, L	L	3

The results showed that alternative 4 (A_4) had a high (H) aggregation value by considering distance, utilization, logistics cost, transportation conditions, and traffic. Thus, by selecting alternative 4, the transportation route is optimal. The proposed method can choose the optimal transportation route.

4.3 The Average Uncertainty Trust Value

Using the uncertainty trust scale, we measure the expert's trust level, as shown in Eq. (6). In this study, we used only nine criteria. The UC_{avg} is entered on the uncertainty trust scale to obtain the expert's trust level. The expert's decision results are declared to be reliable if the average value (CT_{avg}) is above 1.5.

We used only nine trust criteria. Five transportation entrepreneurs assessed each expert. The assessment process was completed by filling out the. Table 5 presents the results of the questionnaire tabulation.

Table 5. The average of uncertainty trust value

Code	Criteria of uncertainty trust	E1	E2	E3	E4	E5
C1	Honesty	2	2	1.7	1.7	1.3
C2	Credibility	2	1.5	2	1.5	2
C3	Competence	2	1.5	1.5	2	1.3
C4	Goodwill	1.5	2	2	1.5	1.5
C5	Predictability	2	1.5	2	2	1.5
C6	Transparency	1.5	2	1.5	1.5	1.2
C7	Commitment	2	2	1.5	2	1.3
C8	Respect	2	2	1.5	1.7	1.5
C9	Communication skills	1.8	1.2	2	2	1.4

Of all the experts, only Expert 5 (E_5) had a score below 1.5. This means that, in general, the average value of trust behavior (CT_{avg}) is 1.7, so that the decisions of the group of experts can be trusted at a moderate level, which means that statistically, these results can still be used for measurement (Fig. 2).

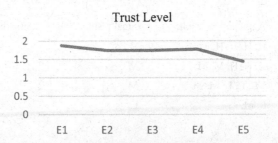

Fig. 2. The average uncertainty trust level

Using a trust scale, the decision results were reliable. It showed that the uncertainty trust model proposed was successfully used to measure the trust level given to decision-makers. The trust value was 1.7, indicating reliability in agri-food logistic decision making.

5   Evaluation and Comparison

Decision-making results show that alternative 4 (A_4) has a high (H) aggregation value. Therefore, A_4 is the optimal alternative. We will review the alternative route 4 (A_4) discussion results-based criteria weights given by the expert. To measure the uncertainty trust level of all experts, we used the uncertainty trust scale so that the results would be more reliable.

Kanuganti *et al.* (2017) and Dubey *et al.* (2014) state that one of the popular methods for selecting alternatives with various criteria is the AHP method [12, 20]. In this study, we compared our proposed method with the AHP. Of the five criteria we suggest, namely distance, utilization, logistic cost, transportation condition, and traffic, the AHP method also determines the importance of criteria (Table 6).

Table 6. Matrix of eigenvalues from alternatives with the weighting criteria

Alternative	Distance	Utilization	Logistic cost	Condition[a]	Traffic
A_1	0.1267	0.1817	0.3410	0.5339	0.1430
A_2	0.1929	0.1365	0.1430	0.1459	0.1267
A_3	0.6140	0.6103	0.4403	0.2544	0.4403
A_4	0.0664	0.0715	0.0757	0.0658	0.0757
Eigenvalue	0.1812	0.1943	0.2619	0.0773	0.2851

[a] Transportation Condition

5.1   Comparison with AHP

The AHP also calculates the comparison and weighing of the decision criteria and sub-criteria. Now it is time to compare all decision alternatives concerning each decision sub-criterion. After evaluating all the decision alternatives concerning the decision sub-criteria, the calculation of the weights for each decision element in the AHP is complete. Based on the priority weights of each alternative (Table 7), we can rank each alternative.

Table 7. The combined weight of criteria for each alternative using AHP.

	Distance	Utilization	Logistic cost	Condition[a]	Traffic	Criteria weights
A_1	0.0872	0.0809	0.1294	0.0638	0.0664	0.4277
A_2	0.0230	0.0155	0.0693	0.0233	0.0548	0.1859
A_3	0.2314	0.1401	0.1901	0.0824	0.1063	0.7503
A_4	0.0051	0.0036	0.0253	0.0079	0.0212	0.0632

[a] Transportation Condition

The composite weights for A_1, A_2, A_3 and A_4 were 0.4277, 0.1859, 0.7503 and 0.0632 respectively. According to the AHP, the best alternative was A_3. This is in contrast to the new method proposed in this study. The AHP did not perform expert aggregation. Our proposed method aggregates criteria and aggregates the experts. Thus, the result of the optimal alternative route has the highest aggregation value (H). AHP also does not measure the level of expert trust as a group decision-maker. In this study, we extend the AHP method by adding the unique characteristics of agri-food logistics, and propose four alternative logistics routes with five conflicting criteria.

In our new method, we measure the expert trust level and reliability of the results of their decisions. The advantage of our new method is that we can infer the uncertainty trust level of the experts. In our proposed model, the uncertainty trust model was successfully used to measure the uncertainty trust level of group decision-makers with a trust value of 1.7, indicates reliability. The results of expert decisions using our method can measure the trust level, where the AHP cannot do this.

5.2 Research Limitation

This new model was developed to measure the level of uncertainty in the confidence of many experts after making agri-food logistics decisions.

6 Conclusions and Recommendations

The complexity and uncertainty in determining routes require the participation of multiple experts. Based on the assessments of the five experts, we propose four alternatives with five criteria. The conflicting criteria suggested were distance, utilization, logistic costs, transportation conditions, and traffic. The study concluded that alternative 4 (A_4) with a high (H) aggregation value was the best route. In addition, model validation shows that this new method has advantages over the AHP. The uncertainty trust model proposed was successfully used to measure the level of trust given to decision-makers in the most optimal transportation route selection for potato commodities. The trust value was 1.7, indicating that the inference results were reliable in agri-food logistic.

There are several methods, and each method will conclude one optimal solution. In the proposed model, the optimality is determined by aggregating the value. Other models may use a different definition of optimality. Therefore, further research needs to better measure the validity of the new method, especially regarding the concept of optimality. This study only measures the level of expert trust for a particular decision. Therefore, for further research, it is necessary to consider calculating the average trust level of many experts in the process of obtaining results before making decisions.

References

1. Liu, K. F.-R., Kuo, J.-Y., Yeh, K., Chen, C.-W., Liang, H.-H., Sun, Y.-H.: Using fuzzy logic to generate conditional probabilities in Bayesian belief networks: a case study of ecological assessment. Int. J. Environ. Sci. Technol. **12**(3), 871–884 (2013). https://doi.org/10.1007/s13 762-013-0459-x

2. Kharouf, H., Lund, D.J., Sekhon, H.: Building trust by signaling trustworthiness in service retail. J. Serv. Mark. **28**(5), 361–373 (2014). https://doi.org/10.1108/JSM-01-2013-0005

3. Salehi-Abari, A., White, T.: The relationship of trust, demand, and utility: be more trustworthy, then I will buy more. In: PST 2010: 2010 8th International Conference on Privacy, Security and Trust, pp. 72–79 (2010). https://doi.org/10.1109/PST.2010.5593256

4. Gelei, A., Dobos, I.: Mutual trustworthiness as a governance mechanism in business relationships—a dyadic data analysis. Acta Oecon. **66**(4), 661–684 (2016). https://doi.org/10.1556/032.2016.66.4.5

5. Chen, S., Dhillon, G.: Interpreting dimensions of consumer trust in E-commerce. Inf. Technol. Manag. **4**(2/3), 303–318 (2003). https://doi.org/10.1023/A:1022962631249

6. Pomponi, F., Fratocchi, L., Tafuri, S.R.: Trust development and horizontal collaboration in logistics: a theory based evolutionary framework. Supply Chain Manag. **20**(1), 83–97 (2015). https://doi.org/10.1108/SCM-02-2014-0078

7. Marimin, M., Adhi, W., Darmawan, M.A.: Decision support system for natural rubber supply chain management performance measurement: a sustainable balanced scorecard approach. Int. J. Supply Chain Manag. **6**(2), 60–74 (2017)

8. Yusianto, R., Marimin, M., Suprihatin, S., Hardjomidjojo, H.: Spatial analysis for crop land suitability evaluation : a case study of potatoes cultivation in Wonosobo, Indonesia. In: International Seminar on Application for Technology of Information and Communication (iSemantic), vol. 1, no. 1, pp. 313–319 (2020). https://doi.org/10.1109/iSemantic50169.2020.9234284

9. Yusianto, R., Sundana, S., Marimin, M., Djatna, T.: Method and mapping of trust and trustworthiness in agroindustry logistic and supply chain: a systematic review. Int. J. Supply Chain Manag. **9**(1), 397–410 (2020)

10. Breite, R., Aramo-Immonen, H.: Trust-related dynamics in the supply chain relationship. J. Purch. Supply Manag. **9**(1), 207–216 (2017)

11. Ding, J.F.: Applying an integrated fuzzy MCDM method to select hub location for global shipping carrier-based logistics service providers. WSEAS Trans. Inf. Sci. Appl. **10**(2), 47–57 (2013)

12. Kanuganti, S., Agarwala, R., Dutta, B., Bhanegaonkar, P.N., Pratap, A., Singh, A.K.S.: Road safety analysis using multi criteria approach: a case study in India. Transp. Res. Procedia **25**(7), 4649–4661 (2017). https://doi.org/10.1016/j.trpro.2017.05.299

13. Xia, L., Monroe, K.B., Cox, J.L.: The price is unfair! A conceptual framework of price fairness perceptions. J. Mark. **68**(4), 1–15 (2004). https://doi.org/10.1509/jmkg.68.4.1.42733

14. Hossain, S.A., Ouzrout, Y.: Trust model simulation for supply chain management. In: Proceeding of the 15th International Conference on Computer and Information Technology, no. 12, pp. 376–383 (2012). https://doi.org/10.1109/ICCITechn.2012.6509744

15. Marimin, M., Umano, M., Hatono, I., Tamura, H.: Non-numeric method for pairwise fuzzy group-decision analysis. J. Intell. Fuzzy Syst. **5**(3), 257–269 (1997). https://doi.org/10.3233/IFS-1997-5307

16. Kumar, A., et al.: A review of multi criteria decision making (MCDM) towards sustainable renewable energy development. Renew. Sustain. Energy Rev. **69**(10), 596–609 (2017). https://doi.org/10.1016/j.rser.2016.11.191

17. Rehman, A., Hussain, M., Farooq, A., Akram, M.: Consensus-based multi-person decision making with incomplete fuzzy preference relations using product transitivity. Mathematics **7**(185), 1–13 (2019). https://doi.org/10.3390/math7020185

18. Ferdousi, T., Akhtar, M.A., Ullah, A.M.M.S.: Knowledge-based systems algorithms for fuzzy multi expert multi criteria decision making (ME-MCDM). Knowl.-Based Syst. **24**(3), 367–377 (2011). https://doi.org/10.1016/j.knosys.2010.10.006

19. Yusianto, R., Marimin, M., Suprihatin, S., Hardjomidjojo, H.: The potato crop: management, production, and food security. In: The Potato Crop Management, Production, and Food Security, no. 1, pp. 305–335 (2021)
20. Dubey, S.K., Mishra, D., Arkatkar, S.S., Singh, A.P., Sarkar, A.K.: Route choice modelling using fuzzy logic and adaptive neuro-fuzzy. Mod. Traffic Transp. Res. 2(4), 11–19 (2014)

Investigating the Predictive Power of Google Trend and Real Price Indexes in Forecasting the Inflation Volatility

Kittawit Autchariyapanitkul[1], Terdthiti Chitkasame[2], Namchok Chimprang[3], and Chaiwat Klinlampu[3(✉)]

[1] Faculty of Economics, Maejo University, Chiang Mai 50290, Thailand
[2] Faculty of Economics, Chiang Mai University, Chiang Mai 50200, Thailand
[3] Center of Excellence in Econometrics, Faculty of Economics, Chiang Mai University, Chiang Mai, Thailand
chaiwat_klinlampu@cmu.ac.th

Abstract. The goal of this study is to examine the predictive power of real price indexes and Google Trend in forecasting the inflation volatility in three nations (the USA, Japan, and the UK). The AIC, BIC, and RMSE are used to select the best GARCH-type models with the most appropriate predictors. The overall result shows that the GARCH model with the skew-student distribution is the most effective model in capturing the inflation volatility. Furthermore, this study reveals that the commodity price index is the strongest predictor variable of the inflation volatility. We also find that the financial crisis and health crisis decisively affect the inflation volatility in the United States of America and Japan.

Keywords: GARCH-type models · Inflation · Predictive power · Volatility forecasting

1 Introduction

Inflation is the increased rate in prices over the period. It might be one of the most familiar words in economics and it has plunged entire countries into long periods of insecurity. As a result, the costs of inflation include unfriendly changes in an adverse effect on the level of economic activity including income distribution and resource allocation [1, 2]. Inflation influences a wide range of financial and economic activities; thus, it motivates economists to develop strategies for effectively forecasting inflation. Inflation practically affects everyone in the economy, including banking institutions, stockbrokers, and corporate finance officials. Additionally, the Chief Financial Officer of the company will make the right decisions on a project, or a banker will make exactly financial decisions on a loan and asset management, preventing their financial problem (bankruptcy) and enhancing their profit [3]. If these different economic stakeholders can accurately forecast inflation rates, they will have well preparation of their financial and economic plans. In short, increasing the forecasting accuracy of inflation will help

K. Honda et al. (Eds.): IUKM 2022, LNAI 13199, pp. 355–367, 2022.
https://doi.org/10.1007/978-3-030-98018-4_29

economic agents in dealing better with the numerous interactive economic components in a business environment in which inflation matters [4, 5].

Generally, the inflation rate is measured by the growth of the consumer price index (CPI). We note that the CPI represents the change in the goods and service's price that can alter the financial burden of consumers. For example, when the CPI increases, the consumers' purchasing power will be decreased [6–8]. The CPI has been focused by numerous economic sectors including investors, private firms, consumers, commercial and central banks as it is a significant index for tracking costs of living and it also affects a country's interest rate, which directly affects investment returns and borrowing costs.

According to Fig. 1, the global inflation has exhibited a high fluctuation along 1960–2020. We can observe that the inflation rate has exhibited a large swing during 1980, 1990, 2000, and 2009 which corresponded to the global financial crisis (i, ii, iii, and iv). However, we observe a large drop of inflation rate in 1984, 2001, and 2020 coinciding with the time of the health crisis. The periods of a–d correspond to the pandemic period when health crisis spread across a large region and worldwide and affected a large number of people. It is well known, and the recent COVID-19 outbreak contributing to a great volatility in many aspects and consumer behavior is not an exception [9]. Observably, both the financial and health crises have had a significant impact on inflation movements. Moreover, crisis is not the only factor causing the high fluctuation of inflation. The literature has revealed that there are several factors for instance, conventional cost push, demand pull, monetary policies, interest rate, etc. causing the volatility of inflation. However, our research mainly focusses on commodity prices, house prices, oil prices impact on inflation volatility.

In this study, we consider two types of informative predictors consisting of real price indexes and Google Trend (GT) to forecast the inflation volatility. According to previous studies, the inflation fluctuations are caused by many real price indexes as an index is derived from the weighted average of the consumer goods and services basket prices [10]. Various real price indexes are confirmed to affect the volatility of the CPI, for example the house prices [11], commodity prices [12], and oil prices [13]. In

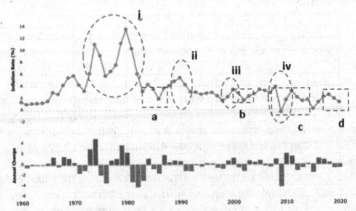

Fig. 1. The global inflation during 1960–2020 and global financial crisis (i–iv) and pandemic crisis (a–d).

addition, Google Trend is now being used as a predictor in various models to improve forecasting accuracy in response to changes in people's behavior in society [14, 15], like forecasting the stock price movements based on stock market interest behavior [16–18]. Furthermore, Guzman [19] confirmed that Google Trend was one of the predictors employed for lowering inflation predictions error.

In the methodological perspective, the Generalized autoregressive conditional heteroskedasticity (GARCH) model [20] has been normally applied for capturing the time series volatility of data. There were many studies using the GARCH model to predict the CPI [21–25]. However, Risteski, Sadoghi and Davcev [26], Yao and Zhang [27]; and Tarkhamtham, Yamaka, and Maneejuk [28] mentioned that the volatility forecasting performance of the GARCH model can be improved by adding additional information from exogenous variables. They revealed that the GARCH model with informative predictor provides better volatility forecast compared to the original GARCH models for both in-sample and out-of-sample investigations.

Note: (1) Global financial crisis: the (i) the oil crisis during 1970–1980, the periods (ii) and (iii) are the dotcom crisis and the Asian financial crisis, respectively and (iv) the Hamburger crisis during 2010.

(2) Pandemic crisis: (a) is presented with the HIV/AIDS pandemic mainly during 1981–1990. (b) the SAR pandemic in 2003, (c) Influenza pandemic during 2009-2010 and (d) the Covid-19 pandemic from 2020 until now.

Source: World Bank, 2021.

The goal of this study is to examine the predictive ability of real price indexes and Google Trend for the CPI of the United States of America, Japan, and the United Kingdom which are the top three developed economies. The Covid-19 crisis is also concerned as another predictor variable of the model. Several different GARCH models (GARCH, GJR-GARCH, and EGARCH) are considered to forecast the volatility of the CPI. Our paper differs from the existing studies in that it assesses the predictive power of real price indexes and Google Trend as well as COVID-19. This is the first attempt ever to investigate the forecasting power of these variables.

The rest of this paper is outlined as follows. Sections 2 and 3 detail the methodology and data used in this paper. Section 4 describes the empirical results and analysis. Section 5 concludes this paper.

2 Methodology

2.1 Generalized Autoregressive Conditional Heteroscedasticity (GARCH)

Time series analysis often assumes that the variance of the data is constant (Homoscedasticity). In fact, the variance of the data depends on the historical error (Heteroscedasticity). Most research works estimated the mean and variance of the serial data by the GARCH model, proposed by Bollerslev [29]. Generally, the GARCH result has the ability to capture the volatility of the serial variable. This study thus applies the GARCH (1, 1) model to identify the volatility. The GARCH formula has the following equation.

$$Y_{i,t} = \mu_{i,t} + \varepsilon_{i,t}, \tag{1}$$

$$\varepsilon_{i,t} = \eta_{i,t}\sqrt{\sigma_{i,t}}, \tag{2}$$

$$\sigma_{i,t}^2 = \omega + \alpha_{i,t}\varepsilon_{i,t-i}^2 + \beta_{i,t}\sigma_{t-1}^2 + \theta X_t, \tag{3}$$

At time t and for country i, where $\varepsilon_{i,t}$ is the error of CPI, μ_t is the constant term, ε_t is the error term, $\eta_{i,t}$ is white noise (an i.i.d. of standard innovation), $\sigma_{i,t}^2$ is the volatility of CPI, and $\alpha_{i,t}, \beta_{i,t}$ are the coefficient parameters representing the ARCH and GARCH effect, respectively. As the predictor (X_i) is considered for volatility forecasing, we include it in the variance Eq. (3).

2.2 Glosten, Jagannathan and Runkle (GJR)-GARCH

The GJR-GARCH model of Glosten et al. [30] is also employed in this study. This model uses indicators function I to capture the positive and negative shocks on the conditional variance asymmetrically. The GJR-GARCH (1, 1) model is defined as.

$$\sigma_{i,t}^2 = \omega + (\alpha_{i,t} + \gamma_{i,t}I_{t-1})\varepsilon_{i,t-1}^2 + \beta_{i,t}\sigma_{i,t-1}^2 + \theta X_i \tag{4}$$

where

$$1_{i,t-i} = \begin{cases} 0 \ if \ \varepsilon_{i,t-i} \geq 0, \\ 1 \ if \ \varepsilon_{i,t-i} < 0. \end{cases}$$

where $\gamma_{i,i}$ is the leverage effect, $\varepsilon_{i,t} = \eta_{i,t}\sqrt{\sigma_{i,t}}$ and $\gamma_{i,t}, \omega_{i,t}, \alpha_{i,t}, \beta_{i,t} \geq 0$, the stationary condition is when $\gamma + \alpha \leq 0$ and $\sigma_t \sim F(\cdot)$ is sequence of independent random variable or innovation. Hence, the $\varepsilon_{i,t-j}$ is the error term of CPI in lag j.

2.3 Exponential GARCH (EGARCH)

The exponential GARCH (EGACH) model was provided by Nelson [31]. It uses the natural logarithmic value of conditional variance to estimate the exponential function of GARCH. This is one of the models able to capture the asymmetric effects referred to as the leverage effects in time series data. The illustration of this model can capture both negative and positive shocks of the same magnitude having an unequal destabilizing effect and remove restrictions on parameters [32]. The EGARCH (1, 1) model can be written as:

$$\log\left(\sigma_{i,t}^2\right) = \omega + \alpha_{i,t}\varepsilon_{i,t-1} + \gamma_{i,t}\left(|\varepsilon_{i,t-1}| - E|\varepsilon_{i,t-1}|\right) + \beta_{i,t}\log\left(\sigma_{i,t-1}^2\right) + \theta X_i, \tag{5}$$

where the conditional variance $(\sigma_{i,t}^2)$ represents the inflation volatility. the coefficient α_i captures the sign effect and $\gamma_{i,t}$ the size effect of the asymmetry. Positive estimates of the volatility are guaranteed due to working on the log variance. There are no restrictions on $\gamma_{i,t}, \omega_{i,t}, \alpha_{i,t}$ and $\beta_{i,t}$ but to maintain stability $\beta_{i,t}$ must be positive and less than one.

3   Data Description

Based on the literature review, several predictors of INF volatility are considered in this investigation. These predictors include the house price index (HOU), the oil price index (OIL), the commodity price index (COM), and a Google Trend (GT) on three keywords: "House prices (G_HOU)", "Oil prices (G_OIL)", and "Commodity prices (G_COM)" (the index data from Google Trends will be de-normalized). The monthly data is collected covering January 2004 to September 2021. We also consider the two dummies representing health and financial crises as another two control variables in our forecasting analysis. The financial crisis (FIC) is represented by the subprime mortgage crisis and the European sovereign debt crisis, while the health crisis (HEC) is represented by H1N1 pandemic in 2009, Ebola outbreak in 2014, and the COVID-19 pandemic in 2020–2021. All data are obtained from https://www.ceicdata.com and https://trends.goo gle.com/trends. The data have been transformed into growth formulas.

Table 1. Descriptive statistics of the USA dataset

Description	USA								
	INF	HOU	COM	G_HOU	G_COM	G_OIL	OIL	FIC	HEC
Mean	1.976	0.002	0.001	−0.002	−0.005	0.005	0.004	0.340	0.491
Median	2.000	0.002	0.002	0.014	−0.026	0.000	0.018	0.000	0.000
Maximum	4.453	0.009	0.028	0.615	1.459	1.996	0.469	1.000	1.000
Minimum	0.603	−0.005	−0.044	−0.556	−1.355	−1.592	−0.555	0.000	0.000
Std. Dev	0.515	0.002	0.009	0.169	0.367	0.425	0.110	0.475	0.501
Skewness	0.913	−0.150	−0.845	−0.010	0.392	0.865	−1.232	0.677	0.038
Kurtosis	7.934	4.850	6.560	4.806	5.207	8.057	9.966	1.459	1.001
Jarque-Bera	244.445	31.038	137.204	28.829	48.459	252.338	482.257	37.192	35.333
Probability	0	0	0	0	0	0	0	0	0
MBF Unit-root test	0.000	0.000	0.000	0.000	0.000	0.000	0.000	0.000	0.000

The descriptive statistics of all series are provided in Tables 1, 2 and 3. The mean value of CPI is approximately 2.00 except only for Japanese CPI which is close to zero. For the predictor variables, their means and skewness are close to zero while their kurtosis are mostly greater than 3, indicating that the distribution of these predictors are leptokurtosis and fat-tailed. This is to say, the predictor variables not normally distributed. Lastly, the unit root test is conducted to investigate the stationarity of the data, and the result shows that the data are strongly stationary. We note the statistic inference of the unit root test is based on Minimum Bayes factor (MBF) [33].

This study considers 14 patterns of predictor sets to forecast the inflation volatility. The specifications of the GARCH-type models are presented as follows:

Model 1: GARCH type-models with house price index
Model 2: GARCH type-models with commodity price index

Table 2. Descriptive statistics of the Japan dataset

Description	Japan								
	INF	HOU	COM	G_HOU	G_COM	G_OIL	OIL	FIC	HEC
Mean	−0.023	0.000	0.003	−0.001	−0.004	−0.009	0.004	0.340	0.491
Median	−0.196	0.000	0.005	0.000	-0.006	0.000	0.018	0.000	0.000
Maximum	2.748	0.005	0.044	1.656	3.932	3.284	0.469	1.000	1.000
Minimum	−1.683	−0.002	−0.098	−1.913	−4.710	−3.516	−0.555	0.000	0.000
Std. Dev	0.898	0.001	0.019	0.385	0.766	0.729	0.110	0.475	0.501
Skewness	1.307	2.508	−1.734	−0.243	−0.033	0.156	−1.232	0.677	0.038
Kurtosis	5.144	18.135	9.495	7.784	17.937	8.421	9.966	1.459	1.001
Jarque-Bera	100.948	2245.821	478.779	204.234	1970.847	260.467	482.257	37.192	35.333
Probability	0	0	0	0	0	0	0	0	0
MBF Unit-root test	0.000	0.000	0.000	0.000	0.000	0.000	0.000	0.000	0.000

Table 3. Descriptive statistics of the UK dataset

Description	UK								
	INF	HOU	COM	G_HOU	G_COM	G_OIL	OIL	FIC	HEC
Mean	1.965	0.003	0.003	0.008	0.007	0.013	0.004	0.340	0.491
Median	1.885	0.001	0.003	−0.020	0.000	0.000	0.018	0.000	0.000
Maximum	4.056	0.043	0.078	1.036	4.710	1.552	0.469	1.000	1.000
Minimum	0.563	−0.015	−0.125	−0.288	−4.710	−1.232	−0.555	0.000	0.000
Std. Dev	0.740	0.007	0.029	0.172	1.250	0.383	0.110	0.475	0.501
Skewness	0.595	2.501	−0.514	1.888	0.009	0.510	−1.232	0.677	0.038
Kurtosis	2.900	13.290	5.265	9.881	6.625	5.371	9.966	1.459	1.001
Jarque-Bera	12.580	1156.324	54.642	544.176	116.097	58.857	482.257	37.192	35.333
Probability	0	0	0	0	0	0	0	0	0
MBF Unit-root test	0.002	0.000	0.000	0.000	0.000	0.000	0.000	0.000	0.000

Model 3: GARCH type-models with oil price index

Model 4: GARCH type-models with house price index and commodity price index

Model 5: GARCH type-models with house price index and oil price index

Model 6: GARCH type-models with commodity price index and oil price index

Model 7: GARCH type-models with house price index, commodity price index and oil price index

Model 8: GARCH type-models with GT "House prices"

Model 9: GARCH type-models with GT "Commodity prices"

Model 10: GARCH type-models with GT "Oil prices"

Model 11: GARCH type-models with GT "Oil prices" and "Commodity prices"

Model 12: GARCH type-models with GT "House prices" and "Commodity prices"
Model 13: GARCH type-models with GT "House prices" and "Oil prices"
Model 14: GARCH type-models with GT "House prices", "Oil prices" and "Commodity prices"

To evaluate the performance of the various GARCH-type models with different sets of predictors, we use three loss functions to measure the forecasting error: Akaike information criterion (AIC), Bayesian information criterion (BIC) and Root Mean Square Error (RMSE).

4 Results

4.1 Model Selection

This study estimates the inflation volatility equation using three different GARCH (1,1) models (GARCH, EGARCH and GJR-GARCH) under different innovation distributions, namely normal (NORM), student-t (STD) and skewed-student-t (SSTD) distributions. To compare the performance of these GARCH-type models, RMSE, AIC and BIC are used.

Firstly, we have investigated the suitable model that precisely describes inflation volatility only for each country relying on the lowest value of AIC and BIC. The result is reported in Table 4, and it shows that the GJR-GARCH model with SSTD-distribution is the best model for capturing the inflation volatility for the USA, whereas the EGARCH model with SSTD-distribution is selected for Japan and the United Kingdom.

Table 4. Model selection.

Type	NORM		STD		SSTD	
	AIC	BIC	AIC	BIC	AIC	BIC
USA						
sGARCH	0.50678	0.57011	0.51767	0.59683	0.35912	0.45411
eGARCH	0.49734	0.54484	0.50823	0.57156	0.34937	0.42853
gjrGARCH	0.51483	0.59399	0.52624	0.62124	**0.30312**	**0.41395**
Japan						
sGARCH	1.51667	1.58000	1.53272	1.61188	1.26830	1.36330
eGARCH	1.50724	1.55474	1.52328	1.58661	**1.25873**	**1.33789**
gjrGARCH	1.52551	1.60468	1.54179	1.63679	1.58389	1.69473
UK						
sGARCH	1.58163	1.64496	1.58917	1.66833	1.50194	1.59693
eGARCH	1.57213	1.61963	1.57965	1.64299	**1.48232**	**1.57149**
gjrGARCH	1.59078	1.66994	1.59827	1.69327	1.49423	1.59506

Note: The bold number indicates the best fit GARCH-type model

Table 5. Evaluation of predictive power of various set of predictors.

Type	Model	USA			Japan			UK		
		NORM	STD	SSTD	NORM	STD	SSTD	NORM	STD	SSTD
sGARCH	1	4.46819	4.46727	4.42738	1.53564	1.53561	1.53329	4.96627	4.96821	4.96511
eGARCH		4.46805	4.46713	4.42723	1.53557	1.53557	**1.53327**	4.96596	4.96794	**4.96481**
gjrGARCH		4.47117	4.46983	**4.39585**	1.53606	1.53594	1.66373	4.96985	4.97208	4.99355
sGARCH	2	4.46769	4.46669	4.42577	1.52761	1.52758	1.52547	4.96636	4.96831	4.96581
eGARCH		4.46755	4.46656	4.42561	1.52754	1.52753	**1.52545**	4.96605	4.96805	**4.96551**
gjrGARCH		4.47086	4.46942	**4.39226**	1.52803	1.52792	1.67919	4.96995	4.97220	4.99435
sGARCH	3	4.47080	4.46985	4.42950	1.52603	1.52599	1.52375	4.96674	4.96869	4.96586
eGARCH		4.47066	4.46972	4.42934	1.52595	1.52595	**1.52372**	4.96643	4.96843	**4.96556**
gjrGARCH		4.47386	4.47249	**4.39665**	1.52643	1.52632	1.66920	4.97032	4.97257	4.99432
sGARCH	4	4.46716	4.46623	4.42627	1.52809	1.52806	1.52559	4.96567	4.96760	4.96495
eGARCH		4.46702	4.46609	4.42611	1.52802	1.52802	**1.52557**	4.96536	4.96734	**4.96465**
gjrGARCH		4.47016	4.46881	**4.39468**	1.52849	1.52837	1.68206	4.96926	4.97149	4.99352
sGARCH	5	4.46984	4.46898	4.42999	1.52682	1.52678	1.52421	4.96615	4.96809	4.96506
eGARCH		4.46970	4.46885	4.42984	1.52674	1.52674	**1.52418**	4.96584	4.96782	**4.96476**
gjrGARCH		4.47266	4.47142	**4.39978**	1.52719	1.52708	1.67142	4.96974	4.97197	4.99353
sGARCH	6	4.46797	4.46704	4.42698	1.52561	1.52558	1.52360	4.96636	4.96831	4.96581
eGARCH		4.46783	4.46690	4.42682	1.52554	1.52554	<u>1.52357</u>	4.96605	4.96805	**4.96550**
gjrGARCH		4.47101	4.46965	**4.39478**	1.52605	1.52592	1.67746	4.96995	4.97220	4.99434
sGARCH	7	4.46744	4.46660	4.42779	1.52616	1.52613	1.52379	4.96567	4.96760	4.96494
eGARCH		4.46730	4.46646	4.42764	1.52609	1.52609	**1.52377**	4.96536	4.96734	**4.96464**
gjrGARCH		4.47026	4.46902	**4.39801**	1.52656	1.52645	1.68029	4.96927	4.97149	4.99351
sGARCH	8	4.46986	4.46888	4.42810	1.53508	1.53504	1.53310	4.96685	4.96881	4.96602
eGARCH		4.46973	4.46875	4.42794	1.53501	1.53500	**1.53308**	4.96655	4.96855	**4.96571**
gjrGARCH		4.47301	4.47159	**4.39458**	1.53552	1.53540	1.66124	4.97044	4.97269	4.99448
sGARCH	9	4.46976	4.46877	4.42796	1.53496	1.53493	1.53296	4.96683	4.96879	4.96589
eGARCH		4.46962	4.46864	4.42781	1.53489	1.53489	<u>1.53294</u>	4.96653	4.96853	**4.96559**
gjrGARCH		4.47290	4.47148	<u>**4.39447**</u>	1.53541	1.53528	1.66132	4.97042	4.97267	4.99432
sGARCH	10	4.46975	4.46877	4.42795	1.53499	1.53496	1.53296	4.96724	4.96919	4.96625
eGARCH		4.46962	4.46864	4.42780	1.53492	1.53492	**1.53295**	4.96693	4.96893	**4.96595**
gjrGARCH		4.47290	4.47147	**4.39448**	1.53544	1.53531	1.66130	4.97082	4.97307	4.99466
sGARCH	11	4.46975	4.46877	4.42795	1.53499	1.53495	1.53296	4.96734	4.96930	4.96634
eGARCH		4.46962	4.46863	4.42780	1.53492	1.53491	**1.53295**	4.96703	4.96904	**4.96604**
gjrGARCH		4.47290	4.47147	**4.39448**	1.53543	1.53531	1.66135	4.97092	4.97317	4.99474
sGARCH	12	4.46991	4.46893	4.42812	1.53517	1.53514	1.53315	4.96737	4.96933	4.96650
eGARCH		4.46977	4.46879	4.42796	1.53510	1.53510	**1.53313**	4.96707	4.96907	**4.96620**
gjrGARCH		4.47306	4.47163	**4.39461**	1.53562	1.53549	1.66125	4.97096	4.97321	4.99494
sGARCH	13	4.46988	4.46890	4.42812	1.53507	1.53503	1.53309	4.96685	4.96880	4.96600

(*continued*)

Table 5. (*continued*)

Type	Model	USA			Japan			UK		
		NORM	STD	SSTD	NORM	STD	SSTD	NORM	STD	SSTD
eGARCH		4.46975	4.46877	4.42796	1.53500	1.53499	**1.53307**	4.96654	4.96854	**4.96570**
gjrGARCH		4.47303	4.47161	**4.39460**	1.53551	1.53539	1.66128	4.97044	4.97268	4.99446
sGARCH	14	4.46991	4.46893	4.42813	1.53516	1.53512	1.53313	4.96747	4.96943	4.96658
eGARCH		4.46978	4.46880	4.42797	1.53509	1.53508	**1.53311**	4.96717	4.96917	**4.96628**
gjrGARCH		4.47306	4.47164	**4.39461**	1.53560	1.53548	1.66129	4.97106	4.97331	4.99502

Then, we use the best GARCH-type models obtained from the previous step to forecast the inflation volatility with include regressors as different model. The RMSE criterion is applied to evaluate the predictive power of each set of predictors. This RMSE is used as a high-standard statistical metric to measure prediction model's performance [34]. According to Table 5, Model 2 presents the lowest value of RMSE (4.39226) for the USA. This indicates that the commodity price index has the highest power to forecast the inflation volatility of the USA. For Japan and the United Kingdom, model 9 and model 7 present the highest forecasting performance, implying that Google Trend on "Commodity prices" and the combination of house price index, commodity price index and oil price index have the highest power for inflation volatility prediction for Japan and the United Kingdom, respectively.

4.2 The Impact of Predictor on the Inflation Volatility

After selecting the most suitable volatility forecasting model in Sect. 4.1, we provide the estimation results of the estimated coefficients of the best forecasting models. In this section, we show the best forecasting models predicted in 2 cases: 1. the real price indexes (Table 6) and 2. the best forecasting models predicted by Google trends (Table 7). According to Table 5, we can conclude that Model 2, Model 6, and Model 7 are the best specification model for USA, Japan, and UK.

Model 2 reveals that the commodity price index has a positive effect on inflation volatility. We also find that the occurrence of a health crisis appears to raise higher inflation volatility when compared to financial crisis. In contrast to the USA, the study in Japan confirms that commodity price index, oil price index, and financial crisis have a negative effect on the inflation volatility, while the presence of a health crisis has a positive impact on the inflation volatility.

Table 7 reports the estimation results of the best model incorporating the Google Trend variables. It can be seen that Model 9 performs the best prediction model for all countries, indicating that Google Trend on "Commodity prices" performs the best in predicting the inflation volatility. Although the Google Trend shows weak evidence supporting the inflation volatility of the USA, Japan, and the UK, we find that financial and health crises remain a strong predictor of the inflation volatility of the USA and Japan.

To illustrate the prediction accuracy of our best fit models presented in this section, we plot the 1-day ahead forecasts of inflation volatility and realized volatility in Figs. 2

Table 6. Estimation result of the best predictive model for each country

Coefficient	USA	Japan	UK
	gjrGARCH_SSTD	eGARCH_SSTD	eGARCH_SSTD
	Model 2	Model 6	Model 7
(Intercept)	0.26153***	0.67345***	0.69036***
	[0.04222]	[0.06630]	[0.04806]
House price index			4.69417
			[4.02965]
Commodity price index	6.66753**	−6.24877**	−0.22836
	[2.89636]	[2.63084]	[0.98259]
Oil price index		−0.35566	−0.01803
		[0.45095]	[0.25507]
Financial crisis	0.21600***	−0.33658***	−0.02251
	[0.05371]	[0.08381]	[0.05618]
Health crisis	0.30136***	0.54060***	0.00191
	[0.05084]	[0.07878]	[0.05644]

Note: *** denote significance level of MBF, by 0.0001–0.01 MBF is decisive evidence. [] denotes the standard error

Table 7. Estimation results of the best predictive model for each country in Google Trend variable group.

Coefficient	USA	Japan	UK
	GJR-GARCH_SSTD	EGARCH_SSTD	EGARCH_SSTD
	Model 9	Model 9	Model 9
(Intercept)	0.26914***	0.64174***	0.71278***
	[0.04263]	[0.06723]	[0.04399]
House prices (Google Trend)			
Commodity prices (Google Trend)	0.02578	0.01116	−0.00077
	[0.06844]	[0.05174]	[0.02074]
Oil prices (Google Trend)		−0.35566	
Financial crisis	0.22104***	−0.31633***	−0.02163
	[0.05433]	[0.08570]	[0.05608]
Health crisis	0.30168***	0.54977***	−0.01870
	[0.05150]	[0.08119]	[0.05313]

Note: *** denote significance level of MBF, by 0.0001–0.01 MBF is decisive evidence. [] denotes the standard error

and 3, respectively, for the best GARCH-type models in Tables 6 and 7. The black line represents the realized volatility, the colored line represents the forecasting line.

According to Fig. 2, it is found that the forecast value is moving in the same direction as the realized volatility. However, it has a high difference in some periods, especially during the global financial crisis (2009), for the USA and Japan. On the other hand, our forecasting model is not well predicting the inflation volatility of the UK. The possible reason is that the real commodity price index has probably no effect on the UK's inflation volatility forecasts (see Table 6). Considering the predictive power of Google Trend in predicting the volatility in Fig. 3. We find that the models with Google Trend are not well predicting the inflation volatility for all countries.

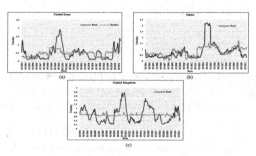

Fig. 2. The forecast inflation volatility based on real price indexes for three countries. (a) the United States of America (b) Japan (c) the United Kingdom. (Color figure online)

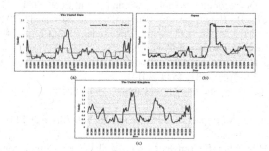

Fig. 3. The forecast inflation volatility based on Google Trend for three countries. (a) the United States of America (b) Japan (c) the United Kingdom (Color figure online)

5   Conclusion

This study focuses on investigating the predictive power of real price indexes and Google Trend keywords in forecasting the inflation volatility for three developed countries (the USA, Japan, and the UK). Several GARCH-type models are also considered in this study as a volatility forecasting model. Overall, our findings show that the Google Trend is not a good predictor of the inflation volatility. However, we find that the real price index, in particular the commodity price index, is the best predictor for all countries. The different models suitable for different countries may arise from the different economic structures that affect the price and cost of goods. In all three countries, commodity prices remain the main factor affecting inflation in all countries, while in Japan and the UK, the cost structure of energy, especially oil prices, is an additional influence on inflation. However, the study results have an interesting part. In addition to energy costs and commodity prices, home prices reflect inflation in the UK. This section is still an interesting point and should be studied in the future. Moreover, we find that financial and health crises have a significant impact on the inflation volatility. Consequently, this study suggests that commodity price index, financial and health crises are the three factors responsible for the inflation volatility. This result is consistent with the literature. Webb [35], Alan Garner [36] and Furlong and Ingenito [37] revealed that there exists a strong relationship between commodity price and the inflation.

References

1. Morag, A.: For an inflation-proof economy. Am. Econ. Rev. **52**(1), 177–185 (1962)
2. Öner, C.: Inflation: prices on the rise. International Monetary Fund (2012)
3. Aiken, M.: Using a neural network to forecast inflation. Ind. Manag. Data Syst. **99**, 296–301 (1999)
4. Mishkin, F.S.: Inflation dynamics. Int. Financ. **10**(3), 317–334 (2007)
5. Fisher, J.D., Liu, C.T., Zhou, R.: When can we forecast inflation? Econ. Perspect.-Federal Reserve Bank of Chic. **26**(1), 32–44 (2002)
6. Stock, J.H., Watson, M.W.: A probability model of the coincident economic indicators (1988)
7. Bryan, M.F., Cecchetti, S.G.: The consumer price index as a measure of inflation (1993)
8. Svensson, L.E.: Open-economy inflation targeting. J. Int. Econ. **50**(1), 155–183 (2000)
9. Reinsdorf, M.: COVID-19 and the CPI: is inflation underestimated? (2020)
10. Bureau of Labor Statistics (BLS). Consumer Price Index (2021). https://www.bls.gov/cpi/questions-and-answers.htm. Accessed 15 Oct 2021
11. Zou, G.L., Chau, K.W.: Determinants and sustainability of house prices: the case of Shanghai China. Sustainability **7**(4), 4524–4548 (2015)
12. Browne, F., Cronin, D.: Commodity prices, money and inflation. J. Econ. Bus. **62**(4), 331–345 (2010)
13. Naurin, A., Qayyum, A.: Impact of Oil Price and Its Volatility on CPI of Pakistan: Bivariate EGARCH Model (2016)
14. Jun, S.P., Yoo, H.S., Choi, S.: Ten years of research change using Google Trends: from the perspective of big data utilizations and applications. Technol. Forecast. Soc. Chang. **130**, 69–87 (2018)
15. Medeiros, M.C., Pires, H.F.: The proper use of Google Trends in forecasting models (2021). arXiv preprint arXiv:2104.03065
16. Loughlin, C., Harnisch, E.: The viability of StockTwits and Google Trends to predict the stock market (2013). StockTwits.com
17. Hu, H., Tang, L., Zhang, S., Wang, H.: Predicting the direction of stock markets using optimized neural networks with Google Trends. Neurocomputing **285**, 188–195 (2018)
18. Huang, M.Y., Rojas, R.R., Convery, P.D.: Forecasting stock market movements using Google Trend searches. Empir. Econ. **59**(6), 2821–2839 (2020)
19. Guzman, G.: Internet search behavior as an economic forecasting tool: the case of inflation expectations. J. Econ. Soc. Meas. **36**(3), 119–167 (2011)
20. Bollerslev, T.: Generalized autoregressive conditional heteroskedasticity. J. Econ. **31**, 307–327 (1986)

21. Lee, J.: Food and energy prices in core inflation. Econ. Bull. **29**, 847–860 (2009)
22. Omotosho, B.S., Doguwa, S.I.: Understanding the dynamics of inflation volatility in Nigeria: a GARCH perspective (2012), working paper
23. Waziri, O.I.O.E.I.: Modeling monthly inflation rate volatility, using generalized autoregressive conditionally heteroscedastic (GARCH) models: evidence from Nigeria. Aust. J. Basic Appl. Sci. **7**(7), 991–998 (2013)
24. Molebatsi, K., Raboloko, M.: Time series modelling of inflation in Botswana using monthly consumer price indices. Int. J. Econ. Financ. **8**(3), 15 (2016)
25. Abbas Rizvi, S.K., Naqvi, B., Bordes, C., Mirza, N.: Inflation volatility: an Asian perspective. Economic research-Ekonomska istraživanja **27**(1), 280–303 (2014)
26. Nyoni, T.: Predicting CPI in Panama., University of Zimbabwe – Munich Personal RePEc Archive (MPRA), Paper No. 92419 (2019)
27. Risteski, D., Sadoghi, A., Davcev, D.: Improving predicting power of EGARCH models for financial time series volatility by using google trend. In: Proceedings of 2013 International Conference on Frontiers of Energy, Environmental Materials and Civil Engineering. Shangai, China (2013)
28. Yao, T., Zhang, Y.J.: Forecasting crude oil prices with the Google index. Energy Procedia **105**, 3772–3776 (2017)
29. Tarkhamtham, P., Yamaka, W., Maneejuk, P.: Forecasting volatility of oil prices via Google Trend: LASSO approach. In: Ngoc Thach, N., Kreinovich, V., Trung, N.D. (eds.) Data Science for Financial Econometrics. SCI, vol. 898, pp. 459–471. Springer, Cham (2021). https://doi.org/10.1007/978-3-030-48853-6_32
30. Bollerslev, T.: Modelling the coherence in short-run nominal exchange rates: a multivariate generalized ARCH model. Rev. Econ. Stat. **72**, 498–505 (1990)
31. Glosten, L.R., Jagannathan, R., Runkle, D.E.: On the relation between the expected value and the volatility of the nominal excess return on stocks. J. Financ. **48**(5), 1779–1801 (1993)
32. Nelson, D.B.: Conditional heteroskedasticity in asset returns: a new approach. Econometrica: J. Econom. Soc. 59, 347–370 (1991)
33. Duan, J., Gauthier, G., Simonato, J., Sasseville, C.: Approximating the GJR-GARCH and EGARCH option pricing models analytically. J. Comput. Financ. **9**(3), 41 (2006)
34. Maneejuk, P., Yamaka, W.: Significance test for linear regression: how to test without P-values? J. Appl. Stat. **48**(5), 827–845 (2021)
35. Willmott, C.J., Matsuura, K.: Advantages of the mean absolute error (MAE) over the root mean square error (RMSE) in assessing average model performance. Clim. Res. **30**(1), 79–82 (2005)
36. Webb, R.H.: Commodity prices as predictors of aggregate price change. FRB Richmond Econ. Rev. **74**(6), 3–11 (1988)
37. Furlong, F., Ingenito, R.: Commodity prices and inflation. Economic Review-Federal Reserve Bank of San Francisco, pp. 27–47 (1996)
38. Garner, C.A.: Policy options to improve the US standard of living. Econ. Rev. **73**(Nov), 3–17 (1988)

Price Volatility Dependence Structure Change Among Agricultural Commodity Futures Due to Extreme Event: An Analysis with the Vine Copula

Konnika Palason[1], Tanapol Rattanasamakarn[1], and Roengchai Tansuchat[2(✉)]

[1] Faculty of Economics, Chiang Mai University, Chiang Mai, Thailand
[2] Center of Excellence in Econometrics, Faculty of Economics,
Chiang Mai University, Chiang Mai, Thailand
roengchai.tan@cmu.ac.th

Abstract. Since the COVID-19 spreads, global food prices have continued to rise and become more volatile because of food security panic, global food supply chain disruption, and unfavorable weather conditions for cultivation. This paper aims to study and compare the dependence structure in price volatility among agricultural commodity futures before and during the COVID-19 pandemic, with different vine copulas, namely the R-vine, C-vine, and D-vine. The daily closing prices of the agricultural commodity futures are used in the investigation, including Corn, Wheat, Oat, Soybean, Rice, Sugar, Coffee, Cocoa, and Orange, traded in the Chicago Board of Trade (CBOT) from January 2016 to July 2021. The conditional volatilities were estimated using the best fit GARCH model with the student-t distribution. The empirical results highlight the dependence structures captured by the C-vine, D-vine, and R-vine copula-based models before and during the COVID-19 pandemic. Although the C-vine copula structures of the two different periods are unchanged, the details of the copula family in such a structure differ. In the case of D-vine and R-vine copulas, the details of the copula families and their vine structures of two different periods are significantly different, meaning that COVID-19 impacts the price volatility dependence structure among the agricultural commodity futures examined. Based on the AIC, the most appropriate dependence structure for pre-COVID-19 period is the C-vine copula, while the during-COVID-19 period is the D-vine copula. The dependence structure of agricultural commodity futures prices can be used in other risk analysis and management methods such as value at risk (VaR), portfolio optimization, and hedging.

Keywords: Dependence structures · Agricultural commodity futures · Vine copula · COVID-19

1 Introduction

Agricultural commodities are agricultural products, of which more than half are used for producing food for both humans, animals, and bioenergy. The common agricultural

© Springer Nature Switzerland AG 2022
K. Honda et al. (Eds.): IUKM 2022, LNAI 13199, pp. 368–378, 2022.
https://doi.org/10.1007/978-3-030-98018-4_30

commodities are corn, wheat, oat, soybean, rice, sugar, coffee, cocoa, and orange. There are two kinds of agricultural commodity trading: spot (or cash) and futures trade. The spot trade is a trade that takes place immediately or within a few days at the current market price. The futures trade is a trade that is not an actual exchange of goods but by an agricultural futures contract, which is an agreement to buy or sell a particular commodity at a future date, at a specific price and amount of the commodity at the time of the agreement.

The agricultural commodity trading price depends on many internal factors such as demand-supply from the producers and the buyers, outstanding position, pre-delivery period, current market price [1–9]. In addition, the uncontrollable external factors are seasonality, climate-changing, and natural disasters. Other external factors are political turmoil or severe epidemic such as COVID-19, which contribute to agriculture and food supply disruptions. These factors also contribute to agricultural commodity price volatility.

Volatility in agricultural prices, known as commodity risk, is the risk of fluctuations in commodity prices. The commodity risk is the main reason for having the commodity futures exchange for hedging against price fluctuations [10]. The futures market also provides an opportunity for investors and speculators to make profits. Although the linear regression analysis of commodity prices can explain the correlation, it is inappropriate to explain the unbalanced dependencies. To date, numerous empirical studies have shown the importance of using several econometric methods to examine the dependence between commodity prices. For example, studies were undertaken on the dependence structure in the agricultural commodity prices [11–13], and interdependence structure between agricultural commodity and oil prices by using the copula model [14–18]. In particular, the copula can be extended to higher dimensions and provide flexible measurements to capture the asymmetric dependence between commodities. Just and Łuczak evaluated the conditional dependency structure in commodity futures markets using the copula-GARCH model [19]. The Vine copula offers better flexibility than the standard copula models. It allows the creation of a model of complex dependency structures that may be analyzed as a tree structure [20]. In addition, from literature review found that Copula-GARCH could be applied to the co-movement or dependence of agricultural commodity. For example, Xinyu Yuan studied Co-Movement among Different Agricultural Commodity Markets of agricultural products using Copula-GARCH [21]. Giot employs stochastic volatility models to analyze the spillover of speculation and volatility between agricultural commodity and crude oil markets [22]. The copula-GARCH approach is useful in investigating the dependence or the co-movement of different series [23, 24].

Currently, the world is facing the COVID-19 pandemic. COVID-19 affects all economic sectors through supply and demand, especially the agricultural sector [25, 26]. According to the Food and Agriculture Organization (FAO), there has been downward pressure on agricultural prices since the early stage of the pandemic [27]. However, as the epidemic continues, the government of each country has set preventive measures to control the spread of COVID-19, including a curfew, border suspension, lockdowns, movement restriction, and social distancing, resulting in labor shortages in the agricultural sector, which indirectly cause the reduction of agricultural production [28–34]. In addition, the measures lead to a degree of food security panic among consumers even in a short-term period. Consequently, agricultural commodities are more expensive [35, 36]. According to a study by Varshney, the coronavirus situation affects the prices of different agricultural commodities differently [37]. Moreover, the news-based COVID-19 sentiment has also affected agricultural prices and price volatility [38].

The literature review revealed that the agricultural commodity markets have different dependence structures; for example, a study by Yamaka et al. [12] in 2018 found that the dependence between commodity futures has different structures. Agricultural commodities appear to move together, with joint movements varying over time [19]. Consequently, the research questions are how the dependence structures of the agricultural commodities volatility look like and how they differ in the pre-COVID-19 time and during the COVID-19 pandemic episode. Therefore, our study used the most famous volatility model, GARCH model, and the different vine-copula models to examine the dependency structure of agricultural commodities composed of nine agricultural commodities (corn, oat, soybean, rice, sugar, coffee, cocoa, orange, and wheat). The rest of the paper is organized as follows: Sect. 2 describes the method, while Sect. 3 presents the information. Section 4 shows the empirical results. Finally, Sect. 5 is the conclusion and discussion.

2   Methods

2.1   GARCH

The Generalized Autoregressive Conditional Heteroskedasticity (GARCH) model is a conditionally heteroskedastic model proposed by Bollerslev (1986) [19]. It has been widely used in financial econometric modeling and analysis since the 1980s because of its ability to capture dynamic volatility and volatility clustering. The ARMA (p,q) GARCH $(1, 1)$ is defined as follows:

$$r_t = \mu + \sum_{i=1}^{p} \phi_i r_{t-i} + \sum_{i=1}^{q} \psi_i \varepsilon_{t-i} + \varepsilon_t \tag{1}$$

$$\varepsilon_t = \sigma_t z_t \tag{2}$$

$$\sigma_t^2 = \varpi + \alpha_1 \varepsilon_{t-1}^2 + \beta_1 \sigma_{t-1}^2 \tag{3}$$

where ε_t is the innovation at time t. z_t is a sequence of *i.i.d.* random variables with mean 0 and variance 1. The restrictions are $\varpi > 0, \alpha_i, \beta_i > 0$ and $\alpha_1 + \beta_1 \leq 1$. The α_1 and β_1 are known as ARCH and GARCH parameters, respectively. However, many studies have indicated that many financial data are non-normally distributed. Instead of the normal distribution, the student's t-distribution is the most popular financial data representation.

2.2   Vine Copula

In 1959, Sklar (1959) [20] introduced The Sklar's theorem. This theorem states that any multivariate joint distribution can be decomposed into two parts: univariate marginal distribution functions and copula. This copula illustrates the dependence structure between the variables. For random variable vector $X = (X_1, \ldots, X_n)'$ with the marginal distribution $F_1, \ldots, F_n$, the Sklar's theorem states that.

$$F(x_1, \ldots, x_n) = C(F_1(x_1), \ldots, F_n(x_n)) \tag{4}$$

for some appropriate n-dimensional copula. The copula from (4) has the expression

$$C(u_1, \ldots, u_n) = \left\{ F_1^{-1}(u_1), \ldots, F_n^{-1}(u_n) \right\} \tag{5}$$

where $F_i^{-1}(u_i)$ is the inverse distribution function of the marginal derived from the ARMA(p,q)-GARCH process. The copula joint density function (f) uses the chain rule with a continuous F with strictly increasing, continuous marginal densities $F_1, \ldots, F_n$.

$$f(x_1, \ldots, x_n) = c_{1 \ldots n}\{F_1(x_1), \ldots, F_n(x_n)\} \cdot f_1(x_1) \ldots f_n(x_n) \tag{6}$$

For conditional density follows that

$$f(x_{n-1}|x_n) = c_{(n-1)n}\{F_{n-1}(x_{n-1}), F_n(x_n)\} \cdot f_{n-1}(x_{n-1}) \tag{7}$$

Consequently, the pair-copula with a conditional marginal density in the general formula can be written as;

$$f(x|v) = c_{xu_j|v_{-j}} \left\{ F(x|v_{-j}), F(u_j|v_{-j}) \right\} \cdot f(x|v_{-j}) \tag{8}$$

Joe (1996) showed that the pair-copula construction involves marginal conditional distributions of the form $F(x|v)$ for every j,

$$F(x|v) = \frac{\partial C_{x,u_j|v_{-j}}\left\{F(x|v_{-j}), F(u_j|v_{-j})\right\}}{\partial F(u_j|v_{-j})} \tag{9}$$

where $C_{ij|k}$ is a bivariate copula distribution function [34].

The Vine copula is a graphical tool for high dimensional data modeling complex dependency patterns from bivariate copulas as building blocks. Bedford and Cooke introduced the vine copula (2001, 2002), denoting it as a regular vine (R-vine), and it can be classified into two cases: C-vine and D-vine. Each model gives a specific way of decomposing the density [39, 40].

For n-dimensional, vine structure can be expressed with $n-1$ trees, and the tree T_j has $n + 1 - j$ nodes and $n - j$ edges which indicate $n - j$ copula density functions. Therefore, the complete decomposition is defined by $n(n-1)/2$ edges.

The density function of D-vine can be written as

$$\prod_{k=1}^{n} f(x_k) \prod_{j=1}^{n-1} \prod_{i=1}^{n-j} c_{i,i+j|i+1,\ldots,i+j-1} \left\{ F(x_i|x_{i+1}, \ldots, x_{i+j-1}), F(x_{i+j}|x_{i+1}, \ldots, x_{i+j-1}) \right\} \tag{10}$$

where index j identifies the trees while i runs over the edges in each tree. The density function of C-vine can be written as

$$\prod_{k=1}^{n} f(x_k) \prod_{j=1}^{n-1} \prod_{i=1}^{n-j} c_{j,j+i|1,\ldots,j-1} \left\{ F(x_j|x_1, \ldots, x_{j-1}), F(x_{j+i}|x_1, \ldots, x_{j-1}) \right\} \tag{11}$$

For R-vine, the structure of R-vine has no uniform like C-vine and D-vine. However, R-vine structure can be the same structure with C-vine and D-vine. R-vine density function is the following

$$f_{1...n}(x) = \prod_{k=1}^{n} f_k(x_k) \prod_{i=1}^{n-1} \prod_{e \in E_i} C_{C_{e,a},C_{e,a}|D_e} \left(F_{C_{e,a}|D_e}(X_{C_{e,a}}|X_{D_e}), F_{C_{e,b}|D_e}(X_{C_{e,a}}|X_{D_e}) \right)$$

(12)

where $x = (x_1, \ldots, x_n)$, $e = \{a, b\}$, $xx_k = \sum_{t=1}^{k}(x_t - \bar{x})$, $yy_k = \sum_{t=1}^{k}(y_t - \bar{y})$, $k = 1, 2, \ldots, N$, and D_e, f_i is the inverse function of F_i.

3 Data

This paper used the futures prices of the agricultural commodities traded in the Chicago Board of Trade (CBOT), including wheat, corn, soybean, rice, oat, sugar, coffee, cocoa, and orange, obtained from Thomson Reuters, for the period running from 1 January 2016 to 31 July 2021. The threshold date of the COVID-19 situation is 30 January 2020. Then, we calculated the agricultural commodities' futures returns from their closing prices based on the continuous compound basis as $r_{i,t} = \ln(P_{i,t}/P_{i,t-1})$, where $P_{i,t}$ and $P_{i,t-1}$ is the futures price of agricultural commodity i at time t and $t-1$, respectively (Fig. 1).

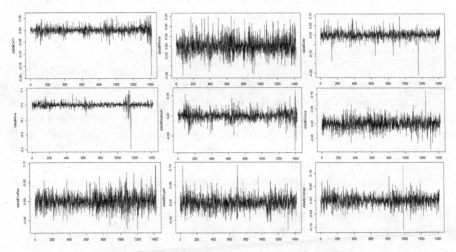

Fig. 1. The daily returns of each agricultural commodity

4 Empirical Results

4.1 Descriptive Statistic

Table 1 shows the descriptive statistics of the agricultural commodity return before COVID-19 and during COVID-19 for wheat, corn, soybean, rice, oat, sugar, coffee, cocoa, and orange futures. The results of the ADF test for the unit root in Table 2 show that all agricultural commodity futures returns for pre-COVID-19 and during COVID-19 periods are stationary.

Table 1. Descriptive statistics of agricultural return

	Before COVID-19						During COVID-19					
	Max.	Min.	Std.	Skew.	Kurt.	JB.	Max.	Min.	Std.	Skew.	Kurt.	JB.
Wheat	0.062	–0.060	0.017	0.271	3.850	42.40	0.054	–0.041	0.017	0.434	3.426	16.4
Corn	0.051	–0.063	0.014	–0.170	5.118	191.90	0.062	–0.191	0.020	–2.102	23.96	7,999
Soybean	0.055	–0.048	0.011	0.039	5.127	189.04	0.064	–0.086	0.013	–0.672	10.25	951.7
Rice	0.072	–0.067	0.014	0.285	5.341	242.18	0.098	–0.300	0.024	–4.584	59.76	57,862
Oat	0.098	–0.140	0.021	–0.105	7.989	1,040	0.055	–0.211	0.019	–3.220	35.54	19,264
Sugar	0.108	–0.053	0.018	0.487	5.717	347.27	0.063	–0.078	0.019	–0.147	3.893	15.5
Coffee	0.067	–0.064	0.017	0.114	3.740	24.98	0.096	–0.076	0.023	0.394	4.467	48.5
Cocoa	0.064	–0.065	0.018	0.011	3.318	4.25	0.115	–0.089	0.020	–0.078	7.017	282.8
Orange	0.131	–0.109	0.020	0.280	6.165	430.88	0.064	–0.074	0.020	–0.199	3.712	11.7

Table 2. ADF unit root test

Variable	Before COVID-19			During COVID-19		
	None	Intercept	Intercept & Trend	None	Intercept	Intercept & Trend
Wheat	–30.868***	–30.856***	–30.848***	–22.307***	–22.303***	–22.283***
Corn	–30.725***	–30.710***	–30.695***	–19.252***	–19.267***	–19.252***
Soybean	–32.988***	–32.972***	–32.976***	–21.029***	–21.177***	–21.154***
Rice	–29.593***	–29.580***	–29.578***	–16.260***	–16.242***	–16.223***
Oat	–31.638***	–31.635***	–31.623***	–19.934***	–19.948***	–20.050***
Sugar	–30.298***	–30.285***	–30.270***	–20.338***	–20.355***	–20.360***
Coffee	–32.342***	–32.326***	–32.310***	–19.787***	–19.814***	–19.953***
Cocoa	–32.843***	–32.830***	–32.859***	–20.646***	–20.624***	–20.599***
Orange	–31.039***	–31.034***	–31.048***	–19.467***	–19.486***	–19.465***

Notes: ***, **, and * denote statistical significance at the 1%, 5%, and 10% levels, respectively

4.2 Vine Copula

This paper considers all types of vine copula, namely R-vine, C-vine, and D-vine. The technique of Dißmann et al. (2013) is applied to construct and estimate the Vine copula structure. The study is divided into two periods: (1) the pre-COVID-19 crisis and (2) during the COVID-19 crisis to determine whether there is a change in the price volatility dependence structure. In the first step, we estimate the best fit of ARMA-GARCH with student's t distribution model for each of the agriculture commodities. In the second step, we transform the estimated conditional volatility to Uniform (0.1) distribution as input data for vine copula estimates. The estimation results of the C-vine, D-vine, and R-vine Copula models are shown in Figs. 2, 3 and 4, respectively. From Fig. 2, the C-vine copula results reveal that corn is a link between the relationships of other commodities in the dependence structure before the COVID-19 crisis and during the COVID-19 crisis. The correlation analysis using the D-vine model (Fig. 3) reveals that the volatility dependence structures in the pre-COVID-19 and during COVID-19 periods are different for virtually all pairs of commodities except for the pairs of rice and orange and the corn and soybean. Meanwhile, the results R-vine model (Fig. 4) show that dependence structures in the pre-COVID-19 time have changed dramatically after the arrival of the COVID-19 epidemic.

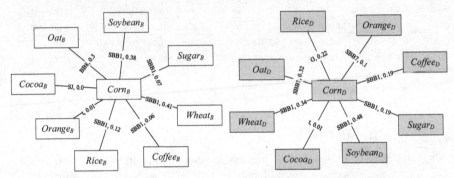

Fig. 2. C-vine Copula tree 1 for Before COVID-19 and During COVID-19

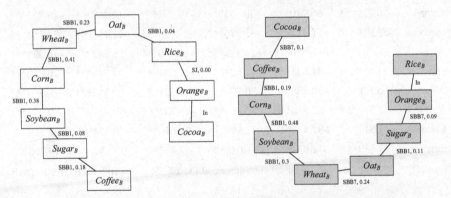

Fig. 3. D-vine Copula tree 1 for Before COVID-19 and During COVID-19

Table 3 presents the Log-Likelihood, AIC, and BIC values from different Vine copulas. For the pre-COVID-19 data, the C-vine model can characterize the relationship between different agricultural commodities better than the D-vine and R-vine types. However, for data during the COVID-19 crisis, the D-vine structure can capture the relationships better than the C-vine and R-vine models.

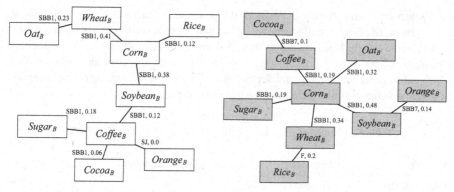

Fig. 4. R-vine Copula tree 1 for Before COVID-19 and During COVID-19

Table 3. Log-likelihood, AIC, and BIC of the vine copulas

	Before COVID-19			During COVID-19		
	Log likelihood	AIC	BIC	Log likelihood	AIC	BIC
C-Vine	3,198.34	−6,284.68	−6,009.79	1,328.96	−2,547.92	−2,325.71
D-Vine	3,074.32	−6,038.63	−5,768.65	1,352.05	−2,598.10	−2,383.96
R-Vine	3,157.86	−6,203.73	−5,928.84	1,339.61	−2,573.23	−2,359.10

Table 4 shows the estimation results of the C-vine copula-based model. In the pre-COVID-19 period, the orders of C-vine structure are corn, wheat, oat, soybean, rice, sugar, coffee, cocoa, and orange, respectively. All pairs of variables have significant co-movement and tail dependence, especially the corn and wheat pair, which possesses the most remarkable dependence.

Table 4. Estimation results using the C-vine copula for pre-COVID-19

Variable	Copula	Parameters 1	Parameters 2	Kendall's tau
Corn & Oat	BB8	2.12*	0.92**	0.30
Corn & Rice	Survival BB1	0.27**	1.00***	0.12
Corn & Cocoa	Survival Joe	1.00***	-	0.00
Corn & Sugar	Survival BB1	0.13*	1.01***	0.07
Corn & Soybean	Survival BB1	1.17*	1.01***	0.38
Corn & Wheat	Survival BB1	1.32*	1.02***	0.41
Corn & Coffee	Survival BB1	0.12*	1.00***	0.06
Orange & Corn	T	0.01**	30.0***	0.01

Table 5 shows that, during COVID-19, the orders of D-vine structure are corn, wheat, oat, soybean, rice, sugar, coffee, cocoa, and orange, respectively. The estimation results of the D-vine copula-based model in Table 5 show that almost all pairs of variables have significant co-movement and tail dependence, especially the corn and soybean pair, which exhibits the most significant dependence. The orange and rice pair is an exception as these two commodities were independent in their price volatility.

Table 5. Estimation results using the D-vine copula during COVID-19

Variable	Copula	Parameters 1	Parameters 2	Kendall's tau
Orange & Rice	Independent	-	-	0.00
Sugar & Orange	Survival BB7	1.00***	0.19*	0.09
Oat & Sugar	Survival BB1	0.23*	1.01***	0.11
Wheat & Oat	Survival BB7	1.01***	0.62**	0.24
Soybean & Wheat	Survival BB1	0.77**	1.03***	0.30
Corn & Soybean	Survival BB1	1.79***	1.02***	0.48
Coffee & Corn	Survival BB1	0.47*	1.00***	0.19
Cocoa & Coffee	Survival BB7	1.00***	0.21*	0.10

5   Conclusion

This study examines the dependence structures of agricultural commodity futures with copulas in the vine class, namely the R-vine, C-vine, and D-vine, in two periods: the pre-COVID-19 and during the COVID-19 pandemic. The empirical results show that the dependence structure or vine structure changed significantly. These findings are consistent with the studies by Yamaka [31] and Yuan that found the dependence structure in agricultural commodities to vary over time [30, 31]. Besides, Sriboonchitta studied the contagion effects of agricultural commodity markets during the 2007–8 food crisis and found that the dependence structure of agricultural commodities changed when the crisis occurred [29], which is consistent with the results of this study that COVID-19 has changed the volatility dependence structure of agricultural commodity futures. The present findings on the change in price volatility dependence structure between different agricultural commodity futures due to such extreme events as COVID-19 can be applied in other risk analysis and management methods such as the Value-at-Risk (VaR), portfolio optimization, and hedging.

Acknowledgments. This research work was partially supported by Chiang Mai University.

References

1. Bessembinder, H., Seguin, P.J.: Price volatility, trading volume, and market depth: Evidence from futures markets. J. Financ. Quant. Anal. **28**(1), 21–39 (1993)

2. Samuelson, P.A.: Proof that properly anticipated prices fluctuate randomly. In: The World Scientific Handbook of Futures Markets, pp. 25–38 (2016)
3. Smit, E.V.M., Nienaber, H.: Futures-trading activity and share price volatility in South Africa. Invest. Anal. J. **26**(44), 51–59 (1997)
4. Daal, E., Farhat, J., Wei, P.P.: Does futures exhibit maturity effect? New evidence from an extensive set of US and foreign futures contracts. Rev. Financ. Econ. **15**(2), 113–128 (2006)
5. Xin, Y., Chen, G.M., Firth, M.: The determinants of price volatility in China's commodity futures markets (2005)
6. Duong, H.N., Kalev, P.S.: The Samuelson hypothesis in futures markets: an analysis using intraday data. J. Bank. Financ. **32**(4), 489–500 (2008)
7. Karali, B., Thurman, W.N.: Components of grain futures price volatility. J. Agric. Resour. Econ. **35**, 167–182 (2010)
8. Lee, N.: Quantile speculative and hedging behaviors in petroleum futures markets. Int. Res. J. Financ. Econ. **53**, 84–99 (2010)
9. Gupta, A., Varma, P.: Impact of futures trading on spot markets: an empirical analysis of rubber in India. East. Econ. J. **42**(3), 373–386 (2016)
10. Just, M., Łuczak, A.: Assessment of conditional dependence structures in commodity futures markets using copula-GARCH models and fuzzy clustering methods. Sustainability **12**(6), 2571 (2020)
11. Brechmann, E.C., Czado, C.: Risk management with high-dimensional vine copulas: an analysis of the Euro Stoxx 50. Stat. Risk Model. **30**(4), 307–342 (2013)
12. Yamaka, W., Phadkantha, R., Sriboonchitta, S.: Modeling dependence of agricultural commodity futures through Markov switching copula with mixture distribution regimes. Thai J. Math. 93–107 (2019)
13. Pennings, J.M., Meulenberg, M.T.: Hedging Risk in Agricultural Futures Markets. In: Wierenga, B., van Tilburg, A., Grunert, K., Steenkamp, J.B.E.M., Wedel, M. (eds.) Agricultural marketing and consumer behavior in a changing world, pp. 125–140. Springer, Boston (1997). https://doi.org/10.1007/978-1-4615-6273-3_7
14. Chen, K.J., Chen, K.H.: Analysis of Energy and Agricultural Commodity Markets with the Policy Mandated: A Vine Copula-based ARMA-EGARCH Model (No. 333-2016-14250) (2016)
15. Liu, X.D., Pan, F., Yuan, L., Chen, Y.W.: The dependence structure between crude oil futures prices and Chinese agricultural commodity futures prices: measurement based on Markov-switching GRG copula. Energy **182**, 999–1012 (2019)
16. Yahya, M., Oglend, A., Dahl, R.E.: Temporal and spectral dependence between crude oil and agricultural commodities: a wavelet-based copula approach. Energy Econ. **80**, 277–296 (2019)
17. Kumar, S., Tiwari, A.K., Raheem, I.D., Hille, E.: Time-varying dependence structure between oil and agricultural commodity markets: a dependence-switching CoVaR copula approach. Resour. Policy **72**, 102049 (2021)
18. Tiwari, A.K., Boachie, M.K., Suleman, M.T., Gupta, R.: Structure dependence between oil and agricultural commodities returns: the role of geopolitical risks. Energy **219**, 119584 (2021)
19. Bollerslev, T.: Generalized autoregressive conditional heteroskedasticity. J. Econom. **31**(3), 307–327 (1986)
20. Sklar, M.: Fonctions de repartition an dimensions et leurs marges. Publ. Inst. Statist. Univ. Paris **8**, 229–231 (1959)
21. Yuan, X., Tang, J., Wong, W.K., Sriboonchitta, S.: Modeling co-movement among different agricultural commodity markets: a Copula-GARCH approach. Sustainability **12**(1), 393 (2020)
22. Giot, P.: The information content of implied volatility in agricultural commodity markets. J. Futures Mark.: Futures Options Other Deriv. Prod. **23**(5), 441–454 (2003)

23. Reboredo, J.C.: Do food and oil prices co-move? Energy Policy **49**, 456–467 (2012)
24. Sriboonchitta, S., Nguyen, H.T., Wiboonpongse, A., Liu, J.: Modeling volatility and dependency of agricultural price and production indices of Thailand: static versus time-varying copulas. Int. J. Approx. Reason. **54**(6), 793–808 (2013)
25. Siche, R.: What is the impact of COVID-19 disease on agriculture? Scientia Agropecuaria **11**(1), 3–6 (2020)
26. Gregorioa, G.B., Ancog, R.C.: Assessing the impact of the COVID-19 pandemic on agricultural production in Southeast Asia: toward transformative change in agricultural food systems. Asian J. Agric. Dev. **17**, 1–13 (2020). (1362-2020-1097)
27. FAO: Food commodities still at risk of coronavirus 'market shock' (2020). https://www.reuters.com/article/us-global-agriculture-outlook-idUSKCN24H19U. Accessed 1 Nov 2021
28. Bakalis, S., et al.: Perspectives from CO+ RE: how COVID-19 changed our food systems and food security paradigms. Curr. Res. Food Sci. **3**, 166 (2020)
29. Aday, S., Aday, M.S.: Impact of COVID-19 on the food supply chain. Food Qual. Saf. **4**(4), 167–180 (2020)
30. Gong, B., Zhang, S., Yuan, L., Chen, K.Z.: A balance act: minimizing economic loss while controlling novel coronavirus pneumonia. J. Chin. Gov. **5**(2), 249–268 (2020)
31. OECD: COVID-19 Crisis Response in ASEAN Member States (2020). https://www.oecd.org/coronavirus/policy-responses/COVID-19-crisis-response-in-asean-member-states-02f828a2/. Accessed 1 Nov 2021
32. Pulubuhu, D.A.T., Unde, A.A., Sumartias, S., Sudarmo, S., Seniwati, S.: The economic impact of COVID-19 outbreak on the agriculture sector. Int. J. Agric. Syst. **8**(1), 57–63 (2020)
33. Pu, M., Zhong, Y.: Rising concerns over agricultural production as COVID-19 spreads: lessons from China. Global Food Secur. **26**, 100409 (2020)
34. Islamaj, E., Mattoo, A., Vashakmadze, E.T.: World Bank East Asia and Pacific economic update, April 2020: East Asia and Pacific in the time of COVID-19, No. 147196, pp. 1–234 (2020). The World Bank
35. Joe, H.: Families of m-variate distributions with given margins and m (m–1)/2 bivariate dependence parameters. Lecture Notes-Monograph Series, pp. 120–141 (1996)
36. Cariappa, A.A., Acharya, K.K., Adhav, C.A., Sendhil, R., Ramasundaram, P.: COVID-19 induced lockdown effects on agricultural commodity prices and consumer behaviour in India–Implications for food loss and waste management. Socio-Econ. Plan. Sci. 101160, 1–23 (2021)
37. Varshney, D., Roy, D., Meenakshi, J.V.: Impact of COVID-19 on agricultural markets: assessing the roles of commodity characteristics, disease caseload and market reforms. Indian Econ. Rev. **55**(1), 83–103 (2020). https://doi.org/10.1007/s41775-020-00095-1
38. Balcilar, M., Sertoglu, K.: The COVID-19 effects on agricultural commodity markets. SSRN, 1–20 (2021). https://papers.ssrn.com/sol3/papers.cfm?abstract_id=3882442. Article ID 3882442
39. Cooke, P.: Regional innovation systems, clusters, and the knowledge economy. Ind. Corp. Chang. **10**(4), 945–974 (2001)
40. Cooke, P.: Knowledge economies: Clusters, learning and cooperative advantage. Routledge, Abingdon (2002)

Author Index

Printed in the United States
by Baker & Taylor Publisher Services